Reliability, Maintainability, and Supportability

WILEY SERIES IN SYSTEMS ENGINEERING AND MANAGEMENT

Andrew P. Sage, Editor

A complete list of the titles in this series appears at the end of this volume.

RELIABILITY, MAINTAINABILITY, AND SUPPORTABILITY

Best Practices for Systems Engineers

Michael Tortorella

WILEY

Published by John Wiley & Sons, Inc., Hoboken, New Jersey
Published simultaneously in Canada

Library of Congress Cataloging-in-Publication Data:

Tortorella, Michael, 1947–
Reliability, maintainability, and supportability : best practices for systems engineers / Michael Tortorella.
pages cm. – (Wiley series in systems engineering and management)
Includes bibliographical references and index.
ISBN 978-1-118-85888-2 (hardback)
1. Reliability (Engineering) I. Title.
TA169.T676 2015
620′.00452–dc23

2014049531

Set in 11/13pt Dante by SPi Publisher Services, Pondicherry, India

Printed in the United States of America

10 9 8 7 6 5 4 3 2 1

1 2015

For Matthew
1982–1999
Lux æterna

Contents

Part III Supportability Engineering

Foreword

PURPOSE AND RATIONALE

Students and professionals have many choices of text and reference books for the sustainability engineering disciplines: reliability, maintainability, and supportability. Available books range from theoretical treatises on the mathematical theory of reliability, applied maintainability and logistics modeling, studies in reliability physics, and books devoted to systems management. But there's still something missing: there is a need for an exposition of the sustainability engineering activities that systems engineers need to carry out, which explains the purposes and benefits of the activities without necessarily explaining how to do them all in detail. This book fills that need.

Several decades of experience in sustainability engineering and management in the telecommunications industry and additional experience in research and teaching have led me to these relevant observations.

1. Few publications in the sustainability disciplines focus on the core systems engineering tasks of creating, managing, and tracking requirements for these disciplines specifically.
2. The small number of degree-granting programs in sustainability engineering means that many systems engineers have no exposure to these ideas until they are assigned to deal with them in the work environment.
3. The gap between what is known and available in the research literature and what is routinely practiced in day-to-day sustainability engineering is large and growing. Many sustainability engineers use oversimplified models and tools to deal with sustainability engineering tasks and consequently miss opportunities to develop more thorough and informative product management and improvement plans at lower cost.
4. Systems engineers, in particular, because of the broad scope of their responsibilities, need support from those with specialized expertise to write good sustainability requirements, understand the results provided

to them by sustainability engineering specialists, and track compliance with stated sustainability requirements. Consequently, they need enough background knowledge in these areas to be good suppliers and customers for the specialist teams.

5. Many software tools essential for executing complex sustainability engineering tasks often (silently) incorporate simplifying assumptions, rely on the user to discern when results are reasonable or not, and do not give the user good insight into what to expect from the tool and what not to expect from the tool.

Sustainability engineering and management is not an obscure, arcane branch of knowledge. It is a human endeavor that can readily be carried out systematically and on the basis of a manageable number of principles. The purpose of this book is to provide that basis for systems engineers in particular. Certainly, few have as much influence on a product's design as do systems engineers. The creation of appropriate sustainability requirements is a key step to developing a system whose realized reliability, maintainability, and supportability meet the needs and desires of the system's customers while promoting success and profit to the vendor. Conversely, incomplete, unfocused, or inappropriate requirements lead to customer dissatisfaction with the system they purchase and use and cost the vendor more in warranty costs, maintenance of an extensive repair business, and lost goodwill. Our purpose here is to provide systems engineers with the principles and tools needed to craft sustainability requirements that make the product or system successful in satisfying the customers' needs and desires for reliability, maintainability, and supportability while keeping costs manageable. Our purpose is also to provide methods and tools systems engineers can use to determine whether sustainability requirements are being met satisfactorily by understanding and analysis of data from field installations. Finally, the book discusses enough quantitative modeling for reliability, maintainability, and supportability to support systems engineers in their engineering, management, validation, and communication tasks.

It is important to note that this book is not intended as a textbook in the mathematical theory of reliability (or the mathematical underpinnings of maintainability or supportability). Rather, our intention is to provide systems engineers with knowledge about the results of these theories so that, while they may sometimes construct needed reliability, maintainability, and supportability models on their own, it is more important that they be able to successfully acquire and use information provided to them by specialist engineers in these disciplines. The customer–supplier model provides a useful context for this interaction:

- Systems engineers act as suppliers in providing specialist engineers with clear and effective reliability, maintainability, and supportability requirements for the product.
- Systems engineers act as customers for the reliability, maintainability, and supportability models, data analysis, and so on, provided by specialist engineering teams during development.

Therefore, systems engineers need a good grasp of the language and concepts used in these areas, while not necessarily needing to be able to carry out extensive modeling or data analysis themselves. While this book is careful to describe the necessary language and concepts correctly and in appropriate contexts, it makes no attempt to provide mathematical proofs for the results cited. References are provided for those interested in pursuing details of the mathematical theory of reliability, but those details are not within the scope or purpose of this book.

GOALS

I hope this book will enable systems engineers to lead the development of systems (which we will interpret broadly in this book as encompassing products and services) whose reliability, maintainability, and supportability meet and exceed the expectations of their customers and provide success and profit to their employers. My intention is that systems engineers will themselves be able to employ, and encourage their sustainability engineering specialists to employ, the best practices discussed here in an orderly, systematic fashion guided by customer needs. I recognize that systems engineers have a very broad range of responsibilities, and it may not be possible for them to deal with every responsibility at equal depth. Therefore, it is important that their sustainability engineering and management responsibilities be supported by as straightforward and systematic a program as possible. I emphasize the thought processes underlying all the activities a systems engineer may have to undertake to ensure successful product or system sustainability. To avoid losing sight of the forest for the trees, we repeatedly return to the basic questions and first principles of the field in all the applications we cover, including hardware products, software-intensive systems, services, and high-consequence systems. My intention in doing this is to help systems engineers choose appropriate methods and tools to accomplish their purposes, and thereby create the most suitable sustainability requirements consistent with fulfilling customer needs and expectations and supplier success.

ORGANIZATION OF THIS BOOK

Every author likes to think that he brings to the reader a uniquely formative experience through the superior organization of topics and methods in his book. If only it were that simple. Success in learning depends primarily on student commitment. I can only try to make that job easier. I hope that the devices I use in this book will fulfill that wish.

- The book is organized into three major divisions, one corresponding to each of reliability, maintainability, and supportability engineering. Within each division, there is material on
    - Requirements development,
    - Quantitative modeling sufficient for understanding, developing, and interpreting requirements,
    - Statistical analysis for checking whether systems in operation meet or do not meet requirements, and
    - Best practices in each of these areas.
- I place a lot of emphasis on correct use of language. As discussed at length in Chapter 1, the language we use in the formal system that constitutes sustainability engineering contains many of the same words we use in ordinary discourse. It is vital to keep in mind which context you are operating in at all times. To help you do this in places where I think there is more than the usual possibility for confusion, I will point out in the text information you need to dispel that confusion. These instances are introduced by the header "Language tip" and they appear in many places in the text.
- This book is primarily for systems engineers whose main concern is the determination and development of appropriate requirements so that designers may fulfill the intent of the customer. Accordingly, the book emphasizes the use of various sustainability engineering methods and techniques in crafting requirements that are
    - Focused on the customers' needs,
    - Unambiguous,
    - Easily understood by the requirements' stakeholders (customers, designers, and management), and
    - Verifiable through collection and analysis of data from system operation.

  The device employed in the book to promote this goal is the frequent interjection of "Requirements tips" that appear when needed and of most benefit.
- An equally important concern of systems engineers is determining when requirements are being met by systems operating in customer environments. Accordingly, a chapter or section in each of the major divisions of the book is devoted to the statistical analyses needed to accomplish this task.
- The title of the book emphasizes "Best Practices." Each chapter concludes with a section summarizing the current best practices for systems engineers concerning the material covered in the chapter.
- Finally, I believe that everything we do is a process, whether we call it that or not. In particular, we should all be mindful that everything we do can be improved. Requirements development and verification are no exception (indeed, this book is no exception, and I welcome suggestions from readers to help make the next version better).

Acknowledgments

I have had the good fortune to work with many talented and caring people who have shared their knowledge and helped shape my thinking in this area. First and foremost, I want to thank the late Norman A. Marlow who gave me the freedom to pursue interesting and important problems in network and service reliability and guided my early career at Bell Laboratories. Elsayed A. Elsayed and Endre Boros promoted my pleasant and fruitful relationship with Rutgers University. Many other colleagues and friends contributed their time and expertise to help me learn more about reliability engineering. These include Susan Albin, Sigmund J. Amster, Lawrence A. Baxter, Michele Boulanger, Chun Kin Chan, Ramon V. Leon, Michael LuValle, and William Q. Meeker. José Ramirez-Marquez helped me pass the torch. Working with Bill Frakes on the SUPER project was an education and a pleasure. A. Blanton Godfrey, Jeffrey H. Hooper, and William V. Robinson provided invaluable management support, and Jon Bankert and Jack Sipress gave me an opportunity to work with the Bell Labs Undersea Cable Laboratory where I learned a great deal of practical reliability engineering. I am grateful to Stevens Institute of Technology for employing me to teach a course in the Systems Engineering program based on these ideas. I am deeply grateful to Chun Kin Chan, Bill Frakes, and D. A. Hoeflin for carefully reading the manuscript and offering excellent suggestions. I have also benefitted from conversations with David Coit, Elsayed Elsayed, Shirish Kher, Mohcene Mezhoudi, Himanshu Pant, William V. Robinson, and Terry Welsher. Outside my immediate circle of colleagues, I thank Harry Ascher, Alessandro Birolini, Ilya Gertsbakh, and William A. Thompson. Learning from their clear and well-constructed books helped me be a better reliability engineer. Laura Madison helped clean up my sometimes too-convoluted prose; her patience in taking on a bigger job than she anticipated is much appreciated. Finally, my thanks to Andrea for her patience with my frequent and extended disappearances into the authorial vortex. To all these and many more too numerous to mention, thank you for helping shape this book. I have tried to learn from your suggestions, but people tell me I am sometimes a stubborn cuss, so there may remain errors in the book, and if so, they are mine alone.

Part I

Reliability Engineering

1

Systems Engineering and the Sustainability Disciplines

1.1 PURPOSE OF THIS BOOK

1.1.1 Systems Engineers Create and Monitor Requirements

The textbook marketplace offers many high-quality books that provide the student, professional, and researcher with many points of view on the sustainability disciplines of reliability engineering, maintainability engineering, and supportability engineering. The point of view we advance here, though, is different from that of other books. This book focuses intently on the roles and responsibilities of the systems engineer in *creating* and *monitoring* the requirements for reliability, maintainability, and supportability that will guide development of products and services that are most likely to satisfy their customers and lead to success for their suppliers. Systems engineers play a pivotal role in this process. Get the requirements wrong and the likelihood of a successful product or service is almost nil. That, coupled with the importance of acting as early as possible in the development process to build in quality and reliability, compels a new emphasis on preparing systems engineers to understand how the sustainability disciplines contribute to product and service success and to enlarge their toolkit to incorporate generation and validation of sustainability requirements that promote greater product and service success. The first major purpose of this book is to provide systems engineers with the

---

Reliability, Maintainability, and Supportability: Best Practices for Systems Engineers, First Edition. Michael Tortorella.

knowledge they need to craft clear, concise, and effective sustainability requirements so that they may fulfill their role of key leader in successful product and service development.

Customers and suppliers also want to know whether requirements are being met by deployed products and services. For example, many telecommunications service providers offer service-level agreements (SLAs) to their larger customers (see Section 8.6). SLAs are usually based on certain service reliability criteria [11, 12]; when these criteria are violated, the customer is offered a full or partial refund for a stated period of service. In addition, many suppliers of commercial and consumer products offer warranties. The cost of servicing the warranty is borne by the supplier. The obvious financial consequences in these examples show why it is important to be able to determine in a systematic way whether and to what degree relevant requirements are likely to be met (in a planning phase) and are being met (in operation). Accordingly, the second major purpose of this book is to provide systems engineers with the concepts, tools, and techniques needed to carry out analyses for determining conformance to quantitative sustainability requirements.

1.1.2 Good Requirements are a Key to Success

Accepting, as we do, that a design faithfully realizing a set of complete and effective requirements will make a product or service that is no more or less than those requirements describe, it is clear that requirements are key contributors to a successful product or service. Accordingly, we need to understand what makes a good requirement. At least two important properties of a good requirement can be immediately discerned:

1. The requirement is written to promote an outcome (product or service property or behavior) that is desired by the customer.
2. The requirement is unambiguous: clear criteria are available to determine whether the requirement is met or not.

Every product or service property or behavior that is needed or desired by the customer for the product or service should be the subject of some requirement(s). There is no other reliable way to ensure that the product or service will have that property or behavior. This is nothing more than a restatement of the idea that if you want something, unless you ask for it specifically, you will only get it by some happy accident. Think of a customer, like a telecommunications service provider, who needs a reliable backup generator to ensure continuity of service during periods when utility power is unavailable. If the customer does not specify the length of time for which the backup generator is required to operate without failure, then the system designer has no guidance about how to specify which backup generator to use and what measures need to be taken to ensure that it operates for the needed period of time. Some backup generator will be chosen, but the reliability of that backup generator may or may not be good enough to meet the customer's need. In this example, "you get what you get" without a clear plan to get, rather, what you need—the result is haphazard rather than systematic. Good requirements are complete (cover all properties and behaviors needed and desired by the customer).

The best way to promote unambiguous requirements is to state them in quantitative terms. Most requirements in the sustainability disciplines involve some quantitative variable. For example, we may wish to limit the amount of time it takes to complete a specified repair. To enforce such a limit, this duration will be the subject of a requirement. In practice, the time it takes to complete a repair is influenced by many factors, including control factors (those that the system designer and operator are able to control) and noise factors (factors that are thought of as "random" and not able to be readily adjusted by the designer or operator).[1] Consequently, it is customary to conceptualize the quantitative variables appearing in requirements as random variables in the sense used in probability theory. That is, the values taken by this variable over the different members of the population of products or service realizations may differ from one to another in unpredictable ways. For instance, the duration of the specified repair in the example will be influenced by factors like the location and ease of access of required spare parts, the location and ease of access of required documentation, how well-trained the repair technicians are, etc. The system designer can influence these factors by appropriate selection of requirements for them; see Part II of this book for maintainability considerations like these. However, the repair duration may also be influenced by factors that the designer cannot control, such as how fatigued the operator may be after having worked an entire shift before beginning the repair, whether the operator has to deal with inclement weather in an outdoor installation, etc. The designer cannot control these "noise factors." For this reason, products or services should be designed to be "robust" against the effects of noise factors. This means that the product or service should be insensitive to variations in the values of the noise factors. The discipline of "robust design" [7, 14] has arisen to make this task systematic. A product or service that is robust in this sense is likely to experience fewer failures, making robust design a valuable tool for the systems engineer and design staff. We return to this idea in more detail in Section 6.8.

It is also important when assessing product or service performance against a set of requirements that statistical ideas be used—monitoring the performance of the product or service in operation generates data for each of the requirements. These data may be a census or only a sample from the population of

[1] This approach to conceptualizing operations in the real world was first introduced by Genichi Taguchi in the 1980s [10], in the context of statistically designed experiments. More broadly interpreted, it offers a useful conceptualization of how much of a given product or service realization may be controlled by requirements and how the design may be arranged, including considerations of how much margin may need to be built into the design, to mitigate the influence that "noise" factors have over the eventual outcome.

installed systems. Treating census data is straightforward; many examples are given in chapter 5 and elsewhere. Sampling data need to be analyzed in a consistent statistical manner that respects the sampling nature of the data collection so that an informative and fair picture of how well the product or service is performing may be obtained. The details of such analyses are discussed in various chapters of this book, so we are not going to dig deeper here, but we will point out, for the first time of many times, that comparisons between performance and requirements are expressed in statistical terms using probabilities, significance levels, and confidence intervals. The nature of real-world operation, especially when we are unable to collect data on anything but a sample of the population of systems in operation, brings with it these uncertainties. Whether it is possible to make absolute judgments about meeting requirements or not also depends on the form in which the requirements are written (see Chapters 3 and 5).

1.1.3 Sustainability Requirements are Important Too

At the most fundamental level, systems engineering exists to promote certain outcomes in product and service development and deployment. These outcomes include customer satisfaction and supplier profitability. The basic tool systems engineers use to carry out this function is to create and monitor requirements for specific product or service properties whose achievement promotes these outcomes. While there are many such properties that matter, this book focuses on those properties connected to reliability, maintainability, and supportability. Before narrowing to that focus, however, we need to discuss the broader context of the systems engineering role in these disciplines.

Promotion of certain key outcomes is a primary systems engineering function. In reliability, for example, understanding of customer needs may indicate that the customer is concerned primarily with the frequency of failures, perhaps because remediation of a failure requires dispatch of a repair crew to a remote or difficult-to-reach location, and the customer wishes to minimize the expense associated with these actions. Therefore the systems engineer creates a requirement for frequency of failures, perhaps something like "The equipment shall not experience failures requiring the dispatch of service personnel more often than once per decade per system." Later in chapter 2, we will see why this requirement is incomplete (it lacks any statement about what conditions are to prevail for this failure frequency limit to be valid), but the key point here is that it is created based on a detailed understanding of the customers' needs and the capabilities that need to be designed into the system to meet those needs.

As with any endeavor that undertakes to reach certain targets, an understanding of the process by which those targets are approached is necessary. This is a fundamental principle of quality engineering in which any effort to design and improve a product or service is based on an understanding of the process by which the product or service is created and used. Here, the systems engineer acts to promote certain outcomes. In the sustainability disciplines, these outcomes represent what is needed of the product or service in reliability, maintainability, and supportability so that the product or service will satisfy its customers and produce a profit for the supplier. To do this effectively, he needs to understand the process used to achieve those outcomes. Then she can determine the key points in that process at which monitoring can be most effective in guiding the process toward its desired output.

1.1.4 Focused Action is Needed to Achieve the Goals Expressed by the Requirements

System or service development often begins with a “wish list” of desirable properties, or “features,” that will attract customers. From this list of features, a set of requirements is created. For purposes of this book, we categorize requirements as attribute requirements and sustainability requirements. Attribute requirements comprise functional, performance, physical, and safety requirements. Sustainability requirements are those pertaining to reliability, maintainability, and supportability that bear on whether a system or service can be developed not only to work satisfactorily when it is new but also to continue to operate satisfactorily for a significant period of time thereafter—enough time so that the system or service creates enough customer satisfaction and supplier profitability to be worthwhile.

Deliberate, focused action must be taken to create a design and realize the system or service that meets requirements. These actions are referred to as “design for x” where x may refer to any of the requirements categories. While this is certainly true for attribute requirements, in this book we emphasize design for reliability, design for maintainability, and design for supportability as key enablers of goal achievement through systematic, repeatable, and science-based actions. Without deliberate attention to design for x, whatever requirement goals may be achieved are achieved only by chance, and the odds of meeting all requirements by chance are slim indeed. In particular, reliability, maintainability, and supportability are sometimes seen by those lacking training in these fields as arcane branches of knowledge whose implementation is beyond the capabilities of most engineers. Our position is emphatically that this is not so. We specifically discuss design for reliability, maintainability, and supportability in Chapters 6, 11, and 13, respectively, from the point of view that the actions constituting these fields are systematic, repeatable, and grounded in sound science, and are readily learned and readily applied by most engineers.

Finally, we point out that almost all, if not all, components of design for x are readily susceptible to quantitative modeling and optimization. For instance, the layout and process flow in a repair facility may be modeled as a stochastic network (see chapter 13) and optimized on that basis so that inefficiencies may be rooted out and speedier, more economical operation is promoted. The decision about whether to engage this greater degree of detail rests largely on the organization's judgment about the balance between prevention costs and external failure costs. In this instance, an optimized repair facility promotes more rapid repair of failures, shorter system outages, and faster turnaround to the customer. This is worth something (even though it may be difficult to quantify); whether it is worth enough to justify the expenditure of scarce, skilled resources to carry out the optimization depends on the organization's quality management approach. In any case, the duration of the improvement is likely to be much longer than the time spent on carrying out the optimization, an argument in favor of the modeling approach. You will see many examples of this approach in the design for *x* chapters, but not every opportunity will be discussed in detail because to do so would require turning this into a book surveying all of operations research. Where an important technique of this kind may be useful but is not covered in this book, appropriate references are provided.

1.2 GOALS

For systems engineers to be able to do these things effectively, they need to reach certain goals. These goals determine the goals of this book.

1. Systems engineers need to know how success is defined for the product or service in development. Two primary indicators of success are profitability for the organization supplying the product or service, and satisfaction on the part of their customers. Throughout, we emphasize the relationships between sustainability requirements, product/service success, and the technical content of sustainability models in helping systems engineers look forward and see how the profitability and customer satisfaction results may play out.
2. Systems engineers rarely will be required to carry out detailed reliability, maintainability, or supportability modeling, but they will almost always receive advice from specialists in these disciplines. They may also
   - subcontract the creation of reliability, maintainability, and/or supportability to teams of experts in those disciplines and
   - be part of a team negotiating sustainability requirements with customers or suppliers.

   Therefore, systems engineers need to know how to be good customers of specialist engineering suppliers and be effective negotiators of sustainability requirements. This requires a minimum level of understanding of some details of reliability, maintainability, and supportability engineering. This book will present this kind of information not with a goal of creating reliability, maintainability, or supportability specialists, but rather with an amount of detail necessary to acquire the understanding needed to be good consumers of specialist information and good negotiators. We aim to give systems engineers the skills needed to ask good questions and understand the answers, particularly with regard to the systems engineer's primary responsibility concerning the creation of suitable requirements for reliability, maintainability, and supportability—those that promote successful product and service development and deployment, satisfied customers, and a profitable business. Experienced sustainability engineers may find that some of the explanations needed to support this goal are already familiar to them and do not bear repeating, but they are included—indeed, emphasized—here to provide systems engineers with the background and understanding they need so that they can be good customers and suppliers in this context.

3. A third goal that is at least as important as others mentioned so far is to promote clarity of language and communication across the community of systems engineering stakeholders: customer representatives, specialist engineers, the product or service development team, management, and executives. The sustainability disciplines are loaded with special terms and the temptation to lapse into jargon is sometimes overwhelming. Nonetheless, we firmly believe that the best ideas are the simple ones, or at least those that can be explained simply and clearly, and this book is written with the promotion of clear, unambiguous, and consistent communication as a primary goal. I once worked for a manager who claimed that at times it was necessary to "vague it up," but my experience has been that "vaguing it up" is more often than not a means for disguising a lack of understanding or playing political games, and rarely does it have its claimed benefits. This book intends to help you express yourself clearly and concisely. You may choose not to do so at times, but at least you will be prepared to do so successfully when needed.
4. In addition to providing a modest introduction to sustainability modeling skills, this book aims to enable systems engineers to employ a systematic and repeatable procedure to determine whether the sustainability requirements they have created are being met by systems and services, both during development and after deployment. This key step in the product or service development process enables management to undertake quality improvement programs based on reliable data and sound analyses. In other words, these requirement verifications are part of the Deming cycle's [9] "check" phase. Preventive action—to maintain good performance—and corrective action—to improve performance—should only be undertaken after a solid understanding of the success or failure of relevant requirements is achieved. This book aims to provide readers with the concepts, frameworks, tools, and techniques needed to efficiently determine the degree to which requirements are being met or not met and form the foundation for management by fact.
5. Sustainability engineering is sometimes practiced by engineers who are not specifically trained in these disciplines. Those fulfilling such a role are expected to make good use of the resources available to them while those resources may be written in language that may be unfamiliar and that may leave a lot of gaps in reasoning because they are intended for experts.

Systems engineers who have a broader background and education may need help filling those gaps. One of the purposes of this book is to present material on the sustainability disciplines carefully and with the needs of the systems engineer in mind. Readers will find discussions intended to fill in these gaps, especially as regards clear use of the language, so that confusion and ambiguity may be avoided. In some cases, experts may find these discussions tedious and/or repetitive, but the detailed, step-by-step discussions are deliberately prepared to help the non-expert rapidly be able to make substantive contributions.

1.3 SCOPE

The sustainability disciplines are reliability engineering, maintainability engineering, and supportability engineering. These disciplines are linked in important ways (see chapter 2) and are important factors in product or service success. To enhance learning about these engineering disciplines, this book emphasizes certain points of view and covers certain topics while omitting others.

The primary point of view expressed in this book is that requirements are a key driver of product or service success. As systems engineers are the developers of requirements, they must be skilled in developing requirements that lead in the right directions. Accordingly, they need enough knowledge about each of the three sustainability disciplines to be able to develop sensible and effective requirements. The scope of this book is dictated primarily by this need.

1.3.1 Reliability Engineering

To be able to accomplish reliability engineering tasks effectively requires two key skills: first (and foremost), understanding how actions taken during product or service design and manufacturing promote (or inhibit) reliability, and second, ability to work with the quantitative aspects of reliability modeling and statistical analysis. Accordingly, the two goals of the reliability engineering part of this book (Part I) are

1. to introduce the reader to design for reliability through learning about failure modes and failure mechanisms, failure causes, and preventive actions, in a variety of electronic and mechanical contexts and
2. to provide enough material on quantitative reliability modeling and statistical analysis so that the reader can understand the implications of writing requirements in various ways and be able to determine when the performance of the product or service conforms to the requirements.

This is not a textbook in reliability physics, software design patterns, or general hardware or software development best practices, so the first goal is approached in rather more general terms. From the examples of failure modes and failure mechanisms given, the reader will be expected to inductively transfer this knowledge to new situations. Neither is this a textbook in the mathematical theory of reliability; many such books of high quality are already available to interested readers and are cited in the references herein. Rather, there is given here enough of quantitative reliability modeling and statistical analysis that readers will see how to speak and write about these correctly, understand the results specialists in this discipline will supply as part of the systems engineering and development process, and create procedures to determine to what degree reliability requirements are being met when the product or service is finally deployed. Reliability engineering specialists may find that the material on reliability modeling for systems engineers presented in chapter 4 may be simultaneously too basic and not complete. It is deliberately presented in basic terms so that it may be accessible to a non-specialist audience. It is not complete in the sense that no proofs of mathematical assertions are given (though plenty of references are provided), but it is comprehensive in the sense that the reliability modeling topics of greatest importance are all covered. Specialists may find some of the foundational discussions useful for refreshing their basic understanding of commonly used techniques.

1.3.2 Maintainability Engineering

As with reliability engineering, and as we will point out again with supportability engineering, systems engineers need to understand how actions taken during product or service design and manufacturing promote (or inhibit) maintainability, and they also need the ability to work with the quantitative aspects of maintainability modeling and statistical analysis. Accordingly, the two goals of the maintainability engineering part of this book (Part II) are

1. to introduce the reader to design for maintainability through learning about the key factors that influence maintainability and
2. to provide enough material on quantitative maintainability modeling and statistical analysis so that the reader can understand the implications of writing maintainability requirements in various ways and be able to determine when the performance of the system conforms to the maintainability requirements.

It is important to realize that maintainability and reliability are not independent. As we will see in detail in chapter 2, decisions about maintainability also have consequences for system reliability. For instance, the architecture you choose for the field-replaceable parts of the system influences the duration of the out-of-service period incident on the failure of such a part. This in turn influences the system availability, a key measure[2] of system reliability. Design for maintainability will consider these implications and help guide the systems

[2] In chapter 3 and thereafter, we will refer to availability as a *reliability figure of merit*.

engineer toward effective maintainability requirements when considering system reliability, maintainability, and cost together. And of course, it is still necessary to assess after deployment the degree to which the maintainability requirements are being met, not only to provide the facts necessary to adjudicate customer claims, but also to provide a factual foundation for management and improvement of system maintainability and of the process by which maintainability requirements are created.

1.3.3 Supportability Engineering

By now, you know what's coming. Like maintainability, supportability is not necessarily an end in itself: because system unavailability is directly proportional to the duration of outages, and poorer supportability increases outage duration, supportability plays a direct role in improving system reliability. Therefore, the importance of proper supportability is not only in its opportunity for decreased system cost but also in its implications for system reliability. The supportability part of this book, Part III, emphasizes the optimal allocation of supportability resources to improve reliability while paying attention to supportability cost. Accordingly, the two goals of the supportability engineering part of this book are

1. to introduce the reader to design for supportability through study of the key factors influencing supportability and
2. to provide enough material on supportability optimization and statistical analysis of supportability data so that the reader can understand the implications of writing requirements in various ways and be able to determine when the performance of the product or service conforms to the supportability requirements.

As noted before about maintainability, it is important to realize that supportability and reliability are not independent. It is possible to create a set of system requirements for reliability and supportability independently of each other, but doing so ignores the synergies that are possible from considering these together. Supportability engineering and design for supportability provide a clear application of optimization techniques that we will introduce in this part of the book.

1.4 AUDIENCE

1.4.1 Who Should Read This Book?

While practicing systems engineers and students of systems engineering are the primary audience for this book, others in the technological systems community too may benefit from it. Customer representatives, who may have yet-unformed ideas about reliability as part of a desired features list, may benefit from understanding how the systems engineering process takes informal, imprecise ideas for desired reliability and makes specific requirements from them and maps these requirements to each desired reliability feature. Reliability, maintainability, and supportability engineering specialists may find new material of interest to them, particularly in the chapters concerning data analysis techniques for comparing field results with requirements. These chapters may also be useful to risk management teams and management in general. Design and development engineers will find organized, systematic treatment of design for reliability, design for maintainability, and design for supportability—key disciplines needed to ensure that sustainability requirements become fulfilled.

1.4.2 Prerequisites

No resource of this kind can hope to be completely self-contained. Readers will need to bring some background in certain subjects to gain the greatest benefit from this book. Foremost among these is a facility with statistical thinking. Almost all the language used in the areas of reliability, maintainability, and supportability engineering is based on a probabilistic and statistical approach. The quantities treated in these disciplines are almost never deterministic and require the language of probability and statistics to deal with properly. While this book does not expect you to be an expert probabilist, some maturity with probability concepts, stochastic processes, and statistical inference is assumed. For probability and stochastic processes, familiarity at about the level of [3] is helpful. For statistics, consider Refs. 2 or 6. To help with concepts or models of this kind that may be unfamiliar, other references are provided with the relevant chapters so that additional explanation may be readily obtained.

In addition, this book assumes a certain familiarity with, and maturity in, quality engineering. Systems engineers are vital contributors to the success of a product or service by crafting the requirements that are needed to drive development of a product or service that will fulfill customers' needs and that customers will find attractive and compelling. We will not dwell on the development of quality engineering methods in this book, but rather will use these concepts and methods when needed. A good introduction may be found in Ref. 13.

1.4.3 Postrequisites

Continuing the thought that no resource of this kind can hope to be completely self-contained, readers also should be aware that there are many places in the book where we discuss things that systems engineers are advised to consider doing (or contracting to have done) but that the details of those things are not given. For instance, in chapter 12, we discuss determining the proper size of an inventory of spare parts as an important part of the systems engineering responsibility in system support. Two approaches to solving this problem are mentioned, one based on minimizing the stockout probability and another based on maximizing the system availability, within budget constraints. However, we do not discuss the details of either of these approaches in the book because they are adequately covered elsewhere, either in other textbooks or in papers in the literature. In either case, references are provided so you can acquire these techniques if you so desire, but they will normally be the province of specialists on the design team. The systems engineer's responsibility is to see that these tasks are attended to, though, in most cases, she will not carry out the tasks herself. The structure of this book largely follows this division of labor. This book emphasizes sustainability engineering tasks that need to be carried out in order to develop a successful system or service without necessarily delving deeply into the operational details of the tasks. Carrying out the tasks will usually be the responsibility of some specialist engineers on the development team, and they will use other resources for their needs.

There are a few exceptions to this rule in this book. These were chosen mainly for pedagogical value or are new approaches to sustainability engineering tasks recommended by the author. For example, in chapter 13, we discuss design optimization of a repair facility as part of design for supportability. While the use of stochastic network flow models for this task is certainly not unheard of, it is unusual enough that a brief introduction to the technique is offered in chapter 13. As always, references are provided for further exploration if needed.

Postrequisites *is* a neologism. We hope it will help you remember our emphasis on the sustainability engineering tasks systems engineers need to make sure are done while in many cases referring to other resources for the details of tasks which are mostly the province of specialist engineers on the development team.

1.5 GETTING STARTED

If you are a working systems engineer, collect a bundle of sustainability requirements that you may be familiar with. As you read through the book, or carry on with a course based on the book, examine the requirements you have collected and see if they conform to the recommendations presented. Understand the similarities and differences using the material presented in the text. Experiment with possible alternatives and improvements. Send feedback to the author.

If you are new to systems engineering, it is our hope that by following the precepts given in this book, you will rapidly mature in the sustainability disciplines to the point where you can be counted on to always create clear and effective sustainability requirements.

1.6 KEY SUCCESS FACTORS FOR SYSTEMS ENGINEERS IN RELIABILITY, MAINTAINABILITY, AND SUPPORTABILITY ENGINEERING

1.6.1 Customer–Supplier Relationships

Setting requirements does not happen in a vacuum. To set requirements effectively, it is important to understand the customer–supplier relationships that are at play in this process.

- The primary customer–supplier relationship to consider is, of course, the customer for the system (or service) purchased from its supplier. This customer is an external customer who has a lot to say about whether the system or service will prove profitable to the vendor. Requirements must flow from a deep understanding of this customer's needs. Systematic procedures are available to elicit these needs and to ensure that any requirements developed can be traced directly to them. These procedures include
  - quality function deployment [1],
  - the "House of Quality" [4], [8], and
  - Kano analysis [5], among others.

  A detailed discussion of these techniques is outside the scope of this book. Our intention is to help you develop effective requirements that truly promote the design and development of a product or service that fully and profitably meets the customer needs once they have been determined by these (or other) techniques. Of course, these procedures can also help in the development of sustainability requirements, and it is recommended that they be used in this development to the extent possible.
- The systems engineer is a supplier to the rest of the design and development team. The product supplied by the systems engineer is the set of requirements for the system (or service). The design and development team, an internal customer, needs clear direction from systems engineering, and the techniques we discuss in this book promote that clarity. A good customer–supplier relationship here includes process management for the requirements development process incorporating a robust feedback mechanism for improvement not only of individual requirements but also of the process by which they are generated.
- The systems engineer is also a supplier to management of information about timeliness of requirements development, appropriateness of requirements as related to customer needs and product or service profitability, scope creep, etc. Management needs in this relationship include clarity and forthrightness. Communication skills form the basis for being able to fulfill these needs well. This book will help you become a more skillful communicator in the sustainability disciplines. We provide many

"language tips" at various places in the text that help clarify communication points so that you can clarify them for your customers.

- The systems engineer is a customer for sustainability engineering specialists on the product or service design and development team. Systems engineers as a rule do not carry out all the modeling and analyses needed to support requirements development, and most often time pressures prevent their doing so except in extraordinary situations. Therefore, the systems engineer needs to learn how to be a good customer for this specialized information. That means having at his/her command at least enough knowledge about reliability, maintainability, and supportability engineering to be able to tell when the results submitted by specialists in these disciplines are reasonable, are products of appropriate modeling and analysis, and form a suitable basis for downstream verification of requirements from data collected during deployment. Accordingly, this book covers some of the basic ideas of quantitative modeling for each of the three disciplines. The intent is not to create specialists in reliability, maintainability, or supportability engineering, but rather to enable systems engineers to be good communicators and good customers of the suppliers of this specialty information.

1.6.2 Language and Clarity of Communication

In any technical discipline, the words we use come from ordinary English but usually carry more precise, restricted meanings. This can be a source of confusion because technical discourse uses the same ordinary English words without necessarily indicating that the precise, restricted meaning is being used. For example, in everyday speech, the word "reliable" usually means something like "able to be depended on to perform duties without fail." From this, the noun "reliability" stems and carries a corresponding meaning. But in reliability engineering, "reliability" has precise technical meanings that are narrower than its meaning in ordinary discourse. The specific definitions of "reliability" used in systems engineering are covered in Chapters 2, 3, and 4. When we talk about reliability with nonspecialists, including managers, we usually intend a wider meaning, something akin to "absence of failure." Leaving aside that "failure" is so far an undefined term (see chapter 2), nonspecialists will almost certainly not intend to use "reliability" as, for example, the probability definition in Section 2.2.5 or the survivor function definition in Section 3.3.2.3, and specialists may sometimes so use it and sometimes not, usually without warning. Finally, the word "reliability" is used as a portmanteau word for a system property that contains within it many possible specific criteria, such as availability, times between failures, repair times, etc., *besides* the probability definition. We refer to these as reliability effectiveness criteria (see chapter 2). Keeping this all straight is an important function of systems engineering and promotes clear and effective communication.

The position taken in this book is that it is the systems engineer who is best positioned to discern language problems of this kind and to sort them out so that all constituencies are clear on what is being said. This is a large and important responsibility. Accordingly, we emphasize learning the technical language of the sustainability disciplines and, while so learning, thinking about ways terms may be misunderstood by important constituencies: executives, managers, team members, and customers. Being conscious of possible misunderstandings helps the systems engineer anticipate and overcome the difficulties his/her various audiences are likely to have and become a great communicator as well as a great engineer.

The value of being able to keep everyone on the same page cannot be overstated. Therefore, we urge systems engineers to learn the different languages spoken by specialists and nonspecialists in the sustainability disciplines. You will find "language tips" in many places in this book where some help may be needed in sorting these out.

1.6.3 Statistical Thinking

The relevant quantities in reliability, maintainability, and supportability are not physical constants. They all come from measurements on populations of the systems to which these disciplines are applied. Consequently, they need to be understood in a statistical context, and it helps to have some familiarity with the basic concepts of statistics. In particular, we place a lot of emphasis on the notion of determining when two statistical quantities are "truly" different, that is, is the difference we observe (between a requirement for some quantity and an estimate of that quantity from operational data) explainable, with high probability, by chance fluctuations in the mechanism generating the data? Or is the difference "real" when we account for the sampling errors involved? Such reasoning is important when comparing the performance of a system against its quantitative requirements: while it is appropriate to respond to a difference that is determined to be significant, it is equally important to know when not to expend resources to make corrections when none may be warranted by the quality of our knowledge about the operational performance involved. These are basic principles of management by fact that apply whenever the quantities involved are statistical in nature, and systems engineers should be able to deal confidently with these matters, and to explain them to other stakeholders who are affected by decisions one way or another.

1.7 ORGANIZING A COURSE USING THIS BOOK

The book is organized so that Parts I, II, and III are mostly self-contained, so a course whose primary emphasis is on either reliability, or maintainability, or supportability can be constructed based on the appropriate part separately. In this case, a one-semester course can be constructed with most of the modeling chapters (3, 4, 8, 11,13) covered in depth. But it would be a mistake to study only the modeling aspects of these disciplines. The real benefit of studying sustainability engineering comes from application of the design for (reliability, maintainability, supportability) principles that are the subject of Chapters 6, 11, and 13. Consistent with the principles of quality engineering, we advocate for application of these principles early in the development process so that prevention costs may be managed and controlled while increasing the chance that the product, system, or service will be successful. As a consequence, a more valuable course could be constructed using the "design-for" chapters as a foundation and using the modeling chapters as supporting material. If a more general overview is desired, this can be done using chapter 2, and parts of Chapters 6, 10, 11, 12, and 13. In the services industry, Chapters 6, 8, and 9 are helpful. For "high-consequence" systems in which the consequences of failure are very serious, perhaps even life-threatening, consider using chapter 7 as a basis with supplementary material from Chapters 3, 4, 5, 8, 11, and 13. An overview course aimed at introducing systems engineers to the sustainability disciplines would draw from all three parts of the book, and to be able to fit this into one semester the modeling chapters can be touched on more lightly.

1.7.1 Examples

This book contains many examples, but not every concept or technique discussed has an example given. For example, the discussion of reliability budgeting in Section 4.7.3 proceeds at a fairly abstract level and does not contain a complete worked-out example. You will find other instances of this in most chapters. This is deliberate: the variety of possible applications is very large, and the author makes no pretense to being familiar with all of them. More importantly, these situations offer the instructor an opportunity to fill in with examples from her own experience and particular field of expertise. Instructors are encouraged to make the most of this opportunity by planning ahead for class discussion of an ample number of applications, drawn from their own experience, of the concepts and techniques presented.

1.7.2 Exercises

Each chapter contains exercises. These are an integral part of the presentation. Some exercises amplify or complete examples introduced in the text. Others give the reader an opportunity to try out some of the ideas and procedures presented in the chapter. Still others are of an advanced nature that may be suitable as research projects. These are marked with an asterisk.

1.7.3 References

Each chapter contains references to supplementary or source material for the ideas in the chapter. Some of the references are to the author's own work which has ranged widely over theoretical and practical aspects of reliability engineering. This field has grown so extensively that citation of all potentially relevant references is an impossible task. The ones chosen aim to provide historical context as well as foundational material and additional amplification for the material in the chapter.

1.8 CHAPTER SUMMARY

This chapter prepares readers to extract maximum value from this book. It tells the aims and scope of the book, but more importantly it tells who may benefit from it and how that benefit may be gained. The book does not aim to turn systems engineers into specialists in the sustainability disciplines, but rather aims to enable systems engineers, who usually do not receive specific training in the sustainability disciplines, to become successful and productive when dealing with that portion of their responsibilities that include reliability, maintainability, and supportability. We emphasize key success factors in this endeavor. These include understanding the customer–supplier relationships at play in systems engineering, clear and proper use of language, and a facility with statistical thinking.

REFERENCES

1. Akao Y. *Quality Function Deployment—Integrating Customer Requirements into Product Design*. New York: Productivity Press (a division of CRC Press); 2004.
2. Berry DA, Lindgren BW. *Statistics: Theory and Methods*. 2nd ed. Belmont: Duxbury Press (Wadsworth); 1996.
3. Chung KL, AitSahia F. *Elementary Probability Theory: With Stochastic Processes and an Introduction to Mathematical Finance*. New York: Springer-Verlag; 2006.
4. Clausing D, Houser J. The house of quality. Harv Bus Rev 1988;66 (3):63–73.
5. Kano N, Seraku N, Takahashi F, Tsuji S. Attractive quality and must-be quality (in Japanese). J Jpn Soc Qual Control 1984;14 (2):39–48.
6. Moore DS, McCabe GP. *Introduction to the Practice of Statistics*. New York: Freeman and Co.; 1993.
7. Park SH, Antony J. *Robust Design for Quality Engineering and Six Sigma*. Singapore: World Scientific; 2008.
8. Park T, Kim K-J. Determination of an optimal set of design requirements using house of quality. J Oper Manage 1998;16 (5):569–581.
9. Scherkenbach WW. *The Deming Route to Quality and Productivity*. Washington: CEE Press Books; 1988.
10. Taguchi G. Quality engineering in Japan. Commun Statist Theory Methods 1985; 14 (11):2785–2801.
11. Tortorella M. Service reliability theory and engineering, I: foundations. Qual Technol Quant Manage 2005;2 (1):1–16.
12. Tortorella M. Service reliability theory and engineering, II: models and examples. Qual Technol Quant Manage 2005;2 (1):17–37.
13. Wadsworth HM, Stephens KS, Godfrey AB. *Modern Methods for Quality Control and Improvement*. New York: John Wiley & Sons, Inc; 2002.
14. Wu Y, Wu A. *Taguchi Methods for Robust Design*. New York: ASME Press; 2000.

2

Reliability Requirements

2.1 WHAT TO EXPECT FROM THIS CHAPTER

This chapter is the foundation for the first third of this book dealing with reliability. The chapter covers various uses of the word "reliability" in ordinary conversation and in its specialized uses in engineering. This prepares the way to study reliability requirements. We explore what makes a good reliability requirement and show how appropriate attention to reliability, maintainability, and supportability can create a virtuous circle of improvement and lower cost. Then we move to a more detailed examination of reliability concepts, including reliability effectiveness criteria and figures of merit. This enables us to review some examples of reliability requirements in four areas: products, flow networks, standing services, and on-demand services. The topic of interpretation of reliability requirements is important for proper comparison of performance with requirements, and some examples of comparisons are given here as a preparation for the more detailed coverage of this topic in Chapter 5. We introduce additional figures of merit and some of the statistical procedures that are covered in more detail in Chapter 5. As with all chapters in this book, this chapter closes with a discussion of best practices in creating reliability requirements and a brief summary of key points.

Reliability, Maintainability, and Supportability: Best Practices for Systems Engineers, First Edition. Michael Tortorella.

2.2 RELIABILITY FOR SYSTEMS ENGINEERS

2.2.1 "Reliability" in Conversation

Most people have a good idea of what "reliability" means in ordinary conversation. Usually, we mean something or someone is reliable if he/she/it can be counted on to do his/her/its job without fail, steadily, for as long as asked. Stated in this way, the meaning of "without fail" is of paramount importance. Usually, in conversation, we take that to mean he/she/it does correctly what he/she/it is supposed to do. This understanding serves us well because as we will see, more precise use of these terms in systems engineering formalizes these ideas, thereby enabling important relationships to be exposed and studied. This chapter is devoted to amplifying the notion of reliability, defining it clearly, and exploring some of the implications of the choices we have made.

2.2.2 "Reliability" in Engineering

Engineering works because its concepts are clearly defined and, very often, quantitative. The reliability engineering framework follows closely from the ordinary sense we have of "reliability" as described earlier: requirements are what he/she/it is "supposed to do"; "failure" is a violation of a requirement, "steadily, for as long as asked" becomes the time period over which failure-free operation is desired. The formal definitions align closely with these ideas.

If you are comfortable with this metaphor, it may help to think of "reliability," in the systems engineering context, as a primitive term in a formal system. That is, in systems engineering we endow "reliability" with a special meaning that is more precise than its meaning in our ordinary day-to-day conversational usage. The next sections are devoted to clarifying these notions.

2.2.3 Foundational Concepts

2.2.3.1 Attribute requirements

The subject matter of systems engineering is *requirements*. Requirements are statements about functions a system[1] or service is supposed to perform and properties that a system is supposed to possess that users and customers may consider necessary or desirable. These include functional, performance, physical, and safety characteristics. Their related requirements will be referred to as "attribute requirements" to distinguish them from sustainability requirements (reliability, maintainability, and supportability), which concern themselves with violations of attribute requirements and correction of those violations. In brief, functional requirements concern what a system is supposed to do; performance requirements concern how efficiently the system does

[1] As a reminder, we interpret the word "system" broadly to encompass tangible products like airplanes and computers as well as less tangible objects like software applications.

them; physical requirements pertain to the appearance of the system in the world, encompassing such things as size and weight; and safety requirements concern the protection of life and limb while the system is used. We may think of these requirements as "static" and the subject matter of *quality* which is concerned with the degree to which these attribute requirements are met by the system or service as designed.

Systems engineers create requirements from a deep understanding of customers' needs and desires and a balance of these with the cost of development to meet them. The appropriateness and completeness of a set of requirements is judged precisely on how well they capture these customer needs and desires, and whether the resulting product, system, or service is profitable to the supplier. Requirements are in turn used by downstream members of the system development team to guide design, testing, validation, and verification, and other development activities so that the end product of the development process is a system, product, or service that fulfills the customers' needs and desires to a degree necessary to ensure its acceptability to the customer and profitability to the supplier. For the purposes of this book, we consider that the system's attribute requirements have been acceptably defined. In practice, this is sometimes not the case; everyone can name examples of products and services that failed in the marketplace because their systems engineers misunderstood the customer and consequently got the requirements incomplete or just plain wrong. However, we postulate the ideal situation so that we can focus on the primary tasks covered in this book, namely learning the principles of sustainability engineering and management, and the creation, evaluation, and tracking of sustainability requirements.

Several tools are available to the systems engineer to help acquire the knowledge needed about customer needs and desires so that good requirements may be developed. These include quality function deployment (QFD) [7], also known as House of Quality [18], and Kano analysis [5]. While alignment with customer needs and desires is absolutely of paramount importance for systems engineers, for good requirements are impossible without it, the details of these techniques are outside the scope of this book even though they are useful in helping to determine not only attribute but also sustainability requirements. Customers and users are interested in the frequency and duration of incidents of violations of attribute requirements, and their needs and desires for the frequency and duration of such incidents of violation are properly the subject of appropriate requirements themselves (these will be the reliability, maintainability, and supportability requirements). These tools are mainly used to develop attribute requirements, but systems engineers may use these tools to develop sustainability requirements also. Once attribute requirements are established, we may consider the question of sustainability—how frequently are the attribute requirements violated, and for how long do such conditions persist? The tools may be used to ascertain customers' needs and desires for system reliability, but it is possibly even more important to get the attribute requirements correct first because a system that does not do what its users want and need it to do will not be improved by its doing those things very reliably. Then reliability requirements can be worked out on a sound basis.

2.2.3.2 *Failures*

Broadly speaking, reliability concerns *failure*. By itself, "failure" is too broad a term to be useful. In this section, we make the meaning of "failure" precise for use in reliability engineering.

Definition: A failure is an action or omission in which one or more system requirements are violated.

We will amplify this concept in the chapters to follows. For now, note that an important implication of defining failure this way is that requirements must be written in such a way that it is possible to discern clearly when they are not being met. This fundamental principle of systems engineering is more readily implemented when requirements are stated in quantitative terms. Fortunately, many concepts of interest in reliability, maintainability, and supportability readily lend themselves to expression in quantitative terms, and many examples will be given throughout the book.

When a failure occurs, the system enters a state in which it does not perform, or it inefficiently performs, one or more of its functions. That is, during this period of time, one or more attribute requirements continue to be violated. We say the system is in a *failed* or *degraded* state. This condition may persist for some time.

Definition: An outage is the period of time following a failure during which the system is in a failed or degraded state.

Throughout this book, we will use "failure" to indicate the change of the system from an operating state to a failed or degraded state, that is, a failure is something that takes place at a particular, distinct instant of time. When a failure occurs, an outage begins, and the outage persists for some length of time until a recovery is completed and the system is returned to normal operation (all attribute requirements are being met). Further discussion may be found near Figure 2.1.

"Hard" failures and "soft" failures

It is sometimes assumed that the content of reliability engineering deals only with failures that look like a complete cessation of system operation, sometimes called "hard failures." This is a common but far from a fruitful point of view. Many violations of system requirements may occur that do not look like complete cessation of system operation. For example, in a transaction processing system, there is often a delay requirement that looks something like "the mean system response time after a user request will not exceed 500 milliseconds under nominal load." This is a performance requirement. Leaving aside for the moment the precise specification of "nominal load," when certain subsystem failures occur,[2] it may be possible for the system to continue

[2] For example, one server (completely) fails out of a bank of seven servers being used to process transaction requests.

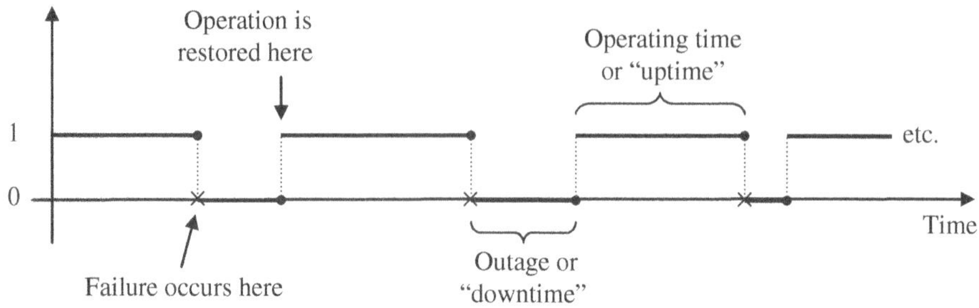

Figure 2.1 *History diagram illustrating failure and outage.*

processing user requests with a mean response time of greater than 500 milliseconds for some requests. The system is still providing service (it has not completely ceased operation), but a system failure has occurred because this requirement is not being met. Users might not be aware that this failure has occurred, particularly if the mean response time remains less than, say, 600 milliseconds; but at some point in the development of the system, the systems engineers decided that keeping user response mean delay below 500 milliseconds was what customers wanted, and crafted a requirement to that effect. Whether, under these circumstances, this was an appropriate requirement is a subject for discussion in the area of developing requirements that faithfully capture the letter and spirit of what the customer wants; for our purposes, here is an example of a requirement that is violated even though the system is still providing some service, albeit at a "degraded" level. Such instances are sometimes referred to as "soft failures." These terms are almost universally recognized in reliability engineering, but prudent practice advises that some effort be spent on ensuring that all parties to the conversation interpret them the same way in any particular case.

Reliability engineering is most effective when the concept of "failure" is not confined to "complete cessation of system operation" but includes violation of some (any) system attribute requirement. For example, the software engineering community has debated the notion of whether safety-related software failures are distinct from other types of failures. Leveson [17] maintains that they are qualitatively different from other kinds of failures. The point of view informing this book is that safety failures, as violations of particular attribute requirements (namely, safety requirements), are failures that can be fruitfully dealt with using the methods of reliability engineering. The key to resolving this disagreement lies in understanding how failures are

- avoided by design engineering actions and
- remedied once they occur.

It is certainly true that the consequences of safety-related failures, which may range all the way to injuries to people and loss of life, may be more serious than the consequences of other kinds of failures. The position adopted in this book is that, nonetheless, safety-related failures are failures whose prevention and remediation still fall within the scope of reliability engineering and management. A safety failure is a violation of a safety requirement,[3] and as such safety failures may be avoided and remedied appropriately by the methods of reliability engineering and management described in this book. It is precisely the application of the principles and practices of reliability engineering described here that promotes effective realization of all attribute requirements, including safety requirements, both at initial system shipment and thereafter throughout the system's useful life.

2.2.4 Reliability Concepts for Systems Engineers

The formal use of the word "reliability" in the systems engineering context encompasses three important aspects of system operation:

- The violation of one or more system requirements,
- The conditions under which the system operates (which may vary from time to time) and prevailing up to and including the time when the violation occurs, and
- The time at which the violation occurs.

We have defined *failure* as the violation of one or more system attribute requirements.[4] Every system attribute requirement presents an opportunity for a failure to occur, that is, each attribute requirement contains within it one or more failure modes, or different ways in which that requirement can be violated and that provide evidence to the user that such a violation (failure) has occurred. For instance, consider a real-time processing system such as an online ticket-selling application. The application may have a delay requirement such as the following: the system response time to a customer request shall not exceed two seconds when the demand is 100 requests per minute or less. At any time, the system response time exceeds two seconds, and the demand is less than 100 requests per minute, a violation of this requirement has occurred. The user can detect a response time of greater than 2 seconds; this is concrete evidence that this requirement has been violated. Only the system operator can detect whether the offered load (demand) was greater than 100 requests per minute when this event occurred. Even if a user can't tell whether this failure has occurred, excessive delay may annoy the user. If 2 seconds is an excessive delay in the sense that it causes user annoyance, or if

[3] Regrettably, in many instances, safety requirements are only implicit. It is preferable to make any such safety requirements explicit so that appropriate attention is drawn to them and effective actions are taken to ensure that they, like all requirements, are met.

[4] We are going to disallow, on infinite regress grounds, the notion of calling a violation of a reliability requirement a failure.

users become annoyed because delays of less than 2 seconds occur too frequently (and user satisfaction studies would be required to establish these propositions), this would provide motivation for a reexamination of the requirement. Important: while this is a *performance* requirement, a violation of the requirement is a failure that is properly within the scope of reliability engineering. The sustainability engineering aspect involves the frequency and duration of violations of this requirement, that is, failures and outages in this particular failure mode.

> **Language tip:** Synonyms for "reliability." It is not unusual to hear terms like "dependability," "longevity," "durability," etc., in discussions concerning system operation. Systems engineers need to be aware that any dictionary will be able to provide common discourse definitions for these terms, but they have no universally accepted meaning in the formal system of reliability engineering. You are of course free to use these terms in reliability engineering provided they are unambiguously defined and used consistently throughout your study and agreed to by all stakeholders. Indeed, because they lack universally accepted meaning, you are free to define them as you may need. "Reliability" has several meanings in ordinary discourse and is defined precisely within the formal system, and this definition is widely accepted, as discussed earlier and elsewhere in this book, and so should be used in this manner without modification. "Dependability" and the like have no universally accepted meaning within the formal system. Sometimes, "dependability" is used as a synonym for "reliability," or for some more-encompassing concept, but the definition is not universally agreed. Be alert for variant meanings: when in doubt about spoken or written uses of words that sound like they want to mean "reliability" but are not universally accepted (which includes the standards in Ref. 16), get confirmation from the speaker or author as to the precise way in which terms are being used. It is easy to get bogged down unless the basic terms and their meanings are clear to all parties to the conversation. The point is that you may choose to use any words you like, subject to the provisos that
>
> - if the word has a precise, universally accepted meaning (in the formal system), then it should always be used with that meaning, and
> - if a word lacks universal agreement about its precise meaning, all parties to the conversation need to agree on the precise meaning of the word as it is being used in the current context.

2.2.4.1 Reliability requirements introduction

With the understanding that reliability deals with violations of the system's functional, performance, physical, or safety requirements, we may also consider that requirements may be written for reliability. Reliability requirements, while distinct from the system's attribute requirements, must necessarily refer to the system's attribute requirements because reliability requirements pertain to violation of the system's attribute requirements.

Reliability requirements exist because the system's customers or users are vitally interested in

- how often do failures occur,
- for how long do the failure-caused out-of-service or degraded-service conditions (outages) persist,
- life cycle costs, and
- what is the impact of failures on the customer and on the business.

While all teams (systems engineering, design/development, operations, etc.) care about all four issues, different teams place different emphasis on each. Typically, designers care mostly about duration (How is a failure event detected? How is a failure event recovered, and how long does this take?) and impact (how can the impact of this failure on users be reduced?), and then frequency and costs. Operations care mostly about frequency (how often do we have to enact manual recoveries?) and duration, then impact and costs. Supply chain care mostly about costs. Systems engineers care mostly about what the customer cares about.

While they may also serve other purposes, the primary reason reliability requirements are needed is to control the *frequency* and *duration* and *impact* of failures and outages. We study in this chapter the methods needed to craft reliability requirements that are successful in this sense.

Example: The Federal Communications Commission (FCC) once defined a Public Switched Telephone Network (PSTN) outage to be reportable if the **P**otential **I**mpacted **U**ser **M**inutes (PIUM) exceeded 900,000. PIUM were defined as the outage duration in minutes times the maximum number of users that could have been affected. The FCC wanted to reduce the number of reported incidents per year. Root cause analysis showed a single, cheap, component was responsible for a disproportionally large number of reportable outages. To meet the objective of reducing the number of reportable outages, one could reduce the frequency of outages by using a more expensive component, reduce the duration of outages by changing operations staffing (a very expensive alternative), or use a second cheap component and split the number of users per component in half. Using the second component and splitting the users per component was the cheapest solution. Note that a systems engineer would point out that while the number of FCC reportable outages would be dramatically reduced, the end user would not benefit at all.

2.2.4.2 Reliability and quality

In quality engineering, *quality* is understood as the degree to which requirements are met. The intent of this definition is to capture a snapshot of the system as designed and produced and when first delivered to the customer. Not meeting one or more requirements at the time of delivery to the customer constitutes diminished quality of the system. In most cases, though, the customer will continue to use the system for some period of time after initial delivery and under conditions that may differ from those prevailing at the time of initial operation. It is possible that a system may meet all its requirements at the time of initial operation, while after the passage of some time or with the application of some different conditions, some requirement(s) may not be met. It is this latter situation that we intend to cover with the term "reliability": reliability includes a time dimension that quality does not. We may summarize this discussion by saying that reliability is the persistence of quality over time when the product, system, or service is operated under the conditions prescribed in the requirements. This commonsense understanding of the word serves as a foundation for our more-precise definition within the formal system that follows.

2.2.5 Definition of Reliability

The definition of "reliability" generally accepted by the engineering community [16] is essentially a distillation of the earlier discussion.

> **Definition:** Reliability is the ability of a system to perform as designed, without failure, in an operational environment, for a stated period of time.

More briefly, reliability is the ability of a system to operate correctly under specified conditions for a specified period of time. This statement contains four key concepts:

1. Ability: A characteristic of the system that encompasses all those properties of the system that enables it to more readily operate with fewer failures and shorter outages. We can learn about reliability by focused engineering activities such as modeling, testing, and analysis of failure data from systems in operation. In particular, creation of reliability effectiveness criteria (Section 2.4.1), quantitative expressions of various features of the abstract ability "reliability," helps systems engineers and development teams undertake effective actions to promote and manage reliability. While "ability" may be considered an abstract property, these measurement opportunities allow it to be measured, managed, and improved as needed.
2. As designed, without failure: As described earlier in detail, this is what is meant by correct operation. All system attribute requirements are satisfied, and the system performs as intended, according to its attribute requirements.
3. Operational environment: As we will see in later chapters, the environment (heat, vibration, offered load, user skill, etc.) in which a system operates has a bearing on whether the system will meet its requirements. It may do so in certain environments but not in others. It is therefore important to specify the environmental conditions under which a system is supposed to operate correctly.
4. Period of time: by contrast with quality, in which we are interested in the correct system operation according to requirements at the time of

completion of manufacture and shipment to the customer, reliability is concerned with continued correct operation of the system as time passes. Therefore, the definition of reliability includes a specification of the period of time over which correct operation is desired.

It is worth repeating the summary that **reliability is the persistence of quality over time when the product, system, or service is operated under the conditions prescribed in the requirements**.

Language tip: We reinforce again the notion that the word "reliability" is used in many senses in ordinary discourse, but it has precise meanings in the formal system we use as the framework for reliability engineering. In later chapters, we will see additional general usages and quantitative definitions for "reliability" when used as an effectiveness criterion, a figure of merit, or a metric for nonmaintained or maintained systems. The use of the same word for different purposes, a common practice in this field and one that must therefore be confronted and rationalized carefully, introduces the possibility of confusion. Systems engineers, who have the most comprehensive understanding of the system as a whole, should develop the skill of detecting the context and meaning of the different uses so that they may be able to use the concepts correctly and explain them to other stakeholders, including suppliers, customers, managers, and executives.

Requirements tip: When constructing a reliability requirement, make sure that the three key elements of the definition are included:

- Definition of correct operation (and therefore, failure) according to the attribute requirement(s) to be covered by the reliability requirement,
- The conditions under which the requirement is supposed to prevail, and
- The period of time over which the requirement is supposed to be fulfilled.

We will note that most often in practice, reliability requirements do not address particular attribute requirement(s) explicitly. In those cases, the only reasonable interpretation is that the reliability requirement is intended to apply to all attribute requirements. If that is not what you intend, revise the wording of the requirement appropriately. To avoid any misunderstanding, even if the reliability requirement is intended to apply to all attribute requirements, it is best to say so explicitly.

Example: You have been assigned to develop reliability requirements for a smartphone. The service offered by the wireless carrier includes a certain number of functions that the smartphone is supposed to perform, such as make voice calls and access the various data services (Internet, GPS, etc.) offered by the carrier. Here is an example of a reliability requirement for the smartphone: "The product will carry out all contracted functions at their nominal performance values when the ambient temperature is between –10°C and 40°C for a period of no less than 10,000 hours." Does this fulfill the definition of a good reliability requirement? Consider the three key elements of the definition as described in the requirements tip earlier.

1. Is "correct operation" defined? The part of the requirement here pertaining to "correct operation" is the phrase "carry out all contracted functions at their nominal performance values." Is the meaning of this phrase clear? Can the systems engineer list the "contracted functions" (i.e., those that the carrier offers and the customer ordered and pays for)? Is an unambiguous "nominal performance value" defined for each contracted function? Does "correct operation" for the smartphone encompass all the functions a customer may expect, even besides those contracted for with the carrier? If the answer to any of these questions is "no," then "correct operation" is not adequately defined for purposes of this requirement.
2. Are the operational conditions specified? Certainly, some operational conditions are specified: the smartphone is required to operate correctly when the ambient temperature is between −10 and 40°C. Is this enough? The requirement leaves all other possible environmental conditions unspecified, and therefore uncontrolled. One way to interpret this omission is to say that proper operation is required under *any* conditions of humidity, vibration, shock, immersion, barometric pressure, etc. This may be perfectly satisfactory if the manufacturer is confident that the smartphone is capable of such operation. However, it is dangerous to leave important conditions tacit or unaddressed, not least because of possible legal difficulties later on if a disagreement arises between supplier and customer. More often, though, a realistic assessment of the smartphone's capabilities would lead the systems engineer to specify a more restricted range of values for each possible environmental variable that is anticipated to be encountered in practice. We explore in Chapter 3 some quantitative models for how reliability is influenced by environmental conditions.
3. Is the period of time specified over which failure-free operation is desired? In this example, a period of "no less than 10,000 hours" is specified. Note that it is not possible to guarantee that every smartphone in the population will operate for more than 10,000 hours without failure. The causes of failure, and the users' modes of operation of the smartphone, are too varied and too numerous to anticipate completely. Consequently, reliability engineers adopt probabilistic and statistical models to help with quantitative characterization of reliability. Under this paradigm, the most one can hope for is an estimate (from operational or test data or from a predictive model) of the probability that the period of failure-free operation is at least 10,000 hours. This is explored more thoroughly in Section 2.6. It might be preferable to write the requirement as "the probability that the smartphone shall operate for a period of at least 10,000 hours is 0.98."

Requirements tip: Operational time and calendar time. In constructing a reliability requirement, consider that the system may not operate continuously. There may be times when the user does not wish to use the system and turns the power off or otherwise causes it to cease operating.[5] Accordingly, when time duration is specified in a reliability requirement, systems engineers and their customers need to be aware whether (cumulative) operating time or calendar time is intended. Most often, reliability requirements are constructed based on operating time. That is, if the requirement does not state whether it applies to operating time or calendar time, the usual assumption is that operating time is intended.[6] It is perfectly legitimate to write a reliability requirement in which the time specified is calendar time, but interpreting and verifying such a requirement requires a way to relate operating time to calendar time (i.e., a quantitative understanding of how the customer is going to use the system). Either is acceptable provided that all parties to the requirement understand which concept of time is being used. The distinction between operating time and calendar time also has ramifications for the design and analysis of warranties because warranties are almost always specified in terms of calendar time. See Section 3.3.7. Also see Chapter 5 for further discussion of warranties and the Exercises in Chapter 9 for some practice relating operating time to calendar time.

Requirements tip: Other markers of aging besides time. In reliability engineering, the word "aging" is used to indicate progression of a system to failure, usually because of the passage of time. However, some systems progress to failure not simply because time passes (or age increases), but by the action of some other insult. For example, an ordinary household wall-mounted light switch is a fairly simple electromechanical system that is unlikely to fail if unused. Cycling the switch on and off introduces mechanical wear on the toggle pivot and on the electrical contacts that makes it more likely to fail if it is operated more often. The number of operations the switch undergoes is a better indicator of progression to failure than is the simple passage of time. Many other nontemporal indicators of progress to failure come up in electrical and mechanical systems: number of compressor on–off cycles, number of pieces worked for a milling machine cutting tool, etc. As with the relation between operating time and calendar time, if it is desirable to express the reliability of the system in terms of time, it is necessary to relate the number of such marker operations to (calendar, usually) time. For this purpose, information about how frequently these operations are initiated is necessary. The system reliability may then be expressed either in terms of time or in terms of the number of operations. The choice can be made by referring to the language the customer uses.

5 It is commonly assumed that when a system is unpowered, it does not age, or accumulate time against any clock measuring time-to-failure. While this may seem a reasonable assumption, it should be checked in each instance. For example, the humidity and salt spray characteristic of marine environments in many cases cause damage to some types of electronics even if no power is applied.

6 Of course, if there is any doubt, state explicitly which is intended.

2.2.5.1 *Many uses of the same word*

The same word, "reliability," is also defined in the engineering literature as "the probability that a system will perform as designed, without failure, in an operational environment, for a stated period of time." So in addition to "reliability" as an abstract quality, we have "reliability" used also as a numerical concept (in the language of Section 2.4.1, this usage of "reliability" is a reliability figure of merit). We see here the first, but not the last, example of reuse of terminology in this field (see also the "reliability function" in Section 3.3.2.2). "Reliability" is also used as a general-purpose word to encompass all the concepts connected with the frequency and duration of failures and outages, as in "reliability engineering" and "design for reliability." Because the same word is used for different purposes, it is important to be able to detect which meaning is in use in any particular instance. Systems engineers are in the best position to help others in this because of their holistic view of the entire system and its development process.

2.2.6 Failure Modes, Failure Mechanisms, and Failure Causes

The occurrence of a failure is by definition a violation of some system requirement(s). How do you tell when this may have happened? Any overt event detectable by a user indicating that a system requirement has been violated is a *failure mode*. For instance, imagine you are driving an internal combustion automobile. The engine suddenly stops rotating and your forward progress ceases. Cessation of forward progress is a failure: presumably, there is a system requirement for the automobile that incorporates the use of the automobile to move from place to place on roads, and cessation of forward progress (unless deliberately initiated by the user, such as through braking to a stop in the course of normal operation), constitutes a violation of this requirement. In this case, the user can readily tell that a failure has occurred. The failure mode is that forward progress ceases.

Once a failure mode is known, reliability engineering may be applied to discern the cause(s) of the failure and apply suitable countermeasures. The *failure mechanisms* are underlying conditions in the system whose occurrence or presence change the system from an operating condition (no failures occur) to a failed condition (one or more failures occur). We can consider these as causes of the failure, but it is best to reserve the phrase *failure cause*(*s*) for the root cause(s), the last answer(s) in the "why?" chain of root cause analysis, because this is where countermeasures are most effectively applied. Root cause analysis is a process of continually asking "why?" whenever a reason is uncovered. Root cause analysis undertaken to uncover failure mechanisms continues until at least enough understanding is reached to be able to apply sensible countermeasures. Root cause analysis is facilitated by the use of Ishikawa or "fishbone" diagrams [26]. Fault tree analysis (Chapter 6) is a formal procedure that uncovers failure mechanisms and failure causes associated with each failure mode. In the automobile example, the proximate cause of cessation of forward progress was that the engine stops rotating. There may be many reasons why the engine may stop rotating, so it is not yet possible to propose an effective countermeasure. Suppose in this case that the reason the engine stopped rotating is that the timing chain has broken. This is certainly a failure mechanism. Is it enough to apply an effective countermeasure? It may depend on the audience. The owner or operator of the automobile could have the engine repaired by having the timing chain, and any other parts that may have been damaged consequentially, replaced. However, if there was a reason the timing chain broke (perhaps the owner did not adhere to the manufacturer's recommended replacement schedule for the timing chain, or did not maintain proper engine lubrication), merely replacing it without correcting the next layer of failure mechanism will cause the failure to happen again (perhaps after some more time has passed). If the audience is the manufacturer of the vehicle, a broken timing chain during testing is an opportunity to learn more about whether the timing chain specified is strong enough to withstand a stated period of "normal" operation. Finally, the failure cause(s) is the end result of the root cause analysis. In the example, a failure cause could be lack of proper engine lubrication, and an effective countermeasure would be: create a reminder scheme (email, text message, postal mail, or in-vehicle messages) to help owners keep to the recommended schedule of oil changes.

The reasoning applied in this example, and in general in the process of determining failure mechanisms and root causes from failure modes, is the same reasoning used in fault tree analysis (Section 6.6.1), a qualitative design for reliability technique that helps make a system more reliable by taking a systematic approach to anticipating and avoiding, or managing, failures. Fault tree analysis is simply a disciplined application of deductive reasoning in a more comprehensive setting. It aims to uncover the root causes of failures so that suitable countermeasures can be applied to prevent the root causes from occurring, thereby preventing the consequences of these root causes from happening as well.

A fruitful analogy can be developed with the language of illness. A failure mode is like the symptoms a person experiences when ill. They are the overt signals that something has gone wrong and the person is no longer healthy. For instance, a person may experience a fever or abnormally high body temperature. This is an overt signal that something has gone awry in the person's body. Usually, a fever indicates that there is an infection somewhere in the person's body. The infection, the condition in the body leading to the fever, is analogous to a failure mechanism. It is an underlying reason for the fever. In turn, the disease causing the infection is analogous to the failure cause. In the language of quality engineering, it is a *root cause*. Medical professionals are trained to interpret symptoms and use them to uncover the underlying cause of the problem, namely, the disease afflicting the patient, and apply a suitable cure if possible. In the same fashion, reliability engineers, teaming with other experts in the system's operation, endeavor to discover the failure mechanisms and root cause(s) associated with each system failure mode so that appropriate countermeasures may be taken.

Later, during the discussion of *design for reliability* in Chapter 6, we will discover that not every potential failure mode need receive corrective action of its root cause(s). There may be some failures that a system provider may choose to allow to remain in the system. As always, this is an economic decision that turns on the systems engineer's deep understanding of the consequences of the failure. For example, it may be judged too expensive to apply countermeasures for a failure mode that occurs very infrequently or has only minor consequences when it does occur. Some failure modes may cause little or no disturbance to a user in some circumstances. For instance, the failure of a power supply bypass capacitor (if it fails "open" rather than "short") in a radio receiver may cause a slight increase in the noise figure of the receiver. If this increase is not enough to cause a violation of the receiver's noise figure requirement, the designer may choose to "settle for" a maximum (positive) number of such capacitor failures over the population of receivers manufactured and over their designated service life rather than employ a higher reliability (and likely more expensive) capacitor in the bypass application. In practice, a probability model will be required to make a sensible decision in this case, because there are likely to be many such power supply bypass capacitors in a single receiver, and while the noise figure increase caused by a single bypass capacitor failure may be tolerable, the noise figure increase caused by several bypass capacitor failures may be intolerable (i.e., beyond the requirement).

A more disciplined approach to these questions is what is meant by balancing the reliability requirements against the economics of the system. The reasoning reviewed above takes us out of the realm of reliability engineering and management and into the world of consequences. Decisions about design for reliability, reliability improvement, and other activities that pertain to the relationship of the system to the world around it are aided by *decision theory*, a statistical method that makes use of knowledge about how the system is used and what the consequences of failure are to users and other stakeholders. Consequences of failure are captured in a mathematical construct called *utility* that is a basic element of a rational decision maker's toolkit. Utility is combined with the probability of various outcomes (in this case, reliability, or the probability of failure in the various failure modes of the system) to build a picture of *risk*. A rational decision maker can use the concept of risk to choose among various alternatives for the system, which means, in this case, which failure modes may be tolerated and which will receive attention to mitigate or eliminate. Full discussion of utility and risk as it pertains to making decisions about reliability is beyond the scope of this book, but Section 6.5.2 is a rudimentary form of this type of reasoning. Interested readers may consult Refs. 1, 20 for a more comprehensive explanation of these ideas and how systems engineers use them to make more informed decisions about system reliability. See also Ref. 6.

2.2.7 The Stress–Strength Model

Much of the early work in reliability engineering was devoted to the discovery of the physical, chemical, thermodynamic, etc., reasons for (hard) failure of tangible items such as bearings, electronic components, and so on. It was found that many reasonable explanations were instances of an abstract model of the interplay between the strength of an item and the stress placed on that item by the environment in which it operates. Failure occurs when the stress offered by the environment exceeds the strength of the item. For instance, suppose a complementary metal-oxide semiconductor (CMOS) integrated circuit may be able to continue to operate properly after an electrostatic discharge (ESD) of no more than 60V. Then an ESD shock in excess of 60V delivered by the environment will cause the integrated circuit to be damaged and no longer function. This body of knowledge gave rise to the notion of the *stress–strength model* in which failure of a device or item was explained in terms of the occurrence of a stress offered by the environment that exceeded the strength of the device or item. See Section 3.3.3.3 for additional discussion of the stress–strength model in populations of devices or items. For now, understand that the stress–strength model is very helpful in the work of determining the failure mechanisms and root causes for failure modes. The stress–strength concept is also used metaphorically in intangible items such as software and services.

2.2.8 The Competing Risk Model

In many cases, there is more than one relevant stress versus strength process unfolding in a single item. This scenario, in which several processes that can cause failure, or failure mechanisms, operate simultaneously, is called *competing risk*. We can think of these processes as internal physical, electrical, chemical, thermodynamic, mechanical, or other mechanisms that act to weaken the item, decreasing its strength and making it more susceptible to a shock of a given size. The time at which failure of the item occurs is the minimum of the times at which each of the individual processes are overcome by a shock of a relevant kind. In a sense, the processes "compete" for the "privilege" of causing the item to fail.

Example: Two such processes that can be at play in a CMOS are oxide breakdown and crack growth. Oxide breakdown is a physical/chemical/electrical process stimulated by electrical potential across the oxide. Crack growth is a cumulative damage mechanism, a mechanical process stimulated by stress relaxation, lattice mismatches, microshocks, vibration, and other mechanical insults. The CMOS device fails if either the oxide is punctured (due to a voltage stress) or a microcrack created during device manufacturing grows to a point where it interrupts a circuit element or via. The time at which failure occurs is the smaller of the two times at which the oxide punctures or the crack grows large enough to interrupt a circuit element or via. This simplified example (there are other processes leading to failure at play in CMOS, including electromigration, hot carrier damage, ion contamination, and others) uses only two failure mechanisms to illustrate the competition idea in a simple setting.

2.3 RELIABILITY, MAINTAINABILITY, AND SUPPORTABILITY ARE MUTUALLY REINFORCING

2.3.1 Introduction

Much of the earlier discussion can be summarized by saying that reliability engineering deals with making products and services free from (or less susceptible to) failures. But failures are inevitable; rare indeed is the product or service that never experiences any failures, and the degree of attention required to ensure that a product, system, or service never fails is justifiable economically only in rare cases.[7] So systems engineers concerned with crafting requirements for the product or service that will satisfy its customers necessarily must also be concerned with how failures will be dealt with.

Almost all technological systems are repaired or otherwise restored to service,[8] rather than discarded, when they fail. The reasons for this are many and varied, but the key point for such "repairable" or "maintainable" systems (see Chapter 4) is that once a failure occurs, it may take some time before the conditions leading to it are corrected and normal operation can resume. It is during this period when the system is out of service (the outage period) that actions intended to return the product or service to normal operation are undertaken. Here is where maintainability and supportability enter the picture. We divide the outage period into two parts: the first part (chronologically) contains all the activities undertaken to prepare to do a repair, and the second part contains all the repair activities. Broadly speaking, this is the distinction between maintainability and supportability: maintainability engineering concerns *execution of* actual repair actions and operations, while supportability engineering concerns *preparation for* repairs. Some of the items covered by maintainability engineering include the designing of planned repair actions and procedures, while supportability engineering would cover operational planning and preparation for repairs. Maintainability engineering would deal with issues like whether a board would be a plug-in only or use screws to retain it in a socket, while support engineering would deal with choices like keeping spares on site versus next-day delivery of spares from a central warehouse. We will return to these definitions in detail in Parts II and III of this book; this introduction will remain rather broad-brush so you can get a sense of the big picture before getting involved in details. Reliability, maintainability, and supportability engineering are sometimes referred to (as they will be here) as the *sustainability disciplines*.

We will see in Section 2.3.2 how improvement in any of reliability, maintainability, or supportability leads to improvement in the other two. But the key reason that maintainability and supportability are treated as separate

[7] See Chapter 7 for further discussion of reliability engineering for high-consequence systems like satellites, nuclear power plants, critical infrastructures, and so on.

[8] For instance, software crashes are often restored by reboot.

disciplines is that during a period of outage, customers are unable to use the failed system. Customers are therefore interested in returning a failed system to service quickly. Dividing service restoration activities into the two periods of preparation for repair and execution of repair enables the creation of specific disciplines addressing each of these with special processes and tools. In short, customers care about maintainability and supportability because doing these things well promotes shorter outage times and higher availability (Section 4.3.3.4). The maintainability and supportability disciplines have arisen because the systematic and effective methods they provide lead to decreasing the amount of time required to carry out repairs or decreasing the length of outages. Customers care about getting their service back quickly, and maintainability and supportability exist to enable this to happen by a systematic, analytical approach to the activities that need to be carried out in order to bring a system back to operating condition. If a system were perfectly reliable and failures never occurred, then maintainability and supportability would not be needed.

> **Language tip:** Failure, outage, failure time, downtime. It is important to define and use each of these terms precisely in your studies because they are used inconsistently in the literature. We have defined failure above, and continue to reserve the term "failure" for any instance of violation of a product or service requirement. In this book, we reserve "failure time" for the time at which a failure occurs (these are the points marked "×" in Figure 2.1) while using "outage" to refer to the entire period during which the failure condition persists (these are the heavy lines on the horizontal axis starting from the points marked "×" and ending at the next large dot). During an outage, one or more system requirements are being violated. The length of the time during which the failure condition persists (i.e., the time-length of the outage) will be called the "outage time" or "downtime" or "duration of the outage." The heavy horizontal lines at level 1 on the diagram are the operating time intervals, or operating times, that is, the time intervals during which the system is operating properly (no requirements are being violated). In our terminology, the phrases "duration of failure" and "failure duration" are not defined because failure refers to something that takes place at a particular instant. It is the outage that consumes a positive amount of time after the failure occurs. We will return to this when we discuss the concept of "time between failures" in Chapter 4.

In the usual reliability modeling paradigm, the illustration in Figure 2.1 is a *sample path* in the system's *reliability process*. The system's reliability process (Section 4.3.2) is usually conceptualized as a two-state stochastic process in which state 1 indicates that the system is operating properly and state 0 indicates that the system is in a failed condition (one or more requirements are violated). The system history diagram is meant to be viewed as a "typical" or "generic" depiction of the alternating periods of proper operation and failure that a product, system, or service normally undergoes. The system reliability process specifications can be customized to accommodate a variety of assumptions about how the system operates and how it is repaired. These details are covered in Section 4.4.

Systems engineers must plan for maintainability and supportability as part of the overall customer satisfaction assurance activity because failures disrupt customer operations and have potential for causing significant customer dissatisfaction even apart from inability to complete the mission. As always, an economic balance is struck between the cost of providing enough support (so that failures are infrequent and of short duration), the cost to the customer of failures themselves, and the cost to the supplier in warranty servicing and lost reputation or goodwill. The cost of providing support is an up-front cost covering all the design-centric activities needed to anticipate and avoid failures as well as to plan and manage the activities needed to restore normal functioning of the product or service after any failure that may occur. And these activities must include planning for what to do when a failure occurs.

It is widely accepted that the least costly approach to achieving any desired degree of reliability is to pay attention to the sustainability disciplines during product/service definition and design. Quality engineering advocates the "1–10–100 rule": fixing a problem that shows up during manufacturing costs 10 times as much as the cost of any design activities needed to prevent occurrence of that problem, while fixing a problem first appearing in the field costs again 10 times as much. While the numbers in this "rule" are not intended to be precise, the different orders of magnitude serve to readily recall the important principle that costs associated with anticipating and preventing failures are almost always repaid tens and hundreds of times over during the life of the product/service by the savings resulting from not having to deal with those failures. See Exercise 10.

However, these costs appear on different budgets. Design-for-reliability costs accrue to the provider of the product or service, while costs associated with failures accrue to the user of the product or service,[9] and these are most often not the same. The provider, therefore, is required to make a convincing total-cost-of-ownership case to the user that whatever (hopefully modest) increase in acquisition cost that may be due to increased attention paid to design for reliability, maintainability, and supportability, will be more than amply repaid in savings over the useful life of the product or service accruing from nonoccurrence of failures. And there are important differences in different markets: the market for defense systems is very different from the market for consumer electronics. Defense procurement officers can be counted on to understand tradeoffs between first cost and recurring costs, while such considerations are not often foremost in the minds of consumers. While the principles in this book are widely applicable, they need not apply everywhere. For many systems and services, design for reliability, maintainability, and supportability may mean the difference between mission accomplishment or mission failure, and even life or death (Chapter 7). For others, the supplier may wish simply to produce a system or service just reliable enough to survive until a next generation is ready for deployment, and full use of all available techniques may not be warranted.

[9] Except for warranty costs (if a warranty is offered) which are borne by the provider.

In light of the foregoing discussion, it might be inferred that varying degrees of reliability may be easily achieved by focusing more or less attention on design activities. As a rule, though, it is a difficult problem to precisely adjust the degree of reliability to be achieved during the life of a system, product, or service according to the degree to which design-for-reliability activities are undertaken. Some reasons for this include

- Lack of precise information regarding the reliability of components and subassemblies of the system, product, or service. This information is usually summarized probabilistically, and some of these estimates may not be very precise at all. In many cases, precision information about the estimates (e.g., confidence intervals, standard errors of estimates, etc.) may not be available at all. As a consequence, the system, product, or service reliability may not be discernible very precisely either.
- Lack of a continuum of choices for reliability of components or subsystems. The literature in the mathematical theory of reliability contains many studies concerning optimal allocation of system reliability to components and subsystems, usually accomplished through a mathematical programming (optimization) model. In practice, there are usually very few (sometimes only one or two) choices for the reliability of a component or subassembly as a function of cost. That is, it is usually impossible to obtain a small incremental increase (or decrease) in the reliability of a component or subassembly by a small incremental increase (or decrease) in its cost.
- The conditions under which the customer will use the system, product, or service may be more or less extreme, or possibly just different, than anticipated.
- In the end, if design for reliability has been implemented properly, most failures will come from unanticipated sources. This introduces another, unquantified uncertainty into any reliability modeling and points up the importance of seemingly intangible assets like experienced design and engineering personnel, robust institutional memory, and a culture that does not punish failure but rather treats it as an opportunity for learning.

In light of these facts, adjusting the reliability of a system, product, or service to meet stated goals is promoted by

- staff who are comfortable with probabilistic and statistical thinking,
- a culture that treats failure as an opportunity to learn for the future, appropriately nurtures institutional memory, and is open to new design for reliability technologies as they appear,
- robust reliability modeling and data analysis capabilities,
- information sources actively maintained, and
- good horizontal and vertical communication throughout the development organization and its customers.

In short, do the best you can with the tools and organization you have, continually improve, and leave margin to allow for inevitable unanticipated failure modes.

2.3.2 Mutual Reinforcement

In this section, we study the proposition that reliability, maintainability, and supportability are mutually reinforcing. In brief, improving any of the three improves the other two as well. Let us see how this works.

- Improving maintainability improves reliability: As we will see in Chapter 4, one of the important figures of merit for reliability of a repairable system is *availability*, essentially the proportion of time for which the system is in proper operating condition (the complete definition is in Chapter 4). Improved maintainability means that the system is easier to fix when it fails: components are more accessible, fewer special tools may be required, repair operations can be carried out quickly, etc. All of these translate into less time required for completing maintenance when a failure occurs. By referring to the history diagram shown in Figure 2.1, you can see that the less time consumed by maintenance actions, the shorter an outage and the more quickly the system is restored to proper functioning. Therefore, the proportion of time when the system is in a properly functioning state increases, so its reliability (as reflected in its availability) is improved.
- Improving supportability improves reliability: Again, this argument rests on the idea that improved supportability means less time spent on supportability actions, like transporting required spare parts from their storage location, diagnosing the correct failure cause(s), locating the correct repair instructions for those cause(s), etc. The same reading of the history diagram of Figure 2.1 with now the shorter period of time required for supportability actions shows that the proportion of time the system is in a properly functioning state is greater when supportability is improved. Again, this means improved system availability.
- Improving supportability improves maintainability: Here we focus on the actions that can be taken to improve the performance of maintenance personnel. Carrying out repairs quickly and effectively requires that diagnosis and location of failure causes be made unambiguously, correct tools and spare parts be located and brought to the work site, and correct repair instructions be provided; in short, doing the supportability things right the first time. This saves maintenance personnel from errors that take additional time to correct and enables them to perform their repair tasks correctly and efficiently the first time. Improving supportability in this way reduces "scrap and rework" in repairs so that maintainability is also improved.
- Improving maintainability improves supportability: Conversely, improving maintainability involves such actions as simplification of repair procedures, minimization of use of special tools, and use of a system architecture that

minimizes the number of different line-replaceable units (LRUs), etc.; then it is easier to provide the preparatory materials and instructions needed to carry out repairs, so important aspects of supportability are improved as well.

- Improving reliability improves both maintainability and supportability: If a system experiences fewer failures, it becomes easier to plan for and execute repairs. Fewer repair facilities and personnel are required, more time is available for training, fewer spare parts are required, etc., all of which reduce the supportability and maintainability burden.

In short, the sustainability disciplines form a virtuous circle. The benefits of this arrangement are best realized when attention is paid to design for reliability, maintainability, and supportability so that the mutually reinforcing consequences may be fully realized. It is easy to destroy this mutual reinforcement by inattention and lack of resources applied early enough in system definition and design. That is why this book discusses design for reliability, design for supportability, and design for maintainability as key disciplines promoting mission accomplishment, customer satisfaction, and profitability. But it is the reliability tail that wags the maintainability and supportability dog. The best reason to be concerned about maintainability and supportability is that they contribute to shorter outage times, and therefore greater reliability.[10] Accordingly, most of the emphasis in this book is on design for reliability. Poor maintainability and supportability can be overcome by having only few failures, but no amount of maintainability and supportability can economically compensate for poor design for reliability and a large number of failures. See Exercise 4.

2.4 THE STRUCTURE OF RELIABILITY REQUIREMENTS

While we have spent some time discussing the relationships among reliability, maintainability, and supportability, this chapter concerns reliability, and so we will here return to studying reliability requirements in detail. Maintainability and supportability requirements are studied further in Chapters 10 and 12, respectively. We begin with a discussion of the general form of quantitative descriptors of reliability.

2.4.1 Reliability Effectiveness Criteria

An effectiveness criterion is a quantitative expression of some system property related to requirements, such as throughput, delay, weight, current draw, etc. Effectiveness criteria serve to direct systems engineering, design, and development attention to those system properties and characteristics that customers

[10] Note that "reliability" is used in its broad sense here.

feel are important and/or desirable. As reliability deals with failures, many reliability effectiveness criteria concern mission completion, frequency of failures, duration of associated outages, and so on. Some examples include

- number of failures per (hour, day, week, month, year, other),
- lifetime of a single-use component,
- proportion of time the system spends in the operating state,
- time between outages, and
- number of replacements of nonrepairable units per (hour, day, week, month, year, other).

It is apparent that there are many possibilities for reliability effectiveness criteria. The above list is certainly not exhaustive. The most widely used reliability effectiveness criteria will be discussed in chapters 3 and 4. The number of reliability effectiveness criteria and the depth of detail the systems engineer chooses to use in requirements depend on the type of system in question and the customer's needs and desires. Some guidance on this matter is provided in Section 2.4.3.1.

In Parts II and III of this book, we will consider maintainability and supportability requirements. The notions of effectiveness criteria and figures of merit apply equally well there. For example, the time required to perform a specified repair operation is an example of a maintainability effectiveness criterion. The cost of required documentation pertaining to system repair is an example of a supportability effectiveness criterion. Many more examples will be seen in the later chapters of this book.

All effectiveness criteria we promote as useful for systems engineering can be considered as *random variables* [4]. This is because systems engineering begins at an early stage of system development. The system does not exist yet, and so we have no way to know what the values of these effectiveness criteria may take when the system is developed and in use. Also, the deployed systems may operate under a wide variety of environmental conditions and different installations of the system may respond to these conditions in different ways. There is no way to tell in advance, with certainty, how many failures (requirements violations) may take place during, say, the first year of operation of a system with a particular serial number. Finally, their values also depend on many factors, some of which cannot be specified precisely (see the discussion of control factors and noise factors in Chapter 1) or for which a good quantitative understanding of how they depend on the factors is lacking. Reliability effectiveness criteria are not physical constants like the speed of light in a vacuum or the specific gravity of mercury. They are properties that usually vary from system to system in unpredictable ways. For example, the number of subassembly failures in a medium-frequency amplitude modulation (MF AM) broadcast transmitter may be 3 in 2014 at WNYC and 5 in the same year (and the same model transmitter) at WKCR.[11] The same transmitter, from

[11] Not a real example; it is used for illustrative purposes only.

the same manufacturer, may be installed in many broadcast stations, and the experience of number of subassembly failures across all these installations may differ—indeed, almost always does differ—from one installation to another. Note that each of the examples cited earlier are like this. For this reason, the reliability engineering community has found it useful to consider reliability effectiveness criteria as random variables in modeling and other quantitative activities.

Because an effectiveness criterion is a random variable, its complete description is its *cumulative distribution function* or *cdf* [4]. The cdf of an effectiveness criterion C is given by $F_C(c)=P\{C\leq c\}$. This is a function of the real variable c, over whatever domain of this variable[12] makes engineering sense, and it takes values in the interval [0, 1]. We discuss cdfs and their properties, including the notion of discretionary variable, in greater depth in Section 3.3.2. For additional perspective, consult Ref. 4. For now, though, it is enough to note that it is only rarely possible to obtain this degree of complete information about an effectiveness criterion. Reliability modeling (Chapters 3 and 4) and analysis of reliability data from laboratory tests, system tests, and deployed systems (Chapter 5) are tools that enable making estimates of the cdf and/or other pertinent quantities related to the effectiveness criterion at various times in the system's life cycle. Section 2.4.2 is devoted to how we work with abbreviations and summaries of parts of the cdf to carry out systems engineering tasks nevertheless and provide useful guidance to those who need to work with the effectiveness criteria.

> **Language tip:** Do not confuse "effectiveness criteria" for reliability, maintainability, or supportability with "(system) measures of effectiveness," which are usually broader quantitative descriptors of some system-level attribute of value to the customer, such as cost per hour of operation or total life-cycle cost. System measures of effectiveness often contain contributions from reliability (or maintainability or supportability) effectiveness criteria or figures of merit, but they are usually broader in scope and intended to provide guidance in other areas such as system economics.

2.4.2 Reliability Figures of Merit

When dealing with a large population of systems, working with the raw random variables can be messy, time-consuming, and ill-suited for clear communication with nonspecialists. In the theory of probability, abbreviations have been developed that allow key properties of random variables to be summarized in briefer terms. The properties that are important for reliability engineering include

- mean or expected value,
- variance and standard deviation, and
- percentiles (including the median).

[12] In this formulation, c is referred to as a *discretionary variable*.

When these more compact descriptions of random variables are used on reliability effectiveness criteria, we call the results *reliability figures of merit*. Definitions for these are given in Section 2.7 and in more detail in chapters 3 and 4.

In the example in Section 2.4.1, we may consider the number of subassembly replacements in the population of *all* installations of transmitters of the same make and model. There may be tens or hundreds of these. It would be in principle possible to list the number of subassembly failures per year in each installation. In practice, such a list is likely to be quite long and hard to use to communicate results to others. The list may be summarized in an *empirical distribution* or *histogram*, simple statistical techniques that enable a briefer summary of complicated or lengthy data. We may also summarize the list (a *census* of the number of subassembly failures in the population) using the mean of the number of such replacements. The number of replacements in the first year of operation is a reliability effectiveness criterion: it takes a (potentially) different value for each installation. The cdf, mean of that random variable, expected number (or mean number) of subassembly replacements in the first year of operation, are all examples of reliability figures of merit.

2.4.3 Quantitative Reliability Requirements Frameworks

Reliability requirements may be written using either reliability effectiveness criteria or reliability figures of merit. Examples of both approaches are encountered in practice. This section discusses each approach in more detail, including some guidance on which to choose for a particular situation.

2.4.3.1 Reliability requirements based on effectiveness criteria

A reliability requirement may take the form of a limit, or bound, on the value some reliability effectiveness criterion is to achieve when a system is deployed. For instance, a reliability requirement may be written: "The number of failures of the system during its first year of operation, under the conditions specified in paragraph x.y.z, shall not exceed two." Note that the requirement pertains to a specific reliability effectiveness criterion, namely, the number of system failures in the first year of operation, and an upper bound is specified, namely, 2. Before proceeding to a more detailed discussion of the basis for the requirement, check that the three important parts of a reliability requirement are included:

1. Is the definition of failure specified?
2. Is the relevant period of time specified?
3. Are the relevant operational conditions specified?

In this requirement, the definition of failure is unspecified, so the requirement is taken to pertain to *all* system failures, that is, any violation of any system (attribute) requirement. The relevant period of time, that is, the first year of operation, is specified. The relevant operational conditions are contained in paragraph x.y.z, and this can be scrutinized for completeness.

A requirement stated in terms of a reliability effectiveness criterion places a limit on the value of that criterion for *every* member of the population to which it applies. It is easy to see whether a particular installation has met the requirement or not: look at the value of the reliability effectiveness criterion achieved by the installation and compare it with the requirement. No statistical analysis is needed for installations from which data can be gathered. To see whether the requirement is met for an installation from which data cannot be gathered, a statistical inference from the data recorded from installations from which data have been gathered can produce an estimate of the probability that an installation chosen at random from the population of all similar installations meets the requirement. The statistical analysis takes into account the possible error that follows from examination of only a sample, rather than the entire population. Several examples are given later in this chapter and in Chapters 3, 10, and 12.

> **Requirements tip:** It is possible to specify, as part of a requirement, the confidence level to be used when estimating from sample data the probability that the requirement is met, but this is not common practice. Setting the value of this confidence level should not be arbitrary but should be guided by utility and risk considerations of downstream reliability economics and decision-making [1, 6, 19, 20] that are important but beyond the scope of this book.

Requirements stated in terms of reliability effectiveness criteria are usable in all engineering situations and are particularly appropriate when

- the population of installed systems will remain small and/or
- individual control of reliability characteristics is critical (as in, e.g., high-consequence systems (Chapter 7)).

2.4.3.2 Reliability requirements based on figures of merit

A reliability requirement may take the form of a limit, or bound, on the values some reliability figure of merit to be achieved when a system is deployed. For instance, a reliability requirement may be written: "The expected number of failures of the system during its first year of operation, under the conditions specified in paragraph x.y.z, shall not exceed two." In this case, the requirement asks for a reliability figure of merit to be limited to the value specified. Because the figure of merit is an abbreviation for the full reliability effectiveness criterion, a requirement written in this form does not control the individual values of the effectiveness criterion for individual system installations. Instead, it attempts to control the figure of merit over the entire population of installed systems. For instance, suppose there are 100 installations of the system covered by the requirement. Then the requirement would be satisfied if in the first year of operation, 99 of the installed systems experienced one subassembly replacement and one system experienced 20 replacements, for then the mean number of replacements is 1.19 which is less than 2. You can easily construct a less extreme example illustrating the same point: when controlling only a figure of merit, as opposed to controlling an effectiveness criterion, you may experience individual installations that fail the number specified in the (figure of merit) requirement while the requirement is being met overall (by the population). Some more material on this topic can be found in Section 2.7.2.1.

If it is possible to take a census of the entire population of installed systems, then checking conformance with the requirement is a simple matter of computing the value of the figure of merit from the data and comparing the result with the requirement. If it is not possible to take a census, but a sample may be obtained, then statistical inference procedures may be employed to assert a probabilistic statement about whether the requirement is met. For relevant procedures for means and population proportions, see Table 5.1.

Requirements stated in terms of reliability figures of merit are usable in all engineering situations but are more suitable when

- the population of installed systems is, or is anticipated to become, large, and/or
- the variation in the values of the figure of merit from system to system is anticipated to be small, and/or
- individual control of reliability characteristics is not critical (as in, for instance, mass-produced consumer entertainment devices).

2.5 EXAMPLES OF RELIABILITY REQUIREMENTS

2.5.1 Reliability Requirements for a Product

Technological societies contain a large number and variety of products used for everything from life-sustaining systems like medical devices, to transport systems like aircraft, railroads, and automobiles, to entertainment products like television receivers. The consequences of failure may differ greatly across these categories, but the basic structure of reliability requirements for a product remains the same while the degree of reliability needed in the different categories may vary. Consider first a simpler product like a kitchen appliance—refrigerator, dishwasher, etc. These are considered simpler in the sense that they have relatively fewer failure modes compared to, say, a fighter aircraft. The operation of the appliance may be continual, as in the refrigerator, or intermittent, as in the dishwasher. The user's expectations for these appliances may be summed up simply as "it works when I want to use it." A reliability requirement consistent with this desire could key on failure-free intervals. For example, a reliability requirement for a home refrigerator could be "The refrigerator will operate without failure for a period of 100,000 hours of continuous operation when the AC line voltage supplied is between 115 and 125 volts and between 58 and 62 Hz and the ambient temperature is between 55°F and 85°F." At this time, we are not concerned with how this requirement was developed, but we want to examine whether the requirement is complete and how it may be interpreted.

Consider first whether it is complete. We have introduced three important components of a reliability requirement: an unambiguous expression of the desired operational behavior, a period of time over which the requirement is to apply, and the operating conditions under which the requirement is to apply. Is each of these present in the example? "Operate without failure" is a definite statement, but a definition of "failure" is not stated. Part of the standard design for reliability process is stepping through each of the attribute requirements of the product to uncover the failure modes in the product; we discuss this further in Section 2.8.1. Because this requirement does not specify which failure mode(s) it covers, it must be assumed to cover all failure modes.[13] The period of time is clear: 100,000 hours[14] of continuous operation. The operational conditions are specified as ranges of supply voltage and frequency and ambient temperature. These seem typical of a consumer kitchen environment. Note that other possible operational environmental variables like humidity, frequency of access, etc., are unspecified. As such, it must be assumed that the requirement is supposed to apply no matter what the values of these unspecified operational conditions may be. Such omissions introduce the possibility of dispute with a customer.

We may alternately consider that the refrigerator may be repairable when it fails. If adopting this point of view, a requirement could be written as a limit on the number of failures over a specified time period. For example, "The number of times the refrigerator fails shall not exceed one over a period of 100,000 hours when then AC line voltage supplied is between 115 and 125 volts and between 58 and 62 Hz, the ambient temperature is between 55°F and 85°F, preventive maintenance is conducted according to the recommended schedule, and repairs are conducted by authorized service personnel."

Either reliability requirement could be appropriate for this product. Some consumers will choose not to repair a failed refrigerator but replace it with a new one instead; for that market, the more appropriate reliability requirement would be the first cited earlier. For those consumers choosing to repair rather than to replace the refrigerator, the latter cited requirement would be more appropriate. Of course, the refrigerator manufacturer could adopt *both* reliability requirements, provided they are consistent.[15]

Reliability requirements for more complicated products like fighter aircraft may differ in degree but are similar in kind. The operator or user is concerned with the product's continued operation without failure throughout some period of time (a mission, for example) or the number of repairs that may be needed per (week, month, year, etc.) to keep the product in a desired operational state. Mission times

[13] It may be desirable to categorize failure modes into more-serious and less-serious categories. For instance, the failure of the interior light bulb in the refrigerator is likely to be viewed with less concern than a failure of the compressor.

[14] A year contains 8,766 hours (to 0 decimal places) so 100,000 hours is about 11.4 years. Most reliability engineers round a year to 10,000 hours for informal use. While this is a useful memory aid, any important reliability engineering exercise should use the more precise figure.

[15] Determination of whether they are consistent is beyond the scope of this chapter and requires the methods discussed in Section 4.4.

may be variable, the list of operational conditions under which the requirement is to apply may be much longer, repairs may be more multifaceted, but these complicating factors do not change the fundamental nature of the reliability requirement: a statement about which failure modes are covered by the requirement, a limit on a reliability effectiveness criterion or a reliability figure of merit, a time period, and the pertinent operational conditions are all needed. Only the degree of detail changes.

Language tip: Many in the engineering community think *reliability* and *availability* are the same thing. Availability is a particular reliability figure of merit applicable to maintainable systems (Section 4.3.3.4), and we will maintain a distinction between reliability and availability throughout this book.

2.5.2 Reliability Requirements for a Flow Network

Many infrastructures essential to societal functioning can be abstractly modeled as flow networks [8]. The ability of the network to deliver the commodity that flows in it without interruption is a critical indicator of the network's value. When considering reliability in flow networks, two features stand out:

1. The reliability and capacity of the elements of the network and
2. The reliability of the promised deliveries the network supports.

By itself, the phrase "network reliability" is ambiguous. Careful examination reveals that users of the phrase "network reliability" usually mean the continued delivery, without interruption, of a desired volume of the commodity supported by the network. This is distinct from the reliability of the network elements considered as individual technological systems themselves. Of course, reliability of the network elements bears strongly on the ability of the network to deliver its commodity in the volume desired and without interruption. Many studies of flows in networks with unreliable elements have recently been undertaken. See Ref. 21 for an introduction.

Network elements may include pipelines, valves, and controllers in a fluid delivery network, transport systems, routers, and billing systems in a telecommunications network, generators, transmission lines, towers, and substations in an electrical power delivery network, trucks, hubs, and routing algorithms in a logistics network, etc. Reliability requirements for these products or systems may be constructed according to the ideas in Section 2.5.1. However, because these elements work together to provide the flow of the commodity in the network, and there are requirements for delivery of certain volumes of the commodity from some originating nodes to some destination nodes, the effect of network element reliability on the reliability of these deliveries must be taken into account using a model for flows in networks with unreliable elements. This is still an active area of research, and many large-scale flow networks can be modeled only approximately. Concepts applicable to this study and some mathematical development of the associated models may be found in Ref. 24.

Flow networks are susceptible to two types of reliability problems that we may label "chronic" and "acute." Chronic reliability problems are network element failures that occur more or less routinely because of component failures, operator errors, software faults, and the "ordinary" vagaries of everyday operation. These problems tend to be isolated and uncorrelated, not long in duration, distributed widely in geography and time over the network, and are viewed as a manageable and inevitable low level of "noise" that must be dealt with. Network operators can plan to mitigate these problems by the kinds of design for reliability processes we recommend and describe here. Acute reliability problems, on the other hand, are more serious, very rare, and involve many neighboring network elements, usually with high correlation across neighboring network elements. Acute reliability problems are often the result of natural disasters (earthquakes, fires, floods, etc.), improperly isolated failures, or deliberate attacks. They tend to be much longer in duration, more serious in their effects on the network flow than the chronic problems, and more difficult to anticipate, plan for, and recover from quickly.

In most cases, it is acceptable to derive network element reliability requirements for chronic reliability problems from flow or delivery reliability requirements using a steady-state model (one which describes stable operation of the network over a long period of time). Acute problems are by their nature not steady-state phenomena and sensible mitigation of these problems relies more on good supportability and maintainability, that is, measures to restore service or flow quickly after a major disruption. This is not to say that sensible measures such as avoiding earthquake-prone areas for locating a nuclear power plant should not be undertaken; recent experience [15] has shown that building a nuclear power plant on an active earthquake zone is a bad idea for many reasons. The consequences of the acute failures caused by an earthquake and tsunami in this example included not only immediate loss of life and serious injuries but also uninhabitability of a wide geographic area for many decades to come. A nuclear power plant is one example of what we call in this book a "high-consequence system." Reliability engineering for high-consequence systems is discussed in detail in Chapter 7.

Example: An (over-)simplified version of a package delivery network. Imagine a logistics carrier who transports goods from city A to city B as in the Figure 2.2.

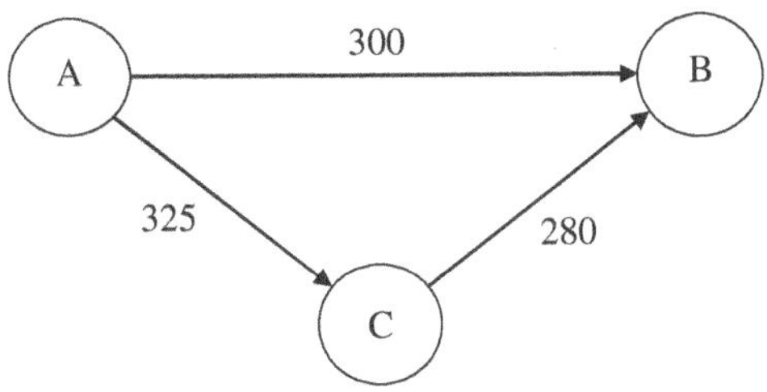

Figure 2.2 *Logistics network example.*

This is a directed network (flow against the direction of the arrows is not permitted) and the capacities of the links are as indicated, in units of packages per day. Suppose that the demand originating at city A is 275 packages per day to be transported to city B and that the reliability requirement for this flow is that the probability of success be at least 0.99. What should the link reliability requirements be so that this flow reliability requirement is satisfied? Let r_{AB}, r_{AC}, and r_{BC} be the probabilities that the indicated links are in a working condition (the links are assumed to be either completely working or completely failed). Then the probability that 275 packages can be transported per day from city A to city B is $r_{AB}+(1-r_{AB})r_{AC}\,r_{BC}$. We then want to find values of r_{AB}, r_{AC}, and r_{BC} so that $r_{AB}+(1-r_{AB})r_{AC}\,r_{BC}\geq 0.99$. There are many values of r_{AB}, r_{AC}, and r_{BC} that make this inequality true, so how do we choose which values to use? One way to choose appropriate values is to incorporate cost into the model. Suppose that the cost for shipping a package from A to B directly is c_1 and the cost for shipping a package from A to B via C is $c_2>c_1$. Then we may write the expected cost for shipping 275 packages per day from A to B as $275c_1r_{AB}+275c_2(1-r_{AB})r_{AC}\,r_{BC}$, and choosing appropriate values for r_{AB}, r_{AC}, and r_{BC} may come from solving the mathematical program

$$\text{Minimize } c_1r_{AB}+c_2\left(1-r_{AB}\right)r_{AC}r_{BC} \text{ subject to } r_{AB}+\left(1-r_{AB}\right)r_{AC}r_{BC}\geq 0.99.$$

This example is very oversimplified: the reliability of the terminals at A, B, and C has not been accounted for; the links may be out of service for more or less than a day and are consequently better modeled by an alternating process (Section 4.3.2), the example becomes more complicated if any of the links have capacity less than 275 per day, the costs are fixed regardless of any other factors such as package weight or size, etc. Nonetheless, the major point of this example is that sensible reliability requirements for elements of a flow network cannot be constructed independently of the flow reliability requirements imposed on the network. The flow reliability requirements are user-oriented requirements, while the network element reliability requirements are of interest mainly to the network operator who presumably has an interest in satisfying user requirements while minimizing the cost of the operation. The influence of network element reliability on flow reliability must be incorporated when constructing requirements for reliability in flow networks.

2.5.3 Reliability Requirements for a Standing Service

Now we begin to draw a distinction between reliability of tangible objects, like products and systems, and reliability of intangible objects, like services. While service reliability is the subject of Chapter 8 of this book, we here introduce some basic ideas and principles that help when devising reliability requirements for services.

First, we distinguish between two types of services: standing services (this section) and on-demand services (Section 2.5.4). A standing service is one that is intended to be always available to users, without interruption. Electric utility power is an example of a standing service. Customers desire utility power to be available at their premises at all times, without interruption. An on-demand service is one which the customer uses intermittently. Each interaction of a customer with the service is a *transaction* that has a defined beginning and end. Internet access is an example of an on-demand service. A user may initiate a web browsing session at a certain time, continue using the Internet for some period of time, and cease doing so at some later time. Each such session constitutes a transaction in the Internet access service. Internet access service need not be present at all times for all customers; the intersection of the customer's transactions with the presence of the service determines the degree of satisfaction the customer may have with the service. If the service is inaccessible only at times when the customer does not try to use it, the customer does not notice whether the service is ever inaccessible.

The foremost distinction between a standing service and an on-demand service is user behavior. In a standing service, the important criterion for reliability is the presence of the service at all times because the user expects or desires that it be present at all times. Electric utility service is being "consumed" at all times by units like refrigerators, life support systems, and other like objects that require continuous, uninterrupted power. So in a standing service, the service consumer requires continuous provision of the service. In a transaction-based service, the user requires the service only occasionally, and understanding how this behavior combines with the service reliability to produce customer satisfaction or dissatisfaction with the service helps when developing reliability requirements for the service.

Because a standing service is supposed to be active all the time, reliability of a standing service is equivalent to the reliability of the infrastructure providing the service. In many important cases (electric power distribution, water distribution, sewage treatment, etc.), this infrastructure is a flow network, and the ideas in Section 2.5.2 apply. A reliability requirement for the service is often stated in terms of accessibility of the service at each customer terminal. For instance, in the electric utility power example, we may write a reliability requirement for power at the meter on the customer's premises (this could look something like: the probability that utility power is present at the customer's meter should be at least 0.999995 at all times for all meters in a stated area); in that case, all the infrastructure of the electric power distribution network, including the customer's drop, is included in the reliability requirement and in any modeling used to relate individual network element reliability to the overall reliability of the service.

2.5.4 Reliability Requirements for an On-Demand Service

The salient characteristic of an on-demand service is that a user from time to time will request service from a service provider, and this interaction lasts for some finite period of time and then is dismissed. Some examples of on-demand services are purchasing gasoline at a filling station, making a voice telephone call, downloading software from the internet, shipping a package, etc. This model is flexible enough that it can accommodate other more abstract scenarios such as use of an application on a personal computer or smartphone.

As with flow networks, there are two important perspectives on reliability on-demand services: that of the user or customer and that of the service provider. The service provider presumably wants to be profitable while providing the user with a good service experience and so must decide what degree of service reliability is compatible with these goals. The service provider is also responsible (either directly or through a repurchase arrangement) for the infrastructure that enables the service. For example, the filling station owner is responsible for storage tanks, pumps, safety systems, billing systems, and other components of the filling station itself. The filling station owner also has to deal with the reliability of the supplier of gasoline: the wholesale purchase of gasoline from the refiner or distributor may be viewed as a transaction in a wholesaling service. Many of the infrastructures used to support service delivery may be conceptualized as flow networks (examples from telecommunications and logistics illustrate this), and similar considerations for reliability apply in flow networks and in on-demand services. Thus, there is the issue of reliability of the service itself (covered extensively in Chapter 8) and the issue of reliability for the elements of the service delivery infrastructure. As with flow networks, these are related. In most normal scenarios, increasing the reliability of the elements of the service delivery infrastructure will improve the reliability of the services carried on it. Quantitative modeling needed to support this activity is described in Refs. 22, 23 and is reviewed in Chapter 8.

The elements of a service delivery infrastructure are technological products or systems for which reliability requirements are considered in Section 2.5.1. The reliability requirements for the service itself are conveniently organized according to the classification described in Chapter 8, which is

- service accessibility,
- service continuity, and
- service release.

Briefly, service accessibility requirements pertain to the ability to set up a transaction when desired by the user, service continuity requirements pertain to the ability to carry on a transaction to its completion without interruption while adhering to relevant quality standards, and service release requirements pertain to the ability to dismiss the transaction when it is complete. These are discussed in greater detail in Chapter 8. An example of a service accessibility requirement for a voice telecommunications service is as follows: the probability that a customer is able to set up a voice call using the service shall be at least 0.99995. This requirement does not specify a time during which it is to apply, so we may conclude that it is intended to apply no matter when the user attempts to initiate a voice call. No other conditions are specified in the requirement, so we may conclude that it is intended to apply under all conditions that may prevail in the network and the user's equipment. If the service provider does not intend either of these broad interpretations, they must include a specification or limitation of time and/or conditions in the requirement. The reliability modeling that would be undertaken to support this requirement must account for the equipment and activities in the service delivery infrastructure that must operate properly in order for a voice call to be set up. More examples of service reliability requirements, including requirements for service continuity and service release, can be found in Section 8.5.1.

2.6 INTERPRETATION OF RELIABILITY REQUIREMENTS

2.6.1 Introduction

It is well understood that requirements in the sustainability disciplines provide key customer satisfier targets for the development team. In addition, they provide a basis for checking whether the system is behaving as intended after deployment. This important function enables the development team to obtain quantitative feedback on their effectiveness and promotes institutional learning from successes and mistakes. In this section, we will introduce a consistent and useful framework for interpreting quantitative requirements that will promote clear and unambiguous guidance (and only as much guidance as is justified by the data gathered) for the development team as well as enable unvarnished understanding of deployed system performance with regard to each of the requirements. This interpretation will be based on the classification of requirements as based on effectiveness criteria or figures of merit (Section 2.4).

Before returning to reliability requirements specifically, we observe that the classification of requirements based on effectiveness criteria and figures of merit applies equally well to supportability and maintainability requirements as it does to reliability requirements. We advocate use of a consistent terminology that makes it easier for systems engineers to accomplish their tasks and communicate important results to key stakeholders, including the development team, management and executives, and customers. Consequently, this section provides a brief introduction to the ideas needed to make useful comparisons between requirements and performance in each of the two categories. In Chapter 5, statistical analyses necessary to carry out this program are described more completely. The introduction given here and the material in Chapter 5 cover the most commonly used practical cases. The statistics of more complicated cases or other custom endeavors are beyond the scope of this book. Those needing additional statistical analyses may consult any of several relevant statistics textbooks, including Refs. 2, 10.

2.6.2 Stakeholders

While many groups—customers, executives and managers, design and development staff, sales forces, and more—have a stake in the successful creation and use of a system, product, or service, two groups are the primary stakeholders in the interpretation of reliability requirements. These are the reliability engineers on the provider side and the reliability engineers on the customer side. Each has unique needs and duties pertaining to reliability requirements. We review these in this section.

2.6.2.1 *Reliability engineers on the provider's team*

When the provider of the system, product, or service develops reliability requirements, the provider has three major relevant interests:

1. The provider needs to convince customers that the reliability requirements meet their needs and that the system, product, or service is capable of meeting the requirements.
2. The design and development team needs to be able to tell whether the design is on track to meeting the reliability requirements when development and manufacturing are complete.
3. The sales and customer service teams need to determine if the system, product, or service is meeting the requirements when the product, system, or service is in operation.

But for very exceptional cases, it is not possible to test a complete product, system, or service for reliability because of the protracted time and large number of samples required.[16] Also, when a product or service is still under development, there may not yet exist finished examples that could be the subjects of a test. The provider's reliability engineering team instead employs reliability modeling to make an estimate of the likely reliability of the product, system, or service on the basis of historical reliability data, the mathematical theory of reliability, and other methods that we will discuss in Chapters 3 and 4. This team needs to be able to compare the results of reliability modeling with the requirement(s). A key point here is that they will choose some reliability model that reflects their understanding of how the system is constructed and how it is maintained, and use this model to compute estimates of the reliability effectiveness criteria and/or figures of merit specified in the requirements. For example, if a reliability requirement for an undersea cable telecommunications system specifies that there shall be no more than three repeater replacements in 25 years of service, the provider of the system needs to choose a reliability model that is capable of estimating the number of repeater replacements in 25 years of service and also reflects to the greatest degree possible the structure of the system, its operations

[16] Nonetheless, accelerated life testing and software reliability growth testing are common, partly because while such testing may not be able to demonstrate reliability, failures that do occur are an indication that the system may contain defects that need to be corrected.

(e.g., how redundancy, if any, is used), and how it is repaired (e.g., by replacing a failed repeater with a new one). Reliability modeling like this is used in the first two items cited in the list provided earlier. The central problem raised by this application is the comparison of the results of reliability modeling with requirements. This problem also arises in planning for warranties.[17]

Once the product, system, or service is in operation, reliability data may be collected. Appropriate analysis of these data enables comparison of real reliability performance results with requirements. This is the central problem faced by the sales and customer service teams. Note that in this case, no reliability modeling of any kind is needed. If a requirement is stated in terms of mean time to first failure, then all one need do is collect data on times to first failure and analyze these directly without regard to how the system is operated. The requirement cares only about the time to the first failure and is agnostic with regard to what model the reliability engineering team may have chosen to demonstrate compliance or how the owner of the system may have chosen to operate it (as long as operation is within the conditions listed in the requirements). It is the responsibility of the provider's reliability engineering team to demonstrate, internally and to customers, that the mean time to the first failure of the product, system, or service meets the requirement, and they will do so using a reliability model that the requirement does not specify. When actual data are available, concerns about reliability modeling do not enter the picture and the data are dealt with directly.

2.6.2.2 *Reliability engineers on the customer's team*

The main interest of the customer or user of the product, system, or service is the same as the third item in the list in Section 2.6.2.1: Is the product, system, or service meeting its reliability requirements while it is in operation? Methods for analyzing reliability data to help answer this question are found in Section 5.1.

2.6.3 Interpretation of Requirements Based on Effectiveness Criteria

When a requirement is written for a reliability effectiveness criterion, the requirement can specify that

- it is to apply to each installation individually, and in this sense it is the most restrictive requirement that can be imposed,
  - for example, each system shall experience no more than three failures during the 25-year service life, or
- it is to apply to some proportion of installations,
  - for example, 95% of systems system shall experience no more than three failures during the 25-year service life.

It may also be of value to consult Section 10.6 for additional insight.

[17] Full treatment of warranty modeling and planning is outside the scope of this book; a comprehensive treatment can be found in Ref. 3.

2.6.3.1 Requirement pertaining to all installations

For instance, the reliability requirement for the refrigerator given in Section 2.5.1 specifies a failure-free period of operation of at least 100,000 hours. As the period of failure-free operation is a random variable, meaning that it may vary in unpredictable ways from one refrigerator installation to another, a requirement stated this way applies to each refrigerator installation by itself. If data are gathered from a particular installation, it is easy to see whether the requirement is met for that installation: either the period of failure-free operation in that installation is greater than 100,000 hours, or it is not.

The situation is more complicated when it is not possible to gather data from a particular installation, but it is still desirable to determine whether that installation meets the requirement. In the absence of data, it is impossible to say with certainty whether this installation meets the requirement or not. But if a sample of failure-free intervals from a population of similar refrigerator installations can be obtained, we are able to make a statement about the probability that the requirement is being met by the refrigerator installations in that population.

Example: The reliability requirement for a refrigerator is as given in Section 2.4.3.1. Suppose that from a sample of 10 refrigerator installations, the following sample of 10 initial failure-free intervals (in hours) was obtained:

TABLE 2.1 Example Failure-Free Intervals

Installation Number	First Failure-Free Interval
1	87,516
2	102,771
3	155,310
4	65,483
5	99,786
6	105,494
7	132,400
8	87,660
9	90,908
10	155,454

First, note that we can state definitely that installations 2, 3, 6, 7, and 10 meet the requirement and the remaining 1, 4, 5, 8, and 9 do not. What is the probability that an installation drawn at random from the population of other installations (besides the ones in the sample) meets the requirement? A "good" estimate of the population proportion is given by the sample proportion, so the estimated proportion of the population of refrigerator installations that meets the requirements is 1/2, which is another way of saying that the probability that a refrigerator drawn at random from this population satisfies the requirement is estimated to be 1/2. This will be made more precise in Chapter 5. To ascertain whether this is a satisfactory result requires consideration of utility and risk questions [1, 6, 20] that are beyond the scope of this book. One may surmise that most customers would probably not find this satisfactory, but a reliable validation of this assertion can only come from the downstream risk analysis of the sort recommended in Ref. 1.

2.6.3.2 Requirement pertaining to a proportion of installations

It is also possible, and often desirable, to write a reliability effectiveness criterion as a limit on the proportion of the installed population that does not meet the specification.[18] For instance, the reliability requirement for the refrigerator given in Section 5.1 may be written instead as "98% of the refrigerators installed will operate without failure for a period of 100,000 hours of continuous operation when the AC line voltage supplied is between 115 and 125 volts and between 58 and 62 Hz and the ambient temperature is between 55°F and 85°F." Now, 2% of the installed population is permitted to have a time to first failure of less than 100,000 hours—how much less is unspecified, so the time to first failure of this 2% may be very short indeed (much the same way as placing a requirement on the mean of an effectiveness criterion allows for possibly large excursions in individual values). If the data in Table 2.1 are a census of the entire installed population, then we conclude that the requirement is not being met. If the data in Table 2.1 are from a sample of the installed population, the sample proportion of installations meeting the requirement is 0.5. In Chapter 5, we will see how the sample size influences the sampling error and the decision about whether the requirement is being met in the larger population of which this is a sample.

2.6.3.3 Repairable systems

A repairable system may fail repeatedly and certain reliability effectiveness criteria like outage time, time between outages, number of failures per month, etc., may assume many values for the same installation. For instance, suppose a particular refrigerator installation experiences three failures and the associated outage times are 1.5 hours, 8 hours, and 4.76 hours (see Figure 2.1). For such systems, the interpretation of a requirement based on an effectiveness criterion is that the requirement applies to each value of the effectiveness criterion generated by operation of the system. In the example, if the requirement is that the outage time shall not exceed 7.5 hours, the system does not meet the requirement because there is one outage time that exceeds 7.5 hours.

2.6.3.4 Conclusion

Some advantages of requirements based on effectiveness criteria are

- simple calculations when a census of the population of installed systems is available,

[18] An argument can be made that this is the only sensible way to write requirements involving effectiveness criteria because verification of these requirements can only estimate the probability that the requirement is met, so you may as well write the requirement in terms of that probability, or proportion of the population.

- straightforward, yes-or-no answer to the question of whether a requirement is met in each system for which data are collected,
- control of the full range of possible values of the effectiveness criterion, and
- easy communication of results in a framework that is easy to explain to all stakeholders.

Some disadvantages of requirements based on effectiveness criteria are

- judgments about conformance can be unstable: analysis of a new set of data from the same system may lead to a different conclusion than the previous analysis, and
- as the number of installed systems becomes large, tracking conformance with requirements can become unwieldy if not properly planned because a comparison is required for each installation individually.

2.6.4 Interpretation of Requirements Based on Figures of Merit

When a requirement is written based on a figure of merit, the requirement can only be interpreted as applying to a population of installed systems.[19] That is because a figure of merit is a summary statistic that is normally intended to summarize the behavior of a (usually large) collection of random variables (values of an effectiveness criterion). A central question in the interpretation of reliability requirements based on figures of merit concerns whether a requirement is intended to apply to only the real population of systems that have actually been built and fielded, or is it intended to apply to a (larger) notional population of all systems of a given type, including those already constructed and those yet to be built? Either interpretation can serve as a basis for a successful enterprise and only slightly different data analyses are needed to support the two cases. In the former case, a census of the installed population yields the easiest analysis.

For purposes of comparing performance with requirements in the case of requirements based on figures of merit, we distinguish two cases, according as a census of the population is available or not.

2.6.4.1 Figures of merit for systems considered as non-repaired

Some reliability effectiveness criteria for non-repairable systems may be used for repairable systems also. For example, reliability effectiveness criteria involving the time to first failure are essentially the same as the criteria for time to failure of a non-repairable system. This section discusses interpretations of requirements built on these effectiveness criteria. Fuller explanation of the practice of using reliability effectiveness criteria for non-repairable systems on repairable systems in found in Section 4.3.4.

[19] We do allow the degenerate case of a population consisting of only a single installation.

A census of the installed population is available

If case data from all the installed systems is available, we may compute the relevant figure of merit from the data (the results of this computation will be called a "metric" in Chapter 5) and simply compare this value against the value in the requirement. For example, suppose now the refrigerator reliability requirement is "The mean time to first failure of the refrigerator shall be no less than 100,000 hours when operated continuously with AC line voltage supplied between 115 and 125 volts and between 58 and 62 Hz, in an ambient temperature between 55°F and 85°F." The requirement is in terms of the mean time to the first failure of the refrigerator, a reliability figure of merit (the time to first failure is a random variable, an effectiveness criterion, and the mean is a measure of the central tendency of that random variable (Section 2.7.2), a figure of merit as defined in Section 2.4.2). Suppose that the 10 refrigerator installations listed in Table 2.1 constitute the entire universe of installed refrigerators of this type. Then Table 2.1 constitutes a census of this population. The mean of the 10 times-to-first-failure in the table is 108,278.2 hours. This is greater than 100,000, so this population of refrigerator installations does satisfy the requirement.

Census of the installed population is not available

Suppose now that the requirement is as in Section 2.6.4.1 but that the 10 installations summarized in Table 2.1 are only a sample from a population of some larger number of refrigerator installations (of the same type). Now it is not possible to determine with certainty whether the requirement is being met in the population because we do not have access to the time-to-first-failure data from any of the other installations. We treat the data from the 10 installations in Table 2.1 as a sample from this population and use the sample data to estimate the population mean. As noted earlier, the sample mean from these data is 108,278.2 hours. The sample standard deviation is 30,035 hours, so the estimator $\hat{\mu}$ (i.e., the sample mean) of the population mean μ is approximately normally distributed with mean 108,278.2 hours and standard deviation $30{,}035/\sqrt{10} = 9{,}497.9$ hours. Then the probability that the population mean is 100,000 hours or less is approximately $\Phi_{(0,1)}(-8{,}278.2/9{,}497.9) = \Phi_{(0,1)}(-0.872) \approx 0.192$. This is an estimate, based on this sample, that the requirement is not being met in this population. Conversely, the data support the contention that the probability that the requirement is being met in this population is approximately 0.808. Note that the probability arising here is due not to a random nature of the population mean (which is fixed, but unknown), but it is due to the variability in the sampling procedure. Another way to put this is that, given this sampling procedure, there is about a one in five chance that this procedure will lead to the conclusion that the requirement is not being met in the population.

2.6.4.2 Figures of merit for repairable systems

A single repairable system may generate many values of a given reliability effectiveness criterion, so a figure of merit may be computed from data from only one system at a time or from data from an aggregate of many systems. There are thus two possible interpretations of a requirement based on a reliability figure of merit for a repairable system: the requirement is considered to be met if it is met on each individual system (so it may be met for some systems and not for others), or it is considered met if it is met in the aggregate (i.e., if the relevant metric computed from all systems in the aggregate satisfies the requirement). For example, suppose the requirement for the refrigerator outage times described in Section 2.6.3 is "The mean outage time for the refrigerator shall not exceed 5 hours." This can be interpreted to mean that the requirement is satisfied if every refrigerator installation in the population has a mean outage time of no more than 5 hours, or it can rather be interpreted to mean that the mean of all outage times over all the installations in the population does not exceed 5 hours. Either interpretation is reasonable, but the first interpretation places tighter control over the possible values the outage times may take in the population and still stay within the requirement. The decision about which interpretation to use rests on an understanding of customer needs as well as an understanding of the amount of variability in outage times that is possible over the population of installed refrigerators. The latter, in turn, devolves from the extent and quality of the design for supportability and design for maintainability performed by the refrigerator supplier.

Again, it is necessary to consider whether a census of the population of installed systems is available. If so, the relevant metric is computed on all the installations in the population, and determining whether the requirement is met is a matter of comparing the computed value from the census with the value in the requirement. If a census is not available, statistical inference must again be used to determine conformance to the requirement, and this conformance will now be expressed in probabilistic terms.

> **Example:** Suppose the requirement states, as given earlier, "The mean outage time for refrigerators of this type shall not exceed 5 hours." Data from outages experienced in eight refrigerator installations, each operated for 100,000 hours, are recorded as Table 2.2.

The second column contains the recorded data while the rightmost two columns are statistics computed from the data (i.e., metrics). Potential interpretations of the requirement are

1. The requirement applies to each installation separately. Then installations numbered 1, 3, 4, and 5 meet the requirement over the stated time period (100,000 hours) and 2, 6, 7, and 8 do not. No statistical inference is required. Given these data, it is possible to estimate the probability that the

TABLE 2.2 Example Refrigerator Outage Times

Installation Number	Outage Times (Hours)	Sample Mean	Sample SD
1	1.5, 8, 4.76	4.75	3.98
2	3.1, 6.5, 4, 7.3	5.23	2.31
3	0.4, 2.25, 9.5	4.05	5.89
4	4.5	4.5	0
5	1.5, 5.5, 6, 7	5.0	2.79
6	4.5, 7.5	6.0	2.25
7	3, 6, 8.75	5.92	3.13
8	4, 7, 9.25, 11	7.81	3.49
Aggregate	All of the above	5.53	2.84

requirement will continue to be met over additional periods of time (i.e., looking ahead after the 100,000 hours over which the data have already been collected, we ask for the probability that the mean from these future data will be less than 5 hours). This reasoning treats the data in hand as a sample from some future stream of data that is not yet visible. For instance, for installation number 1, the current estimate of the mean outage time is 4.75 hours. We ask for the probability that a (future) $\bar{X}$ from installation 1 be less than or equal to 5, assuming that the environment in which the refrigerator is operated does not change. We write

$$P\{\bar{X} \leq 5\} = P\left\{\frac{\bar{X} - 4.75}{3.98\sqrt{3}} \leq \frac{5 - 4.75}{3.98\sqrt{3}} = 0.036\right\}.$$

This is because the distribution of the quantity on the left-hand side is known (approximately). If the number of data points were large, this distribution would be approximately normal. However, because the number of outage times collected from installation 1 is only 3, the distribution is instead approximately a t-distribution with 2 degrees of freedom, so the probability we want is $P\{t_2 \leq 0.036\} \approx 0.51$. We conclude that while the data show that the requirement is now being met for installation 1, the chance that installation 1 will continue to meet the requirement in future (assuming underlying conditions remain the same) is only about 50–50.

2. The requirement applies to all installations, and the data shown are a census of all the installations. That is, in this case, there are only eight installations of this refrigerator, and the table shows the complete record of all outage times from all eight installations. The sample mean of all the outage time data from all eight installations is 5.53 hours, and the requirement is not met. We could again ask for the probability that the requirement may be met after additional time passes, and no computation is required to conclude that this is less than 1/2.

In case a census of the installed population is not available, the requirement applies to all installations and the table shows data from a sample of eight installations. There are more than eight installations, but data are available from only the eight shown. With 24 observations, the sample mean is approximately normally distributed, and

$$P\{\mu \leq 5\} = P\left\{\frac{\mu - 5.53}{2.84\sqrt{24}} \leq \frac{5 - 5.53}{2.84\sqrt{24}} = -0.038\right\} \approx 0.485.$$

Thus the probability that the requirement is being met in the population from which Table 2.2 is a sample is less than 1/2.

2.6.4.3 *Section summary*

This has been a brief introduction to the ideas connected with determining, by studying data from installed systems, whether reliability (or maintainability or supportability) requirements are being met. The purpose of this chapter's discussion is more to show how the varying possible interpretations of reliability requirements color the analysis needed to determine compliance than it is about the comparison methods themselves. The technologies underlying the comparisons needed in reliability engineering—including comparing performance with requirements and comparing reliability predictions with requirements—are discussed further in Chapter 5.

Some advantages of requirements based on figures of merit are

- the framework lends itself more readily to downstream risk analysis,
- judgments about conformance tend to be more stable than when using requirements based on effectiveness criteria, and
- the statistical inference needed to make sense of the data in the framework provides a more nuanced understanding of the system's behavior.

Some disadvantages of requirements based on figures of merit are

- all stakeholders need to be acquainted with the information framework underlying this approach so that appropriate conclusions are reached and communicated and
- slightly more complicated (although easily automated) calculations.

2.6.5 Models and Predictions

So far, we have introduced some ideas useful for comparing the reliability performance of installed systems to reliability requirements. But systems engineers need other kinds of comparisons too. The reliability modeling described in Chapters 3 and 4 produces another kind of estimate of system reliability, one that is based on component reliability estimates, design for reliability activities, and other engineering that takes place during system development. This is commonly called a *reliability prediction*. Comparing the results of this reliability modeling is a way of determining whether the system, in its current state of development, is capable of meeting its reliability requirements once it is installed. Systems engineers have an obvious stake in this determination.

For a series system of components whose life distributions are exponential (see Chapter 3 for definitions), a dispersion characterization may be provided for the parameter of the life distribution of the system (the technique is described in Section "Confidence limits for the parameters of the life distribution of a series system"). This gives a quantitative indication of how much "slop" is present in the system life distribution estimate given the quality of our knowledge about the component life distributions. This information should be used when comparing the results of reliability modeling with either a requirement or with performance inferred from analysis of data from systems in operation. We do not discuss the statistical techniques necessary to do this; however because they are of a more advanced nature than the simple procedures, we introduce them to familiarize systems engineers with this way of thinking. In addition, this technique so far applies only to the limited case of a series system of components having exponential life distributions. Additional research is needed before the same idea can be used with other types of systems. Finally, even though the technique is available for series systems of components having exponential life distributions, a very commonly used model for, for example, printed wiring board assemblies (Section 6.5.1), it is not yet widely used in practice. We can look forward to the day when this use of confidence limit information for reliability models is routine, but that day is somewhat far off in the future at this time.

2.6.6 What Happens When a Requirement is Not Met?

In several of the examples in this section, we concluded that the requirement studied is not being met. This will happen from time to time in real systems. It is important to have a systematic approach to responding to these situations.

First and foremost, understand the strength of the evidence for the conclusion that the requirement is not met. All processes at play in the operation and failure of systems have some degree of statistical fluctuation that is an expected component of their normal operation. The methods shown in this book are intended to help you discern how much the evidence for the conclusions drawn about satisfaction of requirements depends on these fluctuations. This is another way of saying understanding of requirements satisfaction should be managed by fact. If there is a high probability that the results seen are due to chance, given the mechanism that is supposed to be operating, then those results should not be taken seriously as a basis for action until they can be reproduced with more significance. This is akin to the distinction between common causes and special causes in control charting [26]. Explicit control charting may be difficult with reliability requirements because they do not lend themselves to repeated study over different time periods because the time period over which they are intended to apply is usually long. However, for requirements based on shorter time periods, like the maintainability requirements discussed in Section 10.6, explicit control charting is possible and can be effectively used to sort out violations that should be ignored (because they are the result of common causes) and violations that should stimulate further investigation (because they are likely the result of a special cause or causes). See Exercises 5 and 6 to try this on some sample data.

Now assume that you are satisfied, through the statistical analyses recommended here, that a requirement is not met for significant reasons. A sensible next step is to undertake a root cause analysis to determine why the requirement is not met, using the Ishikawa (fishbone) diagram as a tool for guiding the analysis and communicating the results. If the root cause analysis points to a design problem, one should expect that additional failures of the same kind will appear in the population of installed systems. In that case, a review of design for reliability activities undertaken in the system development is called for and changes to the system may be warranted. Changes may be for future versions of the system, or, if the design problem is serious enough, may be retroactively applied to systems already implemented. If the root cause analysis points to randomly occurring failures that seem to have no common origin, a review of the stress–strength interactions possibly at play should reveal appropriate corrective actions. For instance, one of the possible outcomes of the stress–strength review is that the strength distribution in some class of components used in the system was more concentrated on lower values than planned. Another possible outcome is that the system is being used in harsher environments than planned. The root cause analysis will enable implementation of countermeasures based on an understanding of the facts. To help manage the process, it may be desirable to implement a formal improvement program based on the seven-step quality improvement process (QI Story) [25]; see also Ref. 9.

In all cases, it is worth spending some effort to determine whether the unmet requirement is truly not met or if normal statistical fluctuations in the data used for verification are causing it to look like the requirement is not met. It would be naïve to suppose, though, that all customers would be prepared to understand and accept such an analysis. Most customers will insist on attention to the failure they are experiencing now and will not be content to be told that this failure is part of a pattern that is not unusual given the statistics involved. Every failure at a customer location will require attention (even if that attention is just to schedule a repair at a later date; see, for example, Section 10.2.2.1), so this analysis is more for internal use. It helps answer questions about whether extensive redesign efforts are needed (because the pattern of failures seen indicates a special cause at play) or whether the system is perking along normally within the letter and spirit of the reliability requirements, and, while some customers may see some failures, they do not justify major system changes.

2.7 SOME ADDITIONAL FIGURES OF MERIT

The example in Section 2.6.4 was based on the mean time to the first failure of the system. There are many other figures of merit that may be associated with reliability effectiveness criteria and that are useful in creating reliability requirements. We review some of these in this section.

2.7.1 Cumulative Distribution Function

The most complete summary of a random variable is given by its cdf or simply *distribution*. When a random variable X is discrete (takes on only finitely many or countably infinitely many discrete values $x_1, x_2, \ldots$), its distribution function is

$$P\{X = x_i\} = p_i, \quad i = 1, 2, \ldots,$$

where $0 \le p_i \le 1$ and $p_1 + p_2 + \cdots = 1$. For a continuum real-valued random variable X, the distribution of X is

$$P\{X \le x\}, \quad -\infty < x < \infty.$$

There are distributions that have both a discrete and a continuum part. In reliability modeling, these occur most often as descriptions of the lifetimes of switching elements that have a nonzero probability of failure at the moment they are called for (see Section 3.4.5.1 for an example). Additional properties of distributions of lifetime random variables (these are called *life distributions*) are given in Section 3.3.2.3. Many examples of life distributions are considered in Section 3.3.4.

> **Language tip:** In the discrete case, the numbers $\{p_1, p_2, \ldots\}$ are analogous to the density of a continuum random variable. Nevertheless, they are sometimes referred to as the distribution of the random variable. The best way to avoid this confusion is to refer to them as the probability mass function of the random variable. This is accepted terminology, but it is not always common.

The distribution contains all the information about a random variable, so it's no surprise that sometimes it is difficult to get enough information to write down the entire distribution. Fortunately, other briefer summaries are available. The rest of this section discusses some of these.

2.7.2 Measures of Central Tendency

The simplest summary of a random variable is the one that tells where its "center" is. Summaries of this kind are called "measures of central tendency," and there are three in common use:

1. Mean,
2. Median, and
3. Mode.

2.7.2.1 Mean

The mean of a (real) random variable X is the center of gravity of the planar area under the curve of the density (Section 3.3.3.1) of X. It is also called the *expected value* or *expectation* of X. The mean of X is the average value of X. It is computed as a weighted average over all the possible values X may take, each value weighted according to its probability of occurrence. For a discrete random variable X, this computation is

$$\mathrm{E}X = \sum_i x_i P\{X = x_i\} = \sum_i x_i p_i,$$

where the sum is taken over all values x_i (finitely or countably infinitely many) that X may take. For a continuum random variable X, the computation is

$$\mathrm{E}X = \int_\Omega x\, P(dx) = \int_{-\infty}^{\infty} x\, dF_X(x) = \int_{-\infty}^{\infty} x\, f_X(x)\, dx,$$

the next-to-last equality being valid for real-valued random variables (the only ones we shall consider in this book). Here, F_X represents the cdf of X and f_X its density (if it has one); see Sections 3.3.2 and 3.3.3.1). The latter expression shows the "center of gravity" computation using the density.

Example: Suppose the discrete random variable X takes the values 1, 2,..., 10 with probabilities 1/55, 2/55,..., 10/55. Then

$$\mathrm{E}X = \frac{1}{55}\sum_{i=1}^{10} i^2 = 7.$$

But the mean of a variable need not be equal to any of the values of the variable. Suppose Y takes the same values as X but with different probabilities: $P\{Y=i\}=1/20$ for $i=1,\ldots,9$ and $P\{Y=10\}=11/20$. Then

$$\mathrm{E}Y = \frac{1}{20}\sum_{i=1}^{9} i + \frac{110}{20} = \frac{155}{20} = 7.75.$$

Requirements tip: Requirements are very often written as bounds on the figure of merit defined by the mean of some effectiveness criterion. For example, "The mean time between outages shall not be less than 1000 hours." It is important to recognize that controlling the mean of some variable allows for possible wide variation in realized values of that variable. Unless there is good reason to believe that the values of the variable in question will not differ greatly from one another, controlling only the mean allows for possibly large excursions in the realized values of the variable across the population of installed systems. Imagine, for example, that there are two systems deployed, system A and system B, both start in the operating state at time 0, each suffers two failures at the times listed, each outage lasts 1 hour, and the current time is 2001 hours since the start of operation:

1. System A fails at 950 hours and at 2001 hours.
2. System B fails at 100 hours and at 2001 hours.

In System A, the times between outages are 950 and 1050 hours. In System B, the times between outages are 100 hours and 1900 hours. Based on these data, the estimated mean time between outages for System A and System B are both 1000 hours. However, we can reasonably expect that the future failure behavior of these two systems may be quite different. System A exhibits fairly regular behavior, with times between failures (950 and 1050 hours) that are approximately the same. The times between failures for System B (100 hours and 1900 hours) are very different. It can be said that, based on even these sparse data only, we understand more about how System A is likely to behave in future than we do about how System B. The lesson we draw from this example is that, unless you have good reason to expect that the possible values that a variable may take should be close together, controlling only the mean of a variable may leave open the possibility of unduly large excursions from desired behavior. And, of course, gathering and analyzing more data will improve the quality of our knowledge about these two systems.

The concept of "mean" also arises in the statistical analysis of data, such as may be used in verifying conformance to quantitative requirements. Imagine that we have a population of objects whose mean weight, say, is unknown. Perhaps the population is too large, or some members of the population are inaccessible, or for some other reason it is impossible or undesirable to weigh each object in the population (so a census of the weights is not available). Then we may estimate the mean weight in the population by drawing a random sample from the population, weighing each object in the sample, and using standard statistical inference techniques. Let $x_1,\ldots, x_n$ denote n data points, or observations, recorded from some fixed phenomenon (e.g., the weights of the objects in the sample, or the number of failures in the first year of operation of n identical systems, or ...). This set is called a *sample* and consists of what are called in probability theory independent and identically distributed (*iid*) random variables. We are justified in asserting independence[20] if the collection of the observation from any system has no influence on the collection of the observation from any other system. The identical distribution property comes from the fact that all the systems covered by this sample are the same (model, series, manufacturer, etc.). Then the *sample mean* of these data is

$$\bar{X} = \frac{1}{n}\sum_{i=1}^{n} x_i,$$

[20] There is a formal definition of stochastic independence in probability theory [4] that we will suppress in favor of a more informal approach.

which is a simple unweighted average of the observed values. For example, the sample mean of the dataset {38, 55, 27, 10, 88, 41} is 43.167. The sample mean is an *estimator*[21] of the population mean μ (which is inaccessible); when playing this role, it is also denoted by $\hat{\mu}$. Each x_i is a random variable, so the sample mean is a random variable. It is an example of something called a *statistic*, which is nothing more or less than a function of some data. As a random variable, the sample mean has a cdf which is called the *sampling distribution* (*of the sample mean*). In general, it is difficult to compute the sampling distribution explicitly, so we turn to approximations which we discuss in Section 2.7.5. The reason the sample mean is so important is that when comparing the performance of a population of systems against a requirement written as a mean, the distance from the sample mean to the mean specified in the requirement tells something about the probability that the requirement is being satisfied. We have used this reasoning in the examples in Section 2.6.4.

2.7.2.2 Median

The median of a random variable is defined as the 50th percentile of the cdf of the variable. That is, half the values of the variable are less than or equal to the median and half are greater. In symbols, the median of X, denoted m, is any value of the discretionary variable for which $P\{X \le m\} = 0.5$.

In the example from Section 2.7.2.1, the median of X is any value in the interval $(6, 7]$ because $P\{X \le 6\} = 21/55 < 0.5$ and $P\{X \le 7\} = 28/55 > 0.5$.

The sample median of a dataset is the median of the values in the dataset. To compute the sample median, simply place the data in increasing order and find the center value. For the dataset {38, 55, 27, 10, 88, 41} considered in Section 2.7.2.1, the ordered values are {10, 27, 38, 41, 55, 88} and the median is any value between 38 and 41.[22] The sample median is also a statistic and as such has a sampling distribution (which is difficult to compute explicitly for non-normal random variables, so it is usually approximated by simulation).

In applied statistics, the median is sometimes considered a more desirable measure of central tendency than the mean because it is less sensitive to extreme values in the data. Despite this advantage, the median is not often used in engineering requirements. It does share at least one disadvantage with the mean, namely, that controlling only the median leaves open the possibility of large excursions of the variable.

2.7.2.3 Mode

The definition of the mode of a random variable X is different depending on whether X is discrete or continuum. If X is discrete, the mode of X is the value of X that has the largest probability. If X is a continuum random variable, the

[21] Statisticians commonly use the caret ^ over a variable to indicate that an estimator of that variable is being shown.

[22] This peculiarity arises when the number of elements in the data set is even. When the number is $2n-1$, odd, the median is the nth value in the ordered presentation of the data.

mode of X is defined as the value at which the density of X (if it exists) has a maximum. We will not discuss the mode further in this book because it rarely, if ever, appears in any quantitative engineering requirements.

2.7.3 Measures of Dispersion

Measures of central tendency of a random variable's distribution are usually not enough to give high-quality information about the variable. For example, consider two random variables, A and B. A takes on the values 98, 99, 100, 101, and 102 with equal probability (1/5 each), and B takes on the values 0, 50, 100, 150, and 200 with equal probability. The mean and median of A and the mean and median of B are all equal to 100, but you can't help having the feeling that the random variables A and B are quite different in some important sense. Somehow, B is much more spread out, or diffuse, than A. If A and B represented data collected on two different systems, we could reasonably say that we understand the behavior of the system from which A was observed better than we do that of B. At least, you might feel more confident that the next observation from the random phenomenon that produced A is more likely to be near 100 than the next observation from that for B. At least in this sense, the dataset from A provides us with a higher quality of information about the underlying system than does the dataset from B.

2.7.3.1 *Variance*

Fortunately, there is a concise way of expressing the notion of spread-out-ness or diffuseness. This involves the quantity called the *variance* of a random variable. Then variance of a random variable X is a weighted average of the squares of the distances from the mean of X to the values that X may attain. In symbols, when X is discrete,

$$\mathrm{Var}\, X = \sum_i (x_i - \mathrm{E}X)^2\, P\{X = x_i\} = \sum_i (x_i - \mathrm{E}X)^2\, p_i$$

and when X is a continuum variable,

$$\mathrm{Var}\, X = \int_\Omega (x - \mathrm{E}X)^2\, P(dx) = \int_{-\infty}^{\infty} (x - \mathrm{E}X)^2\, dF_X(x) = \int_{-\infty}^{\infty} (x - \mathrm{E}X)^2\, f_X(x)\, dx.$$

You can see that the further away the values of X are from $\mathrm{E}X$, the larger the variance becomes. That is, a large variance is a symptom of a diffuse, or spread-out, distribution, and conversely. A small variance indicates that the possible values of X are clustered near its mean. The variance of a random variable is zero if and only if the variable is equal to a constant with probability 1 (Exercise 12).

For the two random variables A and B discussed earlier in Section 2.7.3, we have $\mathrm{Var}\, A = 2$ and $\mathrm{Var}\, B = 5000$. Remember that both A and B have means equal to 100. Most of the values of A are near 100, while most of the values of B are far from 100. If A and B were datasets from two different populations, we would say that the information provided by A is of higher quality than the information provided by B because we would feel more confident about predicting future values of A than we would about B.

The standard deviation of a random variable is simply the square root of its variance. It is usually denoted by the lower case Greek letter sigma (σ). In the earlier examples, we have $\sigma(A) \approx 1.414$ and $\sigma(B) \approx 70.711$. As with variance, standard deviation tells something about how spread out a random variable (or, equivalently, its cdf) is: a large standard deviation indicates a diffuse, or spread-out, distribution, and conversely. A small standard deviation indicates a random variable whose possible values are clustered near its mean. The standard deviation of a random variable is zero if and only if the variable is equal to a constant with probability 1 (Exercise 12).

Requirements tip: Requirements almost never contain explicit reference to variance or standard deviation. These are usually considered technical issues that are remote from what is trying to be achieved by the requirement. We will see later in Chapter 5 how the notions of variance and standard deviation come into play naturally as part of the process of determining how well a system complies with its requirements.

If $x_1, \ldots, x_n$ denotes a dataset from some phenomenon, we can define the *sample variance* and the *sample standard deviation* for this dataset. The sample variance is defined as

$$\frac{1}{n-1}\sum_{i=1}^{n}(x_i - \bar{X})^2$$

where $\bar{X}$ is the sample mean; it is sometimes denoted by S. The sample standard deviation is the square root of this quantity.[23] The sample variance is an estimator of the population variance, which is inaccessible. The sample variance and the sample standard deviation are statistics, and as such have cdfs, the sampling distribution of the sample variance and the sampling distribution of the sample standard deviation. Again, explicit computation of these sampling distributions is not easy, and we resort to approximations in most practical cases (see Section 2.7.5).

2.7.4 Percentiles

The 100pth percentile of a distribution is the value x_p of the discretionary variable for which $P\{X \le x_p\} = p$. The median is $x_{0.5}$. Other terminology in common use includes quartiles ($x_{0.25}, x_{0.5}, x_{0.75}, x_{1.0}$) and deciles ($x_{0.1}, x_{0.2}, \ldots, x_{1.0}$). As

[23] The denominator is $n-1$ instead of n to provide what statisticians call an *unbiased* estimator of the population variance. See Ref. 7 for more details.

with the median, nonuniqueness is possible. Percentiles are rarely used as figures of merit in requirements even though they provide better control over the range of possible values of the random variable. For example, to require that the mean number of failures in the second year of operation be no greater than 3 leaves open that possibility that, while the requirement is being met, many systems may have more than three failures in the second year of operation and many others may have none at all. If instead we were to require that the 95th percentile number of failures in the second year of operation be no greater than 3, then no more than 5% of the systems installed would have more than 3 failures in that second year of operation if the requirement were being met. Use of percentiles in requirements is not common because determining the percentiles usually requires knowledge of the entire distribution, and computation with percentiles is less straightforward than with, say, means. When pencil-and-paper computations were the norm, these were substantive objections. Simulation modeling provides a convenient way to work with percentiles, even if only modest computing power is available.

2.7.5 The Central Limit Theorem and Confidence Intervals

Probably, the most frequently used figure of merit in sustainability engineering is the mean. Consequently, it is important to have a good grasp of the tools used for working with means. As we saw in the examples from Section 2.6.4, we use the sample mean to infer the population mean in cases where we wish to compare performance with a requirement based on the mean as a figure of merit. This inference rests on two approximations for the sampling distribution of the sample mean, one usable when the number of observations (elements in the sample) is large and the other when it is small. The large-sample approximation is based on the central limit theorem [4] which asserts that the average of a number of independent, identically distributed random variables having finite variance has approximately a standard normal distribution. Formally, if μ is the true (but unknown) population mean,

$$P\left\{\frac{\bar{X}_n - \mu}{\sigma/\sqrt{n}} \le z\right\} = \Phi_{(0,1)}(z)$$

where $\bar{X}_n$ denotes the sample mean from n observations and $\Phi_{(0,1)}$ denotes the standard normal distribution having mean 0 and variance 1. This limit makes it appropriate to use the normal distribution, as earlier, as an approximate distribution for the sample mean when the number of observations is large. In practice, a good rule of thumb is that if there are more than 10–15 observations, the normal approximation is usually acceptable unless the variables involved are very diffuse ("have long tails"), a condition that is not often encountered in run-of-the-mill reliability studies. For smaller datasets, we use instead the fact that the asymptotic distribution of the sample mean is a Student's t distribution with $n-1$ degrees of freedom [10]:

$$P\left\{\frac{\bar{X}_n-\mu}{\sigma/\sqrt{n}}\le z\right\}=t_{(n-1)}(z)$$

where n is the number of elements of the dataset. These facts form the basis for the computations shown in the examples in Section 2.6.4.

We may also express inferences about the population mean in the form of a *confidence interval.* A 100p% two-sided confidence interval ($0<p<1$) for the population mean μ is given by

$$\left[\bar{X}-a\frac{S}{\sqrt{n}},\ \bar{X}+a\frac{S}{\sqrt{n}}\right]$$

where the confidence coefficient a comes from the percentiles of the standard normal distribution when the number of observations is large enough that the normal approximation is appropriate. The confidence intervals most often used are the 90% ($p=0.9$), 95%, and 99%; Table 2.3 gives the corresponding confidence coefficients (based on the normal distribution).

The table includes 68% to show how much of the distribution lies within one standard error ($\pm S/\sqrt{n}$) of its center. When the number of observations is too small for the normal approximation to be suitable, the confidence coefficients are obtained from the Student's t distribution with $n-1$ degrees of freedom (n is the sample size and the coefficients will change with n). For example, when $n=6$, the two-sided confidence coefficients based on the $t_{(5)}$ distribution are 2.01, 2.57, and 4.03 for the 90%, 95%, and 99% confidence intervals, respectively. As n gets large, the Student's t distribution becomes approximately equal to the normal distribution.

A confidence interval expresses the degree of, well, confidence we have about the location of the population mean based on the sample that has been chosen. A 100p% confidence interval represents a conclusion that, if the experiment of choosing a sample from that population were repeated many times, in about 100p% of those repeated samplings, the 100p% confidence

TABLE 2.3 Confidence Coefficients Based on the Normal Distribution

	Confidence Coefficient	
Confidence Level (%)	One-Sided	Two-Sided
68	0.75	1.0
90	1.28	1.645
95	1.645	1.96
99	2.33	2.58

interval that was derived from the data would contain the population mean (note that each time the experiment is repeated, a different sample is obtained, and therefore a different confidence interval results—approximately $100p\%$ of those different confidence intervals would contain the population mean).

Example: Consider the sample of eight refrigerators first encountered in Section 2.6.4.2. Give a 95% confidence interval for the mean outage time in the population of refrigerator installations from which this sample is drawn. From Table 2.2, the sample mean of the 24 outage times recorded is 5.53, and the sample standard deviation is 2.84. We may use the normal approximation because the number of observations is 24 (not 8), so the 95% confidence interval is

$$\left[5.53 - 1.96\frac{2.84}{\sqrt{24}}, 5.53 + 1.96\frac{2.84}{\sqrt{24}}\right] = [4.39, 6.66].$$

This interval contains the requirement (5 hours), but no conclusion is warranted about whether or not the requirement is satisfied. If the interval did *not* contain the requirement, it would be appropriate to assert that the requirement is *not* being met with 95% (or whatever the confidence level was) confidence.

Because of this impreciseness, use of confidence intervals for making inferences about whether a requirement is being met is not recommended. The estimation procedure described in Section 2.6.4 is preferred because it will yield an estimate of the probability that the requirement is being met.

2.8 CURRENT BEST PRACTICES IN DEVELOPING RELIABILITY REQUIREMENTS

One of the premises of this book is that reliability requirements are created through a systematic, repeatable process that may be summarized as follows:

- Catalog the system attribute requirements.
- Determine the failure modes associated with each of the requirements.
- Determine the customer's needs and desires for continued satisfactory operation, considering each failure mode.
- Balance the customers' needs and desires regarding reliability with the economics of developing a system meeting the reliability requirements,
- Create a system reliability budget.
- Document the reliability requirements that result from this analysis.

The remainder of this section will examine each step in detail and offer suggestions about how to accomplish the related tasks.

2.8.1 Determination of Failure Modes

So far, we have established that a failure is a violation of some system attribute (functional, performance, physical, safety) requirement. The failures that customers are concerned about include violations of these and any others that the customer feels are necessary to satisfactory operation throughout the useful life of the product, system, or service. From this perspective, we may catalog the failure modes in the product, system, or service by systematically reviewing the relevant requirements and undertaking analyses to identify the possible failure modes (Section 2.2.6) associated with each requirement. Sometimes, this can be accomplished informally in the systems engineering team if there is enough prior experience with the system or ones like it. However, in totally new systems, or in high-consequence systems (Chapter 7), informal methods may not be enough. In that case, following the same reasoning as in the first steps of a fault tree analysis offers a systematic approach to determining the failure modes associated with a particular requirement. Violation of the requirement is placed as the "top event" in a fault tree, and inductive reasoning is applied to scour the state space of the system for events (configurations of states) that lead to the violation. As a rule, the first layer of the fault tree is a list of the failure modes associated with that requirement. At this stage, it is not necessary to carry out the fault tree analysis further; but when the design for reliability stage is reached, the beginnings of these fault trees can serve as a foundation for the more detailed study that would be then appropriate. Detailed examination of fault tree methods is found in Section 6.6.1.

Example: One of the safety requirements for a home heating system is that it produces no carbon monoxide that can reach living spaces. To identify the failure modes associated with this requirement, set the event "system produces carbon monoxide reaching living spaces" as the top event of a fault tree. Then the fault tree reasoning asks, what are the events in the operation of the system that can cause carbon monoxide to be produced and to reach the living spaces? These events include (i) improper gas/air mixture and (ii) a leak or leaks in the flue. The fault tree reasoning has thus identified two failure modes in the heating system that cause the undesirable event of carbon monoxide reaching the living spaces. At this point, where we are preparing to write a reliability requirement for this safety requirement, there is no need to carry out the fault tree analysis any further. However, when detailed design of the system is begun, it would be appropriate to carry out this fault tree analysis (and others, as necessary) to help determine preventive measures that can be employed to avoid violating this requirement (and others).

2.8.2 Determination of Customer Needs and Desires for Reliability and Economic Balance with Reliability Requirements

Once a list of failure modes for the product, system, or service is in hand, crafting corresponding reliability requirements needs input from customers and users about how often, and for how long, these failure modes can be tolerated.

Users, of course, always want systems that never fail. It is rarely, if ever, possible to reach this goal, and it is usually expensive to approach it closely. A more useful question, then, is "How much reliability is the customer willing to pay for?" Other things being equal, a more reliable system would cost more to develop and manufacture unless appropriate design for reliability techniques are employed early in the system's development. In that case, we can even reach the seemingly paradoxical result that a more reliable system can have a lower development cost than a less reliable system (of the same kind) that is developed using inefficient or ineffective methods. Increased attention to design for reliability means more work in the early stages of a system's design and so the prevention costs for this system could increase. Most system suppliers would translate that into an increase in the price of the system, so systems engineers need a good understanding of, essentially, the customers' elasticity about price and reliability. More sophisticated customers may realize that the additional first cost may be accompanied by a reduction in failure costs. Whether the system supplier can employ this reasoning successfully may depend on how well they understand their markets.

2.8.2.1 Quality function deployment and other formal methods

Among the systems engineer's most important and challenging responsibilities is that of determining customer needs and desires for system operation. Usually, we think of these in terms of the system's functions and other attributes like appearance, weight, etc., but the responsibility extends to sustainability elements as well. We have previously (Section 1.6.1) alluded to quality function deployment (QFD), The "House of Quality," and Kano analysis as structured techniques systems engineers can use to accomplish these tasks. It is not the purpose of this book to teach you how to use any of these methods. Many excellent resources, including textbooks and short courses, both live and online, will help you learn how to apply these successfully. Our purpose here is to make you aware that these methods exist and are suitable not only for requirements pertaining to functionality, appearance, safety, etc., where they are usually advertised, but also for requirements in the sustainability areas.

2.8.2.2 Industry standards

Many industries have developed standard reliability requirements that can be applied directly or can serve as a starting point for new systems, products, or services. Some examples include

- Telecommunications: Globally, the International Electrotechnical Commission (IEC) and International Telecommunications Union (ITU, a part of the ISO) have developed many standards for the reliability of telecommunications equipment and services. In the United States, Telcordia (formerly Bell Communications Research, or Bellcore) has published similar standards. Even if new systems or services are not required to

conform to these standards, they often serve as a starting point for negotiations between suppliers and customers regarding reliability (and other) requirements.
- Defense Acquisition: With the demise of standards under MIL-SPEC reform, the documents formerly known as military standards have been converted to handbooks. A list of handbooks that pertain to reliability in the defense industry can be found at URL [11].
- Electric Power: In the United States, section 215 of the Federal Power Act required the North American Electric Reliability Corporation (NERC) to develop reliability standards that are reviewed by the Federal Power Commission, mandatory, and enforceable. A comprehensive list of reliability standards applicable to the US electric power industry is given on the web page [14].
- Automotive: The Society of Automotive Engineers (SAE) publishes quality, reliability, and durability standards for automobiles [12].
- Aerospace and Commercial Aviation: The SAE also publishes reliability standards for the aerospace industry [13].

2.8.3 Review All Reliability Requirements for Completeness

As noted earlier, all reliability requirements should contain the essential elements that promote clarity and completeness:

- Statement about which requirements violation is being covered,
- Statement of a reliability effectiveness criterion or a reliability figure of merit,
- A quantitative limit or range for that effectiveness criterion or figure of merit,
- The period of time over which the limit or range is to apply, and
- The conditions under which the requirement is to apply.

Often, a reliability requirement is written with no reference to any particular failure mode. In that case, the only reasonable interpretation is that the requirement pertains to *any* failure mode in the system. Systems engineers may find it helpful to formally conduct a cross-functional team review of draft reliability requirements so that the entire development team can contribute to making the requirements better and so that they can also get an early idea about what design for reliability activities may be called for.

2.8.4 Allocation of System Reliability Requirements to System Components

Once system reliability requirements have been established, part of the design responsibility is to assign reliability requirements to the system's constituent components and subassemblies so that these all combine to cause the system to meet its reliability requirements. The result of this allocation is sometimes called a *reliability budget*. The first step in this allocation procedure is to create a system functional decomposition for each major system function to which a reliability requirement is attached. For instance, major functions of a consumer refrigerator include

- keeping the refrigerator compartment within specified temperature limits and
- dispensing ice cubes from an external ice dispenser.

A functional decomposition can be developed for each of these two functions; see Section 3.4.1 for discussion of functional decomposition. The functional decomposition identifies the major subassemblies and components of the system that will need to act together to produce the desired outcome. For instance, keeping the refrigerator compartment within specified temperature limits requires proper operation of the compressor, thermostat, controller, and insulation. A system functional decomposition for this failure mode may include these components or subassemblies together with an understanding of how they function together to keep the temperature stable. This understanding forms the basis for the reliability block diagram (Section 3.4.3) which is then used to evaluate the assignment of reliability requirements to each of the subassemblies so that satisfaction of the overall system-level reliability requirements can be determined. In this example, failure of any of the compressor, thermostat, controller, or insulation results in inability to hold the temperature within range. The compressor, thermostat, controller, and insulation thus constitute an ensemble of single points of failure, or a *series system* in the reliability modeling terminology to be introduced in Section 3.4.4.

Formal methods for assignment of system reliability requirements to constituent parts of the system involve optimal allocation of reliability requirements based on minimizing cost or some other equivalent objective. Some discussion and references are provided in Section 6.6.1.4. Formal methods are not often used in practice, except in high-consequence systems, because they can be time-consuming and they require basic information that is often difficult or impossible to obtain. Instead, more *ad hoc* reliability engineering methods for allocation are used. These are based on approximations to formal methods or on trial-and-error iterations.

Example: Consider the series system of compressor, thermostat, and controller for the refrigeration system. Suppose the refrigerator reliability requirements for the temperature stability is as follows: the probability that the time to the first instance of out-of-range temperature is more than 100,000 hours shall not exceed 0.10.[24] Suppose that the survival probabilities for 100,000 hours for the component parts are as Table 2.4.

[24] We omit the conditions under which this applies as being not germane to the example. But you should have noticed this was missing.

TABLE 2.4 Component Survival Probabilities

Component	100,000-hour Survival Probability
Compressor	p_C
Thermostat	p_T
Controller	p_E
Insulation	1.0

Note that we have assigned the insulation probability 1 of survival past 100,000 hours. This reflects a belief that failures of the insulation are not likely to occur over the service life of the refrigerator.[25] Then the probability that the refrigeration temperature stays within range for at least 100,000 hours is $p_C p_T p_E$. This is to be at least 0.90, and we are now to find values of p_C, p_T, and p_E that satisfy this inequality. A formal method to make this allocation might be as follows: Let $z_A(p_A)$ denote the cost of a unit having the designated survival probability p_A for $A = C, T, E$. Then an allocation based on cost minimization is a solution to the mathematical optimization problem:

Find probabilities p_C, p_T, and p_E
to minimize $z_C(p_C) + z_T(p_T) + z_E(p_E)$
subject to $p_C p_T p_E \geq 0.90$.

If the refrigerator manufacturer's intention is to minimize first cost, then the costs appearing in this problem will be the acquisition costs of the components. If the refrigerator manufacturer chooses instead to minimize the cost to the customer, the costs appearing in the allocation problem will be the repair costs incurred by the customer, including parts and labor. In either case, the challenges in identifying the z-functions are evident. Not least of these is that there may be only one or two choices for each component (although this could be a simplification rather than a challenge). Formal methods remain a good conceptual approach to reliability allocation or budgeting, but more choices are needed before they can become routinely applied. Informally, one way to reason through this problem is as follows. Suppose we roughly judge that the reliability of the thermostat and controller is related to that of the compressor, based on component complexity, past history, or other experience factors. If, for example, we say the reliability of the thermostat is about five times better than that of the compressor, how might that statement be interpreted? Obviously, we can't mean that the survival probability is five times as great, because if the compressor survival probability is 0.90 at some time, then that would make the survival probability for the thermostat 4.50 and this is not a probability. A consistent interpretation can be based on the

[25] Insulation may fail as a consequence of some other failure, such as the refrigerator catching fire; but if the refrigerator catches fire, the owner has bigger problems to worry about than insulation failure.

idea that the probability of failure is one-fifth as small for the thermostat as for the compressor. That is, if the survival probability for the compressor is 0.90 at 100,000 hours, then the probability that the compressor fails before 100,000 hours is 0.10, and one-fifth of that is 0.02, so we may take the survival probability of the thermostat as 0.98 as a consistent interpretation of the desired ratio. Similarly, if we say the reliability of the controller is twice as good as that of the compressor, then the survival probability for the controller is 0.95. Now we return to the general problem: if the survival probability for the compressor is p_C, then (assuming the same ratios) the survival probability for the thermostat is $p_T=1-(1-p_C)/5$ and the survival probability for the controller is $p_E=1-(1-p_C)/2$. These assumptions reduce the allocation problem to a one-dimensional calculus exercise: find the smallest value of p_C (presumably, z_C is a nonincreasing function of p_C) so that

$$p_C\left[1-(1-p_C)/5\right]\left[1-(1-p_C)/2\right]\geq 0.9.$$

While this is not quite as elegant theoretically as are formal methods using mathematical optimization, it does provide some guidance in a situation where the quality of the available information may not be good enough to justify more precise methods.

Regardless how a solution to the problem is approached, allocation of system reliability to the components of the system is an essential step in designing a system so that its reliability requirements will be met, and is considered a best practice in reliability engineering. See also Sections 4.7.3 and 8.7.

2.8.5 Document Reliability Requirements

It hardly rates as exciting news that we are going to recommend careful and systematic documentation of whatever reliability requirements have been created. All members of the development team need access to and understanding of the reliability requirements so that their actions may be guided by clear understanding of the target they are shooting for and so that adjustments may be made on a sound basis if later development indicates that they may be needed. If your organization lacks a process for documentation, it would be a good idea to create one.

2.9 CHAPTER SUMMARY

"Reliability" is used in everyday conversation in a manner that is familiar to most people. The more precise meanings of "reliability" used in engineering are informed by this understanding. The foundational definition of reliability involves the ability of a system to operate properly (according to its attribute requirements) under stated conditions for a stated period of time. Other engineering uses of "reliability" stem from this definition. These include

- the probability that a system operates properly under stated conditions for a stated period of time. This is a reliability figure of merit (Section 2.4.2).
- the survivor function of a component, subsystem, or system: the complement of the life distribution of the component, subsystem, or system (4.3.2).
- as a portmanteau term used when one needs to refer to some aspect of system operation involving frequency and/or duration of failures and outages. In this sense, "reliability" can include availability, failure rate, survivor function, etc.

Reliability requirements are facilitated by understanding reliability effectiveness criteria and figures of merit. These are the pathways by which quantitative concepts are introduced into reliability engineering. Reliability effectiveness criteria are simply quantitative expressions of operation connected with frequency and/or duration of failures and outages and commonly include number of failures per unit time, times between outages, time to first failure, and others. These quantities are usually conceptualized as random variables for reasons discussed in Section 2.4.1. Quantitative descriptors that make these concepts easier to use are called reliability figures of merit, and include the life distribution and survivor function, mean, variance, median, and others.

Reliability requirements should contain reference to

- the failure mode(s) that the requirement is supposed to apply to,
- the reliability effectiveness criterion or figure of merit that is being controlled,
- a quantitative limit or range for that effectiveness criterion or figure of merit,
- the environmental or other operating conditions prevailing under which the requirement is to apply, and
- the interval of time over which the requirement is to apply.

When creating or reviewing reliability requirements, this checklist should be used to ensure that the requirement is complete and unambiguous.

Reliability requirements do not exist in a vacuum. They are intended to drive certain behaviors, and it is important to go back and check after system installation to see whether the requirements are being met. Using the concepts of reliability effectiveness criteria and figures of merit, we study how reliability requirements may be interpreted in light of this need to compare performance with requirements. Some introductory material on this is provided in this chapter as a prelude to more detailed discussion in Chapter 5. A key point is that many comparisons may only be possible in a statistical sense because there is always variability of the values of each reliability effectiveness criterion across the members of the installed base, and, in most cases, data from field reliability performance form only a sample of the installed base and statistical procedures respecting this sampling nature of the activity must be used.

The chapter includes with some more detailed discussion of other figures of merit in common use. Certainly, the mean of some reliability effectiveness criterion is very widely used in reliability requirements, and good understanding of the behaviors that this restricts and allows is an asset to systems engineers. We touch on variance and standard deviation not so much because they are commonly used in requirements (they are not) but because they are important for the statistical procedures used to compare realized reliability with requirements.

2.10 EXERCISES

1. Suppose a device able to withstand a voltage stress of 60 V is operated in an environment where voltage spikes occur at times $T_1, T_2, \ldots$, and have corresponding magnitudes $V_1, V_2, \ldots$. Write an expression for the time at which the device fails. If $\{T_1, T_2, \ldots\}$ forms a homogeneous Poisson process with rate $\lambda > 0$, what is the expected time to failure?
2. It is not uncommon for a system to operate improperly, or not operate at all, when it is used outside the conditions specified in the reliability requirements. Is this a failure? Discuss.
3. In light of the discussion preceding Figure 2.1, what is a reasonable definition for "restoration time?" How have you seen this phrase used in your experience? Is it important that a consistent definition for "restoration time" be universally agreed? What would you recommend as a definition for "restoration time" and what are the advantages and disadvantages of your recommendation?
4. Discuss the relationships between reliability, maintainability, and supportability for the following systems:
   a. A satellite
   b. An undersea cable telecommunications system
   c. A commercial aircraft
   d. A military aircraft
   e. An implantable medical device (e.g., a pacemaker)
   f. A DVD player (consumer product).
5. Is the reliability requirement on the example in Section 2.5.1 based on a reliability effectiveness criterion or on a reliability figure of merit? In a population of 50,000 refrigerators, what is the expected number of refrigerators that do not meet the requirement? Do you have enough information to carry out the computation?
6. Is the reliability requirement on the example in Section 2.5.2 based on a reliability effectiveness criterion or on a reliability figure of merit? Is the requirement complete? Over a period of 1 year, what is the expected number of days in which the requirement is not satisfied? Do you have enough information to carry out the computation? What is a reasonable interpretation of "the probability that the link is in a working condition"? (You may wish to consult Section 4.3.2 for help with this part).

7. Critique the example requirement presented in Section 2.5.3. Is failure well defined? Is the period of time over which the requirement is supposed to apply stated clearly? The conditions under which the requirement is supposed to apply are not stated. What does this mean? Could you improve on the requirement as written?
8. Solve the optimization problem in Section 2.5.2 when $c_1 = \$4.55$ and $c_2 = \$7.12$.
9. Is the electric power utility reliability requirement cited in the example at the end of Section 2.5.3 based on an effectiveness criterion or a figure of merit? Is the requirement complete? A reliability model for the power distribution network may be constructed to compute the steady-state ($t \to \infty$) availability of power at a "typical" customer premises terminal (meter). If the computed value is less than 0.999995, does this provide enough information for you to tell whether the requirement is likely to be met?
10. The "1–10–100 rule." Discuss the intent of the "1–10–100" rule in detail. Who bears the costs at each stage described by the rule? What are the ramifications of the argument that the customer bears most, if not all, of the cost of a failure during use and so it is of no interest to the supplier to do anything about them?
11. Show that the variance of a random variable is zero if and only if the variable is equal to a constant with probability 1.
12. Complete the allocation problem in the example at the end of Section 2.8.4.

REFERENCES

1. Barlow RE, Clarotti CA, Spizzichino F. *Reliability and Decision Making*. Boca Raton: CRC Press; 1993.
2. Berry DA, Lindgren BW. *Statistics: Theory and Methods*. 2nd ed. Belmont: Duxbury Press (Wadsworth); 1996.
3. Blischke WR, Murthy DNP. *Warranty Cost Analysis*. Boca Raton: CRC Press; 1994.
4. Chung KL. *A Course in Probability Theory*. 3rd ed. New York: Springer; 2001.
5. Cohn M. *Agile Estimating and Planning*. New York: Prentice-Hall; 2005.
6. Cui L, Li H, Xu SH. Reliability and risk management. Annal Oper Res 2014;212 (1):1–2.
7. Ficalora JP, Cohen L. *Quality Function Deployment and Six Sigma: A QFD Handbook*. 2nd ed. New York: Prentice-Hall; 2009.
8. Ford LR Jr, Fulkerson DR. *Flows in Networks*. Princeton: Princeton University Press; 1962.
9. Hart CWL, Maher D, Montelongo M. *Florida Power and Light Quality Improvement Story Exercise*. Cambridge: Harvard Business School; 1988.
10. Hoel PG, Port SC, Stone CJ. *Introduction to Statistical Theory*. Boston: Houghton Mifflin; 1971.
11. http://www.rollanet.org/~asemmsd/em-handbook/Resources/ram_r1.html. Accessed November 9, 2014.
12. http://topics.sae.org/qrd/standards/automotive/. Accessed November 9, 2014.

13. http://topics.sae.org/reliability-maintainability-supportability/standards/aerospace/. Accessed November 9, 2014.
14. http://www.nerc.com/pa/Stand/Pages/AllReliabilityStandards.aspx?jurisdiction= United%20States. Accessed November 9, 2014.
15. https://en.wikipedia.org/wiki/Fukushima_Daiichi_nuclear_disaster. Accessed November 9, 2014.
16. Kratz L et al. *Designing and Assessing Supportability in DOD Weapon Systems: A Guide to Increased Reliability and Reduced Logistics Footprint*. Washington, DC: US Department of Defense Memorandum for the Acquisition Community; 2003.
17. Leveson NG. Safety as a system property. Commun ACM 1995;38 (11):146.
18. Madu CN. *House of Quality (QFD) in a Minute*. 2nd ed. Fairfield: Chi Publishers; 2006.
19. National Research Council. *Reliability Issues for DoD Systems: Report of a Workshop*. Washington: The National Academies Press; 2002.
20. Raiffa H, Schlaifer R. *Applied Statistical Decision Theory*. New York: John Wiley and Sons, Inc; 2000.
21. Ramirez-Marquez JE, Coit DW, Tortorella M. A generalized multistate based path vector approach for multistate two-terminal reliability. IIE Trans Reliab 2007;38 (6):477–488.
22. Tortorella M. Service reliability theory and engineering, I: foundations. Qual Technol Quant Manage 2005;2 (1):1–16.
23. Tortorella M. Service reliability theory and engineering, II: models and examples. Qual Technol Quantitative Manage 2005;2 (1):17–37.
24. Tortorella M. Design for network resiliency. In: Cochran JJ, Jr. Cox LA, Keskinocak P, Kharoufeh JP, Smith JC, editors. *Wiley Encyclopedia of Operations Research and Management Science*. Volume 2, Hoboken: John Wiley and Sons, Inc; 2011. p 1364–1381.
25. U. S. Government Accountability Office. *QI Story Tools and Techniques: A Guidebook for Teams*. Bibliogov. Publication no. TQM-92-2; 2013.
26. Wadsworth HM, Stephens KS, Godfrey AB. *Modern Methods for Quality Control and Improvement*. New York: John Wiley & Sons, Inc; 2002.

3

Reliability Modeling for Systems Engineers

Nonmaintained Systems

3.1 WHAT TO EXPECT FROM THIS CHAPTER

It is not the purpose of this book to support your becoming a reliability engineering specialist. As a systems engineer, though, you will be interacting with these specialists both as a supplier and as a customer. You will be supplying reliability requirements that specialist engineers will use to guide their design for reliability work. You will be a customer for information flowing back from reliability engineering specialists regarding how well a design, in its current state, is likely to meet those reliability requirements and whether deployed systems are meeting their reliability requirements. The purpose of this chapter, then, is primarily to support your supplier and customer roles in these interactions. You will need enough facility with the language and concepts of reliability engineering that you will create sensible reliability requirements. Much of this was covered in Chapter 2, and the material covered in this chapter supports and amplifies the concepts introduced there. You will also need enough of this facility to be able to sensibly use the information provided by specialist reliability engineers so that design may be properly guided.

The material in this chapter is designed to support this latter need. What you will find here is chosen so that it reinforces correct use of the concepts and

Reliability, Maintainability, and Supportability: Best Practices for Systems Engineers, First Edition. Michael Tortorella.

language of reliability modeling for nonmaintained systems.[1] It is complete enough that it covers almost all situations you will normally encounter, and if you learn this well you will be able to adapt it to unusual situations as well. While everything here is precise and in a useful order, no attempt is made to provide mathematical rigor with theorems and proofs even though there is a flourishing mathematical theory of reliability [3, 4] that underpins these ideas. If you wish to follow these developments further, many additional references are provided.

3.2 INTRODUCTION

The industrial, medical, and military systems prevalent today are usually very complex and closely coupled, and expensive and time-consuming to develop. For transparent economic and schedule reasons, it is not even remotely realistic to test such systems for reliability. Indeed, to do so would be to fly in the face of the guiding principle of contemporary systems engineering for the sustainability disciplines: design the system from the earliest stages of its development to incorporate features that promote reliable, maintainable, and supportable operation. In short, in preference to a costly and lengthy testing program, or, worse, design scrap-and-rework, take those actions during systems engineering and design that lead to a sustainable, profitable system.

Accepting, then, that testing a complicated system for reliability is not sensible, what can systems engineers and reliability engineers do to ensure that a system meets the reliability needs of its customers? In this book, we advocate strongly for the discipline of *design for reliability*, the discipline that encompasses actions that are taken during systems engineering and design to anticipate and prevent failures. Design for reliability is discussed from this point of view in Chapter 6 where we introduce specific methods such as fault tree analysis (FTA), failure modes, effects, and criticality analysis (FMECA), and others that provide systematic, repeatable techniques that are widely applicable and very effective in anticipating the failures that are possible in the system and deploying suitable countermeasures that prevent those failures from occurring. An important part of the design for reliability process is the ability to project or forecast in quantitative terms the reliability one can expect of the system given the current state of the design. A discipline called *reliability modeling* has been developed to enable these sorts of quantitative projections to be made, even before any of the system may be built (or even prototyped).

Reliability modeling is based on the observation that while the systems we deal with are complex and closely coupled, usually they are made up of a large number of simpler components. Reliability modeling is a process of combining, in suitable mathematical fashion, quantitative information about

[1] The corresponding ideas for maintained systems are covered in Chapter 4.

the reliability of individual components to produce information about the reliability of the complex assembly of those components that is the system in question. It is usually possible to obtain information about the reliability of these simpler components from life testing, fundamental physical principles, and real field experience. Life testing of components is possible because it deals with only one (population of identical) component at a time; complicated interactions with other components are not present, and varying environmental conditions can be applied to characterize the component's reliability in different environments likely to be encountered in operation [16, 62]. Estimation of component reliability from fundamental physical principles is possible in some cases because the physical, chemical, mechanical, and/or electrical mechanisms causing degradation of the component have been identified in many practical classes of components [10, 24]. Component reliability may also be estimated from real operational experience with systems that contain the component provided that the failure that caused a system to be taken out of service for repair can be traced to that specific component [7, 55] (see also Section 5.6). This chapter is devoted to helping you gain an understanding of reliability modeling for nonmaintained systems so that you are equipped to assess whether your reliability requirements are likely to be met as part of an ongoing process throughout the design and development of the system. Reliability models for nonmaintained systems introduced in this chapter form building blocks for the reliability models for maintained systems discussed in Chapter 4.

However, all the reliability modeling you can afford is of little value unless you use what you learned from it to do one (or both) of two things:

1. Improve the reliability of the system if modeling shows that the system in its current configuration is unlikely to meet its reliability requirements.
2. Determine that the reliability requirements originally proposed are too restrictive and may be loosened, possibly creating an opportunity for development cost savings.

Chapter 5 discusses comparison of what is learned from reliability modeling (usually called a "reliability prediction") with the relevant reliability requirement(s). To improve the reliability of the system, additional design for reliability actions must be undertaken or the design for reliability actions already undertaken should be re-examined at greater depth (Chapter 6). The alternative is to decide that the original reliability requirements were more restrictive than they needed to be—but this decision can't really be made without thorough re-examination of the process by which they were created (QFD, House of Quality, Kano analysis, etc., introduced in Section 1.6.1). Without this response, reliability modeling has little value.

Finally, most systems are intended to be repaired when they fail, and by the repair to be restored to service. There are obvious exceptions, of course (viz., satellites, although the example of the Hubble Space Telescope shows that when the stakes are high enough, truly heroic measures will be undertaken to repair even some systems that are traditionally designated as non-repairable). Many reliability effectiveness criteria are appropriate for describing the frequency and duration of failures of a maintainable system (see Section 4.3). The system maintenance concept (see Chapter 10) tells how the system will be restored to service when it fails, and which part(s) of the system are designated as repaired and which parts are not repaired. A reliability model for the system mirrors the system maintenance plan: the model builds up reliability descriptions of the maintained parts of the system from reliability descriptions of their constituent components and subassemblies. All systems contain some components that are not maintainable in the sense that if a system failure is traceable to one such nonmaintainable component, repair of the system is effected by discarding the failed component and replacing it with another (usually new) one. Some systems also contain more complex subassemblies that may be removed and replaced in order to bring a system back to proper operation and that are sufficiently complex and expensive that the removed units are themselves repaired and used as spare parts for later system repairs. See Chapter 11 for more details on this type of operation. Accordingly, Chapters 3 and 4 are structured so that we learn about reliability effectiveness criteria and models for nonmaintainable components first and then we learn how these are combined to form reliability effectiveness criteria and models for the higher level entries in the system maintenance concept—the subassemblies, line-replaceable units, etc., on up to the system as a whole.

3.3 RELIABILITY EFFECTIVENESS CRITERIA AND FIGURES OF MERIT FOR NONMAINTAINED UNITS

3.3.1 Introduction

This section discusses the various ways we describe quantitatively the reliability of a nonmaintained component or system. An object that is not maintained is one that ceases operation permanently when it fails. No repair is performed and a failed nonmaintained component is usually discarded. An object that is not maintained may be a simple, unitary object like a resistor or a ball bearing (these are not repaired because it is physically impossible or economically unreasonable to repair them), or it may be a complicated object like a rocket or satellite (not repaired because they are destroyed when used or are impossible to access). Simple nonmaintained components usually form the constituents of a larger system that may be maintained or not. Most complex systems are maintained to some degree. For example, while failed hardware in a consumer router (for home networking) may not be repairable, the firmware in the router can be restored to its original factory configuration by pressing the reset button. We study reliability effectiveness criteria and figures of merit for nonmaintained items because

- reliability effectiveness criteria and figures of merit are used to describe mission success probabilities for systems that may be maintainable but cannot be maintained while in use (see Section 4.3.4) and
- reliability models for a maintained system are built up from simpler reliability models for the nonmaintained components making up the system.

By contrast, of course, an object that is maintained undergoes some procedure(s) to restore it to functioning operation when it fails; in this case, repeated failures of the same object are possible. The system maintenance concept will tell which part(s) of the system are nonmaintained and which are maintained and will give instructions for restoration of the system to functioning condition when it fails because of the failure of one of the nonmaintained parts of the system (or any other type of failure, for that matter).

Language tip: The concepts presented in Section 3.2 apply to any object that is not maintained, no matter how simple or complicated. We will use the language of "unit" or "component" to describe such objects even though the words "unit" or "component" seem to imply a single, unitary object like a resistor or ball bearing and do not seem to apply to complicated objects like satellites. Nonetheless, the reliability effectiveness criteria we shall describe in Section 3.2 pertain to all such objects, simple or complicated, provided they are, or when they are considered to be, nonmaintained.

Most real engineering systems are maintained: when they experience failure, they are repaired and put back into service. There are, of course, significant exceptions (most notably, satellites) for which repair is not possible at all,[2] and other systems (such as undersea cable telecommunications systems) for which repair is possible but extremely expensive. All systems contain components that are not maintained but instead are replaced when they fail. The replaced component is discarded if it is not repairable, like a surface-mount inductor. Other replaced "components" are more elaborate subassemblies that may be repaired and placed into a spares inventory if it makes economic sense to do this. Reliability models that produce reliability effectiveness criteria for maintained systems are constructed from simpler models for the reliability of their nonmaintained constituent components and subassemblies, and it is these latter models that we study in this chapter.

This is a good time to explore the relationship between failures of parts or components and system failures. A system failure is any instance of not meeting some system requirement. As discussed in Chapter 2, not meeting a system requirement does not necessarily mean that the system has totally ceased operation. Many reliability models are constructed based on the belief that

[2] But even a satellite may have its software rebooted remotely.

system failure is equivalent to total cessation of system operation. The reality is somewhat more complicated. Some system requirements pertain to performance characteristics like throughput, delay, tolerance, etc., that may be measured on a continuum scale. Instances of system operation where some performance characteristic falls outside the range specified in the requirement constitute system failures, even though the system may still be operating, perhaps with some reduced capability. Such failures are indeed within the scope of reliability modeling, and component failures may contribute to these events. This points up the importance of an effective system functional decomposition (Section 3.4.1) as a first step in creating a reliability model and a maintenance plan for each system failure mode. Obviously, any realistic system has too many failure modes for it to be feasible to create a reliability model for every one of them. Some method is required to decide which failure modes to focus attention on; an effective system reliability analysis requires this as a first step.

The key operational characteristic of a nonmaintained item is that when it fails, no attempt is made to repair it, and it is instead discarded (possibly recycled, but whatever disposition it may receive, it is not reused in the original system). The decision about whether any particular component should be considered maintained or nonmaintained is largely an economic one, and is closely connected with the maintenance concept for the system as a whole (see Chapter 10). The always-cited classic example of a nonmaintained unit is the incandescent light bulb (and now we will refer to anything nonmaintained as a unit; this may encompass individual components such as resistors, bearings, hoses, etc., or various assemblies, composed of several components, that are part of a larger system, or in some cases an entire system that is not maintained). When a light bulb burns out and ceases to produce light, it is discarded and the socket that contained it is filled with another, usually new, light bulb.

The repair-or-replace decision is part of the system maintenance concept. In addition to other factors such as accessibility, staff training, etc., which are covered extensively in Chapter 10, this decision has a large economic component. Consider, for example, that it is technically possible to repair a light bulb. Careful removal of the glass envelope from the base, reinstallation of a good filament, and resealing and re-evacuating the bulb are all operations that are easily within contemporary technical capabilities. However, this is never done because it would be a monumentally stupid thing to do from an economic point of view (note that, however, some kinds of expensive ceramic/metal high-power vacuum tubes are sometimes repaired by a process very much like that described here [31]). At this time, raw materials for incandescent bulbs are cheap and plentiful, and the cost of manufacturing a new bulb is measured in pennies. The cost to carry out the repair operations cited would be orders of magnitude greater than the cost of producing a new bulb, and so today this is never done (except possibly for some signally important units like Edison's original bulb which is kept running for historical purposes). There may come a time (and this will probably be an unhappy time) when these raw materials may be scarce and/or expensive, and the consequent increased cost of manufacturing a new bulb may change the discard versus repair equation.[3] But for now, in the decision to characterize a component, unit, assembly, or system as nonmaintained, economics plays a primary role. This reasoning should be very familiar to systems engineers.

Again, the key operational characteristic of a nonmaintained unit is that when it fails, it is discarded. Thus, it can suffer at most one failure. To describe this scenario quantitatively, it is useful to consider the time from start of operation of a new unit until the time the unit fails (assuming continuous, uninterrupted operation). This interval of time is called the *lifetime* of the unit. It can be reasonably represented by the upper case letter L (although this is not obligatory), and is most often thought of as a random variable.

> **Requirements tip:** We have seen that a good reliability requirement must include a specification of the length of time over which the requirement is to apply. When writing these requirements, and undertaking modeling studies to support them, it is important to remember when *operational time* is intended and when *calendar time* is intended. Calendar time refers to elapsed time measured by an ordinary clock and is always greater than or equal to operational time, the period of time during which the object in question is in use. Some systems are intended to be used continuously (most web servers and telecommunications infrastructure equipment are of this nature) while other systems are used only intermittently (an automobile, for instance). Be aware of whether the system you are developing is intended to be used continuously or intermittently, and state reliability requirements accordingly. This matters because equipment is usually considered to be not aging (i.e., accumulating time to failure) when it is not operating.[4] Usually, a model is required to relate operational time to calendar time so that users may anticipate their maintenance and replacement needs based on calendar time that is normally used for operations planning purposes. Some material on relating operational time to calendar time in the context of software products is found in Refs. 33, 46, 47.

3.3.2 The Life Distribution and the Survivor Function

3.3.2.1 Definition of the life distribution

Much discussion has taken place over the choice to model lifetimes as random variables. Suffice it to say that the most satisfactory explanation is that the factors influencing the lifetime of a unit are numerous, not all fully understood, and sometimes not controllable. In a sense, the choice to describe lifetimes as random is a cover for this (inescapable) ignorance [17, 61]. In some rare cases,

[3] Yes, technology has changed, and incandescent light bulbs are now going the way of the buggy whip. But this does not change the lesson of the example.

[4] There are exceptions. The accumulation of corrosion on relay contacts, for example, may take place faster when the relay is not operating than when it is.

it might be possible in principle to identify precisely the lifetime of a particular component. This would involve a deep understanding of the physical, chemical, mechanical, and thermodynamic factors at play in the operation of the component, as well as extremely precise measurements of the geometry, morphology, electrical characteristics, etc., of the component. Even if it were possible in principle to acquire such understanding, it would be prohibitively expensive in practice, and the knowledge obtained about the lifetime of a component A would not be transferrable to any knowledge about the lifetime of a component B from the same population because components A and B are not likely to be identical to the degree necessary to justify not having to perform all the same measurements on component B also. Clearly, this is an impossible situation.

What we do instead is attempt to describe the distribution (in the probabilist's sense) of the lifetimes of a population of "similar" components. For example, imagine a collection of 8 μF, 35-V tantalum electrolytic capacitors in an epoxy-sealed package manufactured by Company C during July 2011. Assuming the manufacturing process at Company C did not change during July 2011, we may reasonably assume that these are "similar" components for the purposes of calling them a "population" in the sense that a statistician would do. Every member of the population has a (different) lifetime that, under specified operating conditions, is fixed but unknown. The difference in lifetimes may be explainable by differences in raw materials, manufacturing process controls, varying environmental conditions in the factory, etc. Instead of trying to ascertain the lifetime of each individual in a deterministic fashion, what we do instead is consider populations of similar components and assign a distribution of the lifetimes (under specified operating conditions) in each population. A distribution of lifetimes for a population is called the *life distribution* for that population. The life distribution is a cumulative distribution function ("cdf"), in the sense that it is used in probability theory, and is often (though this is not obligatory) denoted by the upper case letter F (or sometimes F_L if it is necessary to explicitly call out the pertinent lifetime random variable). Thus, denoting by L the lifetime of a component drawn at random from the population,

$$F(x) = P\{L \le x\} \quad \text{for } x \ge 0 \quad \text{or} \quad F_L(x) = P\{L \le x\} \quad \text{for } x \ge 0.$$

Here, x is a variable that is at your disposal (we will call this a *discretionary variable*). You specify a value of x and the life distribution value at that x is the probability that a unit chosen at random from that population has a lifetime no greater than x, or, in other words, fails at or before time x. For instance, suppose a population of components has a life distribution given by $F(x) = 1 - \exp(-x/1000)$ for $x \ge 0$ measured in hours. Then the probability that a component chosen at random from that population fails at or before 1 year is $F(8766) = 1 - \exp(-8.766) = 0.999844$ which is almost certainty. We will return to this example later to explore some of the other things it has to teach but before we do, here is a picture (Figure 3.1).

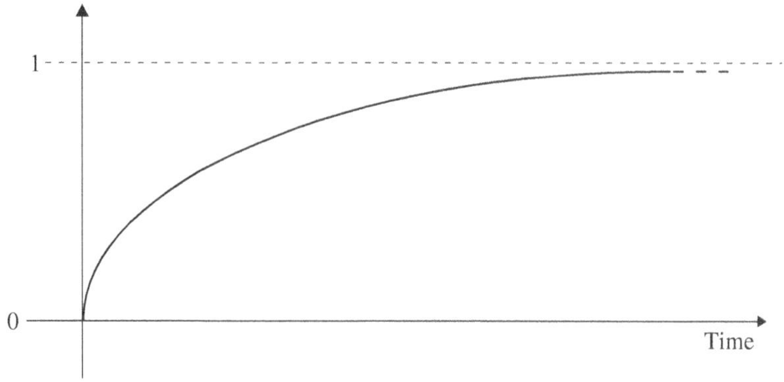

Figure 3.1 Generic life distribution.

The dashes at the end of the curve serve to indicate that the curve continues further to the right. A life distribution need not be continuous (as drawn), and it may have inflection points (not shown), but it is always nondecreasing and continuous from the right (see Section 3.3.2.3).

Example: Suppose the population of tantalum capacitors described earlier has a life distribution given by

$$F(x) = 1 - \exp\left(-\left(\frac{x}{10{,}000}\right)^{1.1}\right) \quad \text{for } x \geq 0 \text{ measured in hours}$$

when operated at 20°C. Suppose 100 capacitors from this population are placed into operation (at 20°C) at a time we will designate by 0. After 1000 hours of uninterrupted operation have passed, what is the expectation and standard deviation of the number of capacitors that will still be working?

Solution: The number of capacitors still working at time x has a binomial distribution with parameters 100 (the number of trials in the experiment) and the probability of survival of one capacitor past time x. For $x = 1000$, this probability is

$$P\{L > 1000\} = 1 - P\{L \leq 1000\} = \exp(-(0.1)^{1.1}) = 0.92364.$$

As the expected value of a binomial random variable with parameters n and p is np, the expected number of capacitors still working after 1000 hours is $100 \times 0.92364 = 92.364$. The variance of a binomial distribution with parameters n and p is $np(1-p)$, which in this case is equal to 7.05292. Consequently, the standard deviation of the number of capacitors still working after 1000 hours is equal to 2.65573.

Requirements tip: We have carried out the computations in this example to five decimal places, which is far more than would be desirable in almost any systems engineering application, solely for the purposes of illustrating the computations. Choose the appropriate number of decimal places whenever a quantity is specified in a requirement. The choice is often dictated by economic factors, practicality of measurement factors, and/or commonsensical factors that indicate how many places is too many for the application contemplated. For instance, specifying the length of a football field, in feet, to two decimal places is too much precision, whereas specifying the dimensions (in inches) of a surface-mount component may require more than two decimal places. Note that the units chosen bear on the decision as well.

3.3.2.2 Definition of the survivor function

The example points to another useful quantity in reliability modeling of non-maintained units, and that is the *survivor function* or *reliability function*. The survivor function is simply the probability that a unit chosen at random from the population is still working ("alive") at time x:

$$S(x) = P\{L > x\} = 1 - P\{L \le x\} = 1 - F(x)$$

and is consequently one minus the life distribution (the complement of the life distribution) at x. Again, a subscript L is sometimes used if it is necessary to avoid ambiguity.

Note that we have consistently pointed out that the discretionary variable x is nonnegative in lifetime applications. This is because, for obvious physical reasons, a life distribution can have no mass to the left of zero. That is, the probability of a negative lifetime is zero. Lifetimes are always nonnegative, so when L is a lifetime random variable, there is no point in asking for $P\{L \le x\}$ when $x < 0$ because $P\{L \le x\} = 0$ whenever $x < 0$.

3.3.2.3 Properties of the life distribution and survivor function

This discussion leads naturally into a discussion of other useful properties of life distributions. We consider four of these:

1. The life distribution is zero for $x < 0$.
2. The life distribution is a nondecreasing function of x for $x \ge 0$.
3. $F(0^-) = 0$ and $F(+\infty) = 1$.
4. The life distribution is continuous from the right and has a limit from the left at every point in $[0, \infty)$.

Return to Figure 3.1 to explore how the generic (continuous) life distribution shown there has these properties. We have indicated in Section 3.3.2.1 how the first property comes about. For the second property, consider that $F(x)$ is the probability that a unit[5] fails before time x. That is, $F(x)$ is the probability that the unit fails in the time interval $[0, x]$. Choose now x_1 and x_2 with $x_1 < x_2$ and

[5] Henceforth, we expect the reader to supply the phrase "drawn at random from the population."

consider $F(x_1)$ and $F(x_2)$. The interval $[0, x_2]$ is larger than (and in fact contains) the interval $[0, x_1]$, so there are more opportunities for the unit to fail in the additional time from x_1 to x_2.[6] Thus $F(x_2)$ must be at least as large as $F(x_1)$, which is property 2. From property 3, $F(0^-)=0$ says that the limit as $x \to 0$ from the left (i.e., through negative values) of the probability that a unit fails immediately upon being placed into operation is zero. $F(+\infty)=1$ says that every unit in the population eventually fails. There are situations in which we may wish to assume $F(0)>0$ (an example is given by a switch that fails to operate when called for) or $F(+\infty)<1$ (an example could be some component that is certain to not fail until after the service life of the system in which it is used is expired). But in most cases, property 3 is used as stated. Finally, the continuity of the life distribution from the right is a consequence of the choice of ≤, rather than <, in the cdf definition of life distribution. An equally satisfactory probability theory can be constructed on the choice of < (and in fact many notable probability textbooks do this), but the convention we have chosen to follow is as above, and in this case the cdf is continuous from the right (in the other case, it is continuous from the left).

Language (and notation) tip: For most of the life distributions in common use in reliability engineering, it is immaterial whether the < sign or the ≤ sign is chosen, because these life distributions are continuous. However, once the choice is made, it is important to continue the current analysis with the same choice throughout for consistency. This only matters when the life distribution has discontinuities (such as the switch life distribution, used in the example in Section 3.4.5.1, which contains a non-zero turn-on failure probability). Even when all life distributions in a study are continuous and it doesn't make any difference to the outcome, it is just sloppy practice to switch between < and ≤ arbitrarily. When working with someone else's analysis, endeavor to determine which choice was made and whether it is consistently applied.

Because the survivor function S is the complement of the life distribution F (i.e., $S=1-F$), the corresponding four properties for the survivor function are

1. The survivor function is one for $x<0$.
2. The survivor function is a nonincreasing function of x for $x\geq 0$.
3. $S(0^-)=1$ and $S(+\infty)=0$.
4. The survivor function is continuous from the left and has a limit from the right at each point in $[0, \infty)$.

Language tip: The survivor function is also sometimes called the reliability function. Recalling our discussions from the Foreword and Chapter 2, the fact that we have just encountered yet another use of the same word

[6] The probabilist would say that $\{\omega \in \Omega : 0 \leq L(\omega) \leq x_2\} = \{\omega \in \Omega : 0 \leq L(\omega) \leq x_1\} \cup \{\omega \in \Omega : x_1 < L(\omega) \leq x_2\}$, that is, there are more elements of the sample space for which the lifetime expires before x_2 than there are for which the lifetime expires before x_1. Most systems engineering studies will never reach this depth, but you need to see this at least once so that if it ever becomes necessary to explain lifetime random variables, you could do so this way if the audience would find it helpful.

"reliability" should strengthen your resolve to master potential confusions inherent in this language and be prepared to clarify for your teammates, customers, and managers another of the many unfortunate language clashes that abound in reliability engineering.

3.3.2.4 *Interpretation of the life distribution and survivor function*

The easiest way to maintain a consistent interpretation of the life distribution and survivor function is to visualize

- the population of components to which they apply and
- the "experiment" of choosing an item from that population at random.[7]

When you make this choice at a certain time (call it t, meaning that you have chosen some time to start a clock and that clock now measures t time units later), the probability that the item chosen is still alive ("working") at that time is given by the value of the survivor function $S(t)$ for that population. Because of the nature of selection at random without replacement, the number of items in the population still alive at time t is a random variable having a binomial distribution. If the initial size of the population is $A < \infty$ and $N(t)$ denotes the (random) number of items still alive at time t, then

$$P\{N(t)=k\}=\binom{A}{k}S(t)^k\left[1-S(t)\right]^{A-k}=\binom{A}{k}S(t)^k\,F(t)^{A-k},\quad k=0,1,\ldots,A.$$

This is a binomial distribution with parameters A and $S(t)$. Its mean is $AS(t)$ and its standard deviation is $\sqrt{AS(t)[1-S(t)]}=\sqrt{AS(t)\,F(t)}$. So the expected proportion of the population that is still alive at time t is $AS(t)/A=S(t)$. As more time passes (t increases), this proportion does not increase.

Similarly, the (random) number of items that have failed by time t (or, to put it another way, the number of items that have failed in the time interval $[0, t]$ from 0 to t) has a binomial distribution with parameters A and $F(t)=1-S(t)$.

> **Language tip:** Note that we have used t and x interchangeably in this section to denote a discretionary variable having the dimensions of time. This is not cause for alarm. It is routinely acceptable provided the definition is clear and the same letter is used consistently throughout each application.

3.3.3 Other Quantities Related to the Life Distribution and Survivor Function

As with cumulative distribution functions in probability, other related quantities enhance our ability to make reliability models. The ones we shall study in this section are the *density* and *hazard rate*.

[7] Choosing at random means that every member of the population has an equal chance of being chosen.

3.3.3.1 Density

Should it happen that the life distribution is absolutely continuous (i.e., can be written as an indefinite integral of some integrable function), that integrable function is called the *density* of the lifetime random variable. So if we can write

$$F(x) = \int_0^x f(u)\, du$$

for some integrable function f, then f is called the density of F. If this is the case, then F is necessarily continuous at every x for which this equation holds. More simply, if F is differentiable on an interval (a, b), then it is absolutely continuous there and $f(x) = F'(x) = dF/dx$ for $x \in (a, b)$. Because of properties 1 and 2 of life distributions, we have $f(x) = 0$ for $x < 0$ and $f(x) \geq 0$ for $x \geq 0$. Most of the life distributions in common use in reliability modeling have densities (see the examples in Section 3.3.4) (Figure 3.2).

> **Example:** Suppose $F(t) = t/(1+t)$ for $t \geq 0$ and $F(t) = 0$ for $t < 0$. Then properties 1, 3, and 4 (Section 3.3.2.3) are readily verified. Also, F is differentiable on $[0, \infty)$ and $F'(t) = 1/(1+t)^2 > 0$ there, so F is increasing (property 2) and $f(t) = 1/(1+t)^2$ is its density. Thus, this F is a life distribution with a density.

3.3.3.2 Interpretation of the density

When the lifetime L has a distribution F that has a density in a neighborhood of a point t, we may write

$$P\{t < L \leq t + \varepsilon\} = F(t+\varepsilon) - F(t) = \varepsilon f(t) + o(\varepsilon) \quad \text{for } \varepsilon \to 0^+.$$

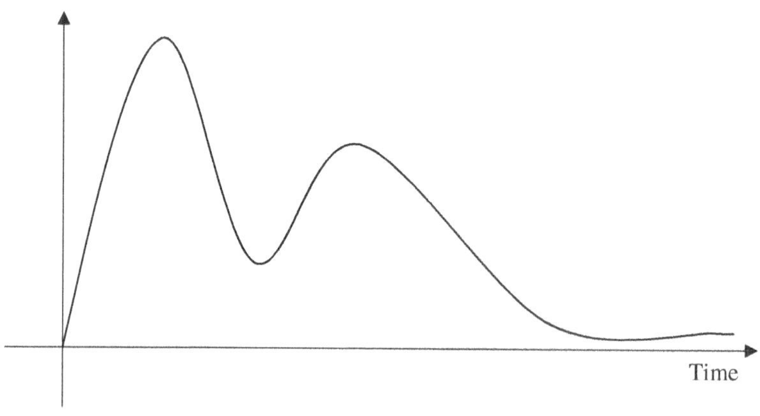

Figure 3.2 *A generic density function.*

That is, for a small positive increment ε, the probability that an item chosen at random from the population fails in the (small) time interval $[t, t+\varepsilon]$ is approximately ε times the value of the density at t. Note that this item may have already failed before time t—there is no requirement that the item be alive at the beginning of this interval. Contrast this with the hazard rate interpretation discussed in Section 3.3.3.5.

3.3.3.3 Return to the stress–strength model

The stress–strength model was introduced in Section 2.2.7 and the example of destruction of a single complementary metal-oxide semiconductor (CMOS) integrated circuit was explained as resulting from a single environmental stress, namely the application of a voltage stress exceeding the strength of the oxide in the device. Here we explore the stress–strength model in a population of devices and an environment that can offer a range of stresses.

Imagine that a population of devices has a range of strengths that is described by a strength density. That is, for some device characteristic V that indicates "strength" (e.g., oxide breakdown voltage), there is a density f_V characterizing that population with respect to that strength variable, or characteristic. That means that we describe the strength of an item drawn at random from the population by a random variable V that has density f_V, and when that item is subjected to a stress greater than V, it fails. Further suppose that the environment offers stresses (on the same scale) described by a random variable S with density g_S. Figure 3.3 shows this relationship graphically. The density of stresses offered by the environment, g_S, and the density of strength in the population of devices, f_V, is shown on the same axes. Figure 3.3 depicts a situation where most of the population strengths are greater than most of the environmental stresses, except for the small area where the two densities overlap. For a stress in this area (a value indicated by the × on the horizontal axis), a device whose strength

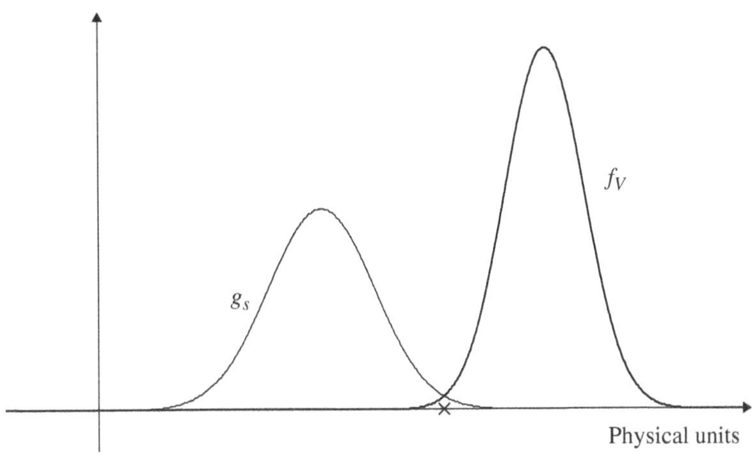

Figure 3.3 *Stress–strength relationship in a population.*

is to the left of this stress (weaker than this stress) will fail. In this picture, this small area indicates that there are few devices in the population whose strength is less than (to the left of) this value. The area under the stress density to the right of the chosen stress value is also small, and this indicates that stresses so large are rarely offered (most stresses are less than this value, or almost all of the stress density lies to the left of this value).

The probability of failure, $P\{S > V\}$, is the probability that a stress chosen at random from the population of stresses (described by the density g_S) exceeds the strength of a device chosen at random from the population of devices whose strength density is f_V. Then the probability of failure of a device drawn at random from that population, when subjected to a stress drawn at random from that environment, is

$$
\begin{aligned}
P\{S > V\} &= \int_0^\infty P\{S > V \mid V = v\} f_V(v)\,dv = \int_0^\infty P\{S > v\} f_V(v)\,dv \\
&= \int_0^\infty [1 - G_S(v)] f_V(v)\,dv = 1 - \int_0^\infty G_S(v) f_V(v)\,dv
\end{aligned}
$$

as long as we assume the environmental stresses are stochastically independent of the population strengths.

Note that neither of these relates to *time* to failure. The distributions (densities) here are both on a scale of some physical property (e.g., volts). To develop this model further to the point where a lifetime distribution could be obtained, it would be necessary to describe the times at which the environment offers stresses of a given size. This could be done with, for example, a compound Poisson process in which at each (random) time an event occurs, a stress of a random magnitude is applied. Some details of this model may be worked out in Exercise 1. A deeper discussion of stress–strength models is found in Ref. 38.

3.3.3.4 Hazard rate or force of mortality

The second related quantity, one that is widely used in modeling the reliability of nonmaintained units, is the *hazard rate*. The hazard rate is customarily denoted by h, and the definition of hazard rate is

$$
h(x) = \lim_{\varepsilon \to 0^+} \frac{1}{\varepsilon} P\{L \le x + \varepsilon \mid L > x\}
$$

when the limit exists. This is the hazard rate of the lifetime random variable L. It is also sometimes spoken of as the hazard rate of the life distribution. Note this definition contains a conditional probability, and, unlike the quantities we have studied so far which are dimensionless, the hazard rate has the dimensions of 1/time (probability is dimensionless and ε has the dimensions of time).

In case F is absolutely continuous at x, the hazard rate may be computed as follows:

$$h(x) = \lim_{\varepsilon \to 0^+} \frac{1}{\varepsilon} P\{L \le x + \varepsilon \mid L > x\} = \lim_{\varepsilon \to 0^+} \frac{1}{\varepsilon} \frac{F(x+\varepsilon) - F(x)}{1 - F(x)}$$

$$= \frac{1}{1 - F(x)} \lim_{\varepsilon \to 0^+} \frac{F(x+\varepsilon) - F(x)}{\varepsilon} = \frac{1}{1 - F(x)} \lim_{\varepsilon \to 0^+} \frac{1}{\varepsilon} \int_x^{x+\varepsilon} f(u)\, du = \frac{f(x)}{1 - F(x)}.$$

If we further assume F is differentiable, the differential equation

$$\frac{F'(x)}{1 - F(x)} = h(x)$$

with initial condition $F(0) = \alpha$ may be solved to yield

$$F(x) = 1 - (1 - \alpha) \exp\left(-\int_0^x h(u)\, du \right).$$

Thus when the life distribution is differentiable, there is a one-to-one correspondence between the life distribution and its hazard rate. Knowing either one enables you to obtain the other. Most often α will be zero, but it is useful to know the expression for life distribution in terms of hazard rate even when $\alpha > 0$. An example of a component whose life distribution is a switch for which the probability of failure when it is called upon to operate is $\alpha > 0$.

3.3.3.5 *Interpretation of the hazard rate*

Return to the definition above to see that

$$P\{L \le x + \varepsilon \mid L > x\} = \varepsilon h(x) + o(\varepsilon)$$

as $\varepsilon \to 0^+$. Imagine for the moment that time is measured in seconds and consider this equation for $\varepsilon = 1$ (second). Then the hazard rate at x is approximately equal to the conditional probability of failure in the next second (i.e., before time $x+1$) given that the unit is currently alive (using time x to represent the current time). So the hazard rate is something like the propensity to fail soon given that you are currently alive. In fact, the concept of hazard rate is lifted directly from demography, the study of lifetimes of human populations, where it is called the *force of mortality*. This description is very apt: the hazard rate, or force of mortality, describes how hard nature is pushing you to die (very) soon when you are alive now.

Example: Let $F(x) = 1 - \exp((-x/\alpha)^{\beta})$ for $x \ge 0$ and $F(x) = 0$ for $x < 0$, where α and β are positive constants. This is readily verified to be a life distribution (Exercise 2). Its particular properties depend on the choice of the constants

α and β which are called parameters. This life distribution is called the Weibull distribution in honor of the Swedish engineer, scientist, and mathematician Ernst Hjalmar Wallodi Weibull (1887–1979). See also Section 3.3.4.3. This distribution has a density

$$f(x) = \frac{\beta}{\alpha}\left(\frac{x}{\alpha}\right)^{\beta-1} \exp\left(-\left(\frac{x}{\alpha}\right)^{\beta}\right)$$

for $x \geq 0$. Consequently, the hazard rate of the Weibull distribution is given by

$$h(x) = \frac{\beta}{\alpha^{\beta}} x^{\beta-1}$$

again for $x \geq 0$.[8] It follows from this expression that the hazard rate of the Weibull distribution may be increasing, decreasing, or constant, depending on the choice of β: if $\beta < 1$, the hazard rate is decreasing, if $\beta > 1$, the hazard rate is increasing, and if $\beta = 1$, the hazard rate is constant. The special case $\beta = 1$ has a long and extensive usage in reliability modeling: it is the *exponential* life distribution $F(x) = 1 - \exp(-(x/\alpha))$ (Section 3.3.4.1). We have seen that the hazard rate of the exponential distribution is constant; it has been shown that this is the only life distribution in continuous time whose hazard rate is constant [34] (the geometric probability mass function $p(x) = (1-\alpha)\alpha^{x-1}$ for $x = 0, 1, 2, \ldots$ and $0 < \alpha < 1$ is a life distribution on a discrete time scale that has a constant hazard rate, and it is the only life distribution in discrete time that is so blessed [35]). We will explore additional properties of the exponential distribution when we discuss more examples in Section 3.3.4.

Finally, contrast the interpretation of hazard rate with the interpretation of density given in Section 3.3.3.2. Owing to the equation

$$P\{t < L \leq t + \varepsilon\} = F(t + \varepsilon) - F(t) = \varepsilon f(t) + o(\varepsilon) \quad \text{for } \varepsilon \to 0^{+},$$

ε times the density at t is approximately equal to the probability that a lifetime falls between t and $t + \varepsilon$, that is, the probability that $L > t$ <u>and</u> $L \leq t + \varepsilon$. Here, we are selecting a unit at random from the population and asking if its lifetime is between t and $t + \varepsilon$. The hazard rate, instead, satisfies

$$P\{L \leq t + \varepsilon \mid L > t\} = \varepsilon h(t) + o(\varepsilon) \quad \text{for } \varepsilon \to 0^{+},$$

which indicates that ε times the hazard rate at time t is approximately equal to the <u>conditional</u> probability that a lifetime expires (at or) before $t + \varepsilon$, <u>given</u>

[8] It is now high time for the reader to be able to supply the domain of existence for the life distributions to be discussed in this book. Henceforth, we shall suppress the "0 for $x < 0$" in all life distribution definitions, asking the reader to be aware that it is now his/her responsibility to fill in this detail, if only tacitly.

that it is greater than t. Here, we are selecting from a restricted portion of the population, namely that set of units whose lifetimes are greater than t (those that are still alive at time t). Selecting a unit at random from those, we ask what is the probability that the lifetime of that unit does not exceed $t+\varepsilon$. In more mathematical terms, this is the difference between $P(A \cap B)$ and $P(A \mid B)$.

Language tip: The hazard rate or force of mortality is almost always called the *failure rate* of the relevant life distribution. This is unfortunate, the more so because it is almost universal, because the word "rate" makes engineers think of "number per unit time," and there is nothing like that going on here (even though the dimensions of the hazard rate are 1 over time). The closest one can come to interpreting "hazard rate" as a rate is as in the following example. Suppose the population of units we are considering initially contains N members and we start all of these operating at an arbitrary time we shall label "zero." At a later time x, the expected number of failed units is $NF(x)$ (where F is the life distribution for this population) and the expected number of units still working is $NS(x)=N(1-F(x))$. One of these still-alive units fails before time $x+1$ with probability approximately equal to $h(x)$.[9] So the hazard rate is like the proportion of the remaining (still-alive) population that is going to fail very soon. This looks like a "rate" when referred to the number of remaining (still-alive or "at-risk") members of the population. Extended discussion of this deplorable situation is available in Ref. 2. See also the "Language Tips" in Section 4.4.2.

Requirements tip: Be very careful when contemplating writing a requirement for "failure rate." Because the phrase can be interpreted in (at least three) different ways in reliability engineering, it is vital that you specify which meaning is intended in the requirement. For this reason, it is probably best to avoid "failure rate" altogether in requirements. Instead, spell out the specific reliability effectiveness criterion intended. For example, "The number of system failures shall not exceed 3 in 25 years of operation under the specified conditions" is preferable to "The system failure rate shall not exceed 1.37×10^{-5} failures per hour during the service life of the system when operated under specified conditions." Indeed, the latter formulation tends to induce one to think that system failures accrue uniformly over time, while the former formulation allows for arbitrary patterns of failure appearance in time, as long as the total number does not exceed 3 in 25 years.

The concept of *cumulative hazard function* will be useful later in the study of certain maintained system models (Section 4.4.2). The cumulative hazard function H is simply the integral of the hazard rate over the time scale:

$$H(t)=\int_0^t h(u)\,du.$$

It is easy to see that $H(t)$ can also be written as $H(t)=-\log S(t)=-\log[1-F(t)]$.

[9] This works best if time is measured in very brief units like nanoseconds.

3.3.4 Some Commonly Used Life Distributions

3.3.4.1 *The exponential distribution*

The lifetime L has an exponential distribution if $P\{L \le x\} = 1 - \exp(-x/\alpha)$ for $x \ge 0$ and $\alpha > 0$. α is called the *parameter* of the distribution. As x has the dimensions of time, so does α because the exponent must be dimensionless. In fact, α is the mean life:

$$\mathrm{E}L = \int_0^\infty t\,dP\{L \le t\} = \int_0^\infty t \exp\left(\frac{-t}{\alpha}\right) dt = \alpha.$$

The exponential distribution has a density, namely $(1/\alpha)\exp(-x/\alpha)$. Consequently, the hazard rate of the exponential distribution is constant and is equal to $1/\alpha$. Note this has the units of 1 per time as it should. The variance of the exponential distribution is

$$\mathrm{Var}(L) = \mathrm{E}L^2 - \alpha^2 = \int_0^\infty t^2 \exp\left(\frac{-t}{\alpha}\right) dt - \alpha^2 = 2\alpha^2 - \alpha^2 = \alpha^2,$$

so its standard deviation is α. The median of the exponential distribution is the value m for which $P\{L \le m\} = 0.5$; solving $\exp(-m/\alpha) = 0.5$ for m yields $m = \alpha \log 2$.

Frequently, the exponential distribution is seen with the parameterization $1 - \exp(-\lambda x)$ for $\lambda > 0$. This is perfectly acceptable; simply replace α by $1/\lambda$ in all the earlier statements.

The exponential distribution is also blessed with a peculiar property called the *memoryless property*. As a consequence of the following computation

$$P\{L > x + a \mid L > x\} = \frac{\exp(-(x + a)/\alpha)}{\exp(-x/\alpha)} = \exp(-a/\alpha) = P\{L > a\},$$

we see that if an item's lifetime L has an exponential distribution, then the probability that the item will fail after the passage of a (additional) units of time is the same no matter how old the item is. That is, if the item is currently x time-units old, then the probability of the item's surviving to time $x + a$ is the same as the probability that a new item survives to time a, regardless what x may be. To get some sense of how peculiar a property this is, consider the purchase of a used flat-screen television. If reliability were your only concern, and the life distribution of the (population of) flat-screen TV(s) were exponential, then you would be willing to pay the same price for a used flat-screen TV of any age as you would for a new one. Of course, there are other factors at play here, and reliability is not your only concern, but the example serves as a caution you should remember when you contemplate using the exponential distribution for the lifetime or a nonrepairable item. The exponential distribution is the only life distribution (in continuous time) that has this property [34].

One reason for the popularity of the exponential distribution in reliability modeling is that it is the limiting life distribution of a series system (Section 3.4.4) of "substantially similar" components [15]. In this context, "substantially similar" has a precise technical meaning which we will defer discussing until Sections 3.4.4.3 and 4.4.5 when a similar result (Grigelionis's theorem [53]) will be seen as relevant to both maintained and nonmaintained systems. Implications for field reliability data collection and analysis are discussed in Chapter 5.

3.3.4.2 The uniform distribution

A random variable L is said to have a uniform distribution on $[a, b], a<b$, if

$$P\{L \le x\} = \frac{x-a}{b-a}, \quad -\infty < a \le x \le b < \infty.$$

If $a \ge 0$, the uniform distribution can be used as a life distribution. In this model, the lifetimes are between a and b with probability 1, and the distribution has the name "uniform" because the probability that a lifetime lies within any subset of $[a, b]$ of total measure τ, say, is $\tau/(b-a)$ regardless where within $[a, b]$ this subset may lie (as long as it lies wholly within $[a, b]$). The density of the uniform distribution is $1/(b-a)$ on $[a, b]$ and zero elsewhere. The expected value of a uniformly distributed lifetime is $(a+b)/2$ and the variance is $(b-a)^2/12$. In other uses of the uniform distribution, a may be negative, but for use as a life distribution a must be nonnegative. See Exercise 6 for the hazard rate of the uniform distribution.

3.3.4.3 The Weibull distribution

The lifetime L has a Weibull distribution if $P\{L \le x\} = 1 - \exp(-(x/\alpha)^{\beta})$ for $x \ge 0$ and $\alpha > 0, \beta > 0$. α and β are the parameters of the distribution. As we saw in the example in Section 3.3.3.4, the Weibull distribution has a density

$$\frac{\beta}{\alpha}\left(\frac{x}{\alpha}\right)^{\beta-1} \exp\left(-\left(\frac{x}{\alpha}\right)^{\beta}\right)$$

and its hazard rate is

$$\frac{\beta}{\alpha^{\beta}} x^{\beta-1}$$

all for $x \ge 0$.

As noted previously, the hazard rate for the Weibull distribution can be increasing, decreasing, or constant, depending on the value of β (Table 3.1).

When $\beta = 1$, the Weibull distribution reduces to the exponential distribution (Section 3.3.4.1). The Weibull distribution with $\beta > 1$ is frequently used to describe the lifetimes in a population of items that may suffer mechanical wear.

TABLE 3.1 Weibull Distribution Hazard Rate

If β is	Then the Weibull hazard rate is
>1	Increasing
=1	Constant
<1	Decreasing

For example, ball bearings normally exhibit wear (decrease of diameter) as they continue to operate.[10] A population of identically sized ball bearings made of the same material, when operated continuously, will accumulate more and more failures due to wear as time increases. That is, failures will begin to accumulate more rapidly the longer the population continues in operation. This phenomenon is labeled "wearout" in reliability engineering, the term being inspired by the concept of mechanical wear such as illustrated in this example. Note that this example treats a nonrepairable item. Any individual ball bearing may suffer at most one failure; the "accumulation of failures" pertains to multiple failures in a population of many bearings, each of which may fail at most once. See Section 3.3.4.8 for additional development of this idea.

Finally, the Weibull distribution is the limiting distribution of the smallest extreme value (i.e., the minimum) of a set of independent, identically distributed random variables [27]. The lifetime of a component under the competing risk model (Section 2.2.8) is a smallest extreme value. This may account for the frequent appearance of the Weibull distribution as a reasonable description of the lifetime of individual components.

3.3.4.4 A life distribution with a "bathtub-shaped" hazard rate

Demographers have determined that the force of mortality in human populations follows a broad U-shaped, or "bathtub-shaped," curve (see Figure 3.4).

The commonly accepted explanation for this shape posits that the decreasing force of mortality in early life comes from infant mortality and the diseases that afflict the young, which, after some period of time, are outgrown and subsequently exert little influence on the population. The increasing force of mortality in late life is due in large part to the finite lifetime of human beings (see Exercise 6), but is also due to what are termed "wearout mechanisms" such as atherosclerosis, loss of telomeres, and others, that promote earlier death. The (approximately) constant force of mortality in mid-life is primarily due to deaths caused by accidents that occur at random times and the rarer occurrence of diseases that strike prematurely in middle age. A similar interpretation obtains in reliability engineering: decreasing force of mortality in the early part of the lifetime in a population of components is explained by the early failure of some components in the population that have manufacturing defects (see Section 3.3.6) that cause them to fail prematurely (such failures are often referred to as "infant mortality failures"). Increasing force of mortality in the later part of the lifetimes is explained by physical and chemical wearout mechanisms

[10] Lubrication greatly slows, but does not stop, this process.

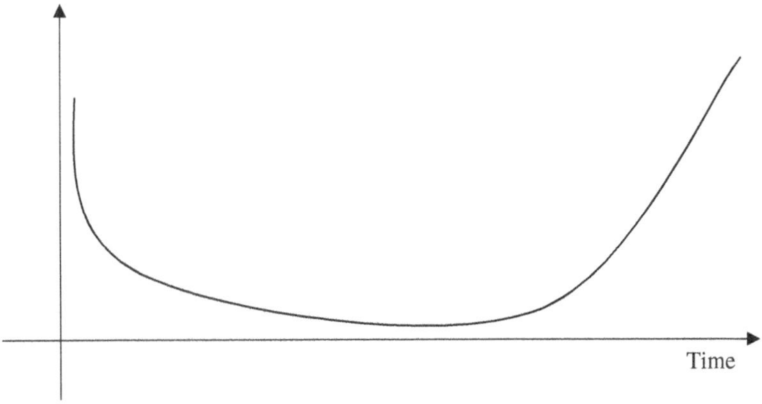

Figure 3.4 *Force of mortality for human populations.*

such as mechanical wear, depletion of reactants, increase of nonradiative recombinations, increase in the number and/or size of oxide pinholes, etc. Indeed, the presence of an increasing hazard rate is often taken as a symptom of the presence of an active wearout failure mode, even if no physical, chemical, or mechanical wearout explanation can be discerned. The constant force of mortality during "useful life" is due primarily to the occurrence at random times of shocks whose stresses exceed the strengths of the components (see Section 3.3.3.3 and Exercise 1).

None of the life distributions discussed elsewhere in this section has a force of mortality with this shape. To develop such a life distribution, we need to employ the method shown in Section 3.3.3.4 in which a life distribution is developed from a hazard rate by the integral formula shown there.

At least one attempt at implementing a practical version of such a life distribution has been made. Holcomb and North [30] introduced a life distribution of this type for electronic components. Their model is a Weibull distribution describing the component's reliability until a time called the crossover time, at which time it changes to an exponential distribution that applies thereafter. That is, the population life distribution is described by a Weibull distribution up until the crossover time, and the (conditional) life distribution of the subset of the population that survives beyond the crossover time is an exponential distribution. This distribution is continuous everywhere and has a density everywhere except at t_c. The hazard rate model is as follows:

$$h(t) = \begin{cases} \lambda_1 t^{-\alpha}, & 0 \le t \le t_c \\ \lambda_L, & t \ge t_c \end{cases}.$$

This model contains four parameters, λ_1, λ_L, t_c, and α. $\lambda_1 > 0$ is the early life hazard rate coefficient and represents the hazard rate of the life distribution at $t=1$ (conventionally, the time unit in this model is hours). $\alpha > 0$ is the early life hazard rate shape parameter; it represents the rate at which the hazard rate decreases until time t_c. At time t_c, the hazard rate becomes equal to a constant $\lambda_L > 0$. The model further imposes the condition that the hazard rate be continuous, so the four parameters are not independent. They are linked by the relation

$$\lambda_1 = \lambda_L t_c^{\alpha}.$$

Note that while this hazard rate model allows for a decreasing hazard rate in early life and a constant hazard rate in "mid-life," the increasing hazard rate characteristic of wearout is not present. This is because it was reasoned that wearout mechanisms in electronic components take so long to appear that the service life of the equipment or system is over before this occurs.[11] Finally, note that in this model the conditional life distribution of components, given that they survive beyond t_c, is exponential with parameter $1/\lambda_L$.

For the life distribution and density corresponding to this hazard rate model, see Exercise 7.

3.3.4.5 The normal (Gaussian) distribution: a special case

A random variable Z has a *standard normal* (or *standard Gaussian*) distribution if

$$P\{Z \le z\} = \frac{1}{\sqrt{2\pi}} \int_{-\infty}^{z} \exp\left(-\frac{x^2}{2}\right) dx := \Phi_{(0,1)}(z), \quad -\infty < z < \infty;$$

the mean of this distribution is 0 and its variance is 1 (this is the definition of "standard" for the normal distribution and the explanation of the subscript on the Φ). The density of this distribution is given by

$$\varphi_{(0,1)}(z) = \frac{1}{\sqrt{2\pi}} e^{-z^2/2}, \quad -\infty < z < \infty.$$

Clearly, evaluating the normal distribution is not a paper-and-pencil exercise. The old-school method is to use the table of standard normal percentiles, which appears in all elementary statistics textbooks; the tables are usually constructed by numerical integration or polynomial approximation [1]. Now, all statistical software and many scientific calculators include a routine for evaluating the standard normal distribution, and many office software programs, such as Microsoft Excel®, also include this capability.

If Z is a standard normal random variable, the random variable $\sigma Z + \mu$ has mean μ and variance σ^2 where $-\infty < \mu < \infty$ (could be negative!) and $\sigma > 0$; the

[11] If considering use of this life distribution, the condition that wearout mechanisms do not become active before the end of the system's service life is over should be verified.

distribution of $\sigma Z+\mu$ is conventionally denoted by $\Phi_{(\mu,\sigma)}$ or simply Φ if μ and σ are clear from the context. Correspondingly, if Z is a normally distributed random variable having mean μ and standard deviation σ, then $(Z-\mu)/\sigma$ has a standard normal distribution. The normal distribution is also called the Gaussian distribution in honor of the great mathematician Carl Friedrich Gauss (1777–1855) who first used it to describe the distribution of errors in statistical observations.

The normal distribution is not a life distribution because it has mass to the left of 0, i.e., it gives positive probability to negative lifetimes. Nonetheless, some studies use a normal distribution with large positive μ and small σ as an *approximate* life distribution because when μ is large positive and σ is small, the probability that the lifetime is negative is quite small and may for some purposes (e.g., computing moments) be neglected. However, the normal distribution is not appropriate for use with many of the important models for the reliability of a maintainable system. For example, the equations of renewal theory (Section 4.4.1) fail for the normal distribution (even if μ is large positive and σ is small).

Some studies make use of a *truncated* normal distribution to avoid the difficulty with negative lifetimes. A truncation of a normal distribution with parameters μ and σ is the conditional distribution of a random variable Y that is normally distributed with mean μ and variance σ^2, conditional on Y belonging to some interval. To use the truncated normal distribution as a life distribution, this interval would be $[0, \infty]$, or the conditioning is on $Y \geq 0$. If we denote by W the lifetime random variable described by this truncated distribution, then

$$P\{W \leq w\} = P\{Y \leq w | Y > 0\} = \frac{\Phi_{(\mu,\sigma)}(w) - \Phi_{(\mu,\sigma)}(0)}{1 - \Phi_{(\mu,\sigma)}(0)} \quad \text{for } w \geq 0$$

and $P\{W \leq w\} = 0$ for $w \leq 0$ so that the truncated normal distribution is a *bona fide* life distribution. Note that the mean and variance of the truncated normal distribution are no longer μ and σ^2. For more details on the truncated normal distribution, see Ref. 28.

3.3.4.6 The lognormal distribution

A lifetime L is said to have a lognormal distribution if the logarithm of the lifetime has a normal distribution. That is,

$$P\{L \leq x\} = P\{\log L \leq \log x\} = \Phi_{(\mu,\sigma)}(\log x), \quad x \geq 0.$$

Note that while $L \geq 0$, log L may have any sign because the logarithms of numbers between 0 and 1 are negative. If Y has a normal distribution, then $L = e^Y$ has a lognormal distribution. If μ and σ are the parameters of the underlying normal distribution, then the mean of the lognormal distribution is $e^{\mu+\sigma^2/2}$ and its variance is $(e^{\sigma^2} - 1)\, e^{2\mu+\sigma^2}$.

The lognormal distribution has been successfully used for modeling repair times of complex equipment [37,51]. Its hazard rate is decreasing as $t \to \infty$, leading to the interpretation that when equipment is complex, repairs are often complicated, and the longer a repair lasts, the less likely that it is that it will be completed soon. For example, times to complete repairs for undersea telecommunications cables that require a repair ship to visit the site of the failure have been postulated to follow a lognormal distribution, but citations in the literature are hard to find.[12]

3.3.4.7 The gamma distribution

The lifetime L has a gamma distribution if

$$P\{L \le x\} = \frac{1}{\alpha^k \Gamma(\nu)} \int_0^x u^{k-1} e^{-u/\alpha} \, du \quad \text{for } 0 \le x < \infty$$

where $\alpha > 0$ and $k > 0$ are the location and shape parameters, respectively, of the distribution, and Γ is the famous gamma function of Euler (Leonhard Euler, 1707–1783), defined by

$$\Gamma(x) = \int_0^\infty u^{x-1} e^{-u} \, du$$

for $x > 0$. The gamma function is perhaps most well-known for being an analytic function that interpolates the factorial function: $\Gamma(n+1) = n!$ whenever n is a positive integer. α is a location parameter and k is a shape parameter (α has the units of time and k is dimensionless); when $k = 1$ the gamma distribution reduces to the exponential distribution (Section 3.3.4.1) with parameter α. The importance of the gamma distribution in reliability modeling lies largely in its property that the gamma distribution with parameters α and n (n an integer) is the distribution[13] of the sum of n independent exponential random variables, each of which has mean α. Actually, more is true: the sum of two independent gamma-distributed random variables with parameters (α_1, ν) and (α_2, ν) again has a gamma distribution with parameters $(\alpha_1 + \alpha_2, \nu)$, and of course this extends to any finite number of summands as long as the shape parameter ν is the same in each. There is a natural connection with the life distribution of a cold-standby redundant system (see Section 3.4.5.2 for further details).

The density of the gamma distribution is given by

$$\frac{1}{\alpha^k \Gamma(k)} x^{k-1} e^{-x/\alpha} \quad \text{for } x > 0.$$

[12] Most commercial organizations are loath to share specific data on operating times, repair times, and so on, in public, stating that these are equivalent to trade secrets.

[13] In queuing theory, these distributions are known as the Erlang distributions (Agner Krarup Erlang, 1878–1929, a pioneering teletraffic engineer).

Consequently, the hazard rate of the gamma distribution is given by

$$\frac{x^{k-1}\, e^{-x/\alpha}}{\int_x^{\infty} u^{k-1}\, e^{-u/\alpha}\, du},$$

again for $x>0$. When $k=1$, this reduces to $1/\alpha$, a constant, as it should because for $k=1$ the gamma distribution is the exponential distribution. The hazard rate is clearly increasing for $k>1$; it is the content of Exercise 3 that the hazard rate is decreasing when $0<k<1$. So the behavior of the gamma distribution is similar to that of the Weibull distribution according to the shape parameters (Table 3.2).

The mean of the gamma distribution is $k\alpha$ and its variance is $k\alpha^2$.

The main importance of the gamma distribution elsewhere comes from its relation to commonly used quantities in statistics that we use in Chapters 2, 5, 10, and 12. The sample variance from a population having a normal distribution has a gamma distribution. Formally, if $X_1, X_2, \ldots, X_n$ are normally distributed random variables with mean 0 and variance σ^2, then $X_1^2 + X_2^2 + \cdots + X_n^2$ has a gamma distribution with parameters $1/2\sigma^2$ and $n/2$. For historical reasons, this distribution when $\sigma=1$ is also called the chi-squared distribution with n degrees of freedom (Karl Pearson, 1857–1936). Other important quantities in statistics have distributions related to the gamma distribution, including student's T-statistic (student was a pseudonym adopted by William Sealy Gosset (1876–1937) to enable him to publish his works over the objections of his employer, the Guinness brewing company), Snedecor's F-statistic (George W. Snedecor, 1871–1974), and Fisher's Z-statistic (Sir Ronald A. Fisher, 1890–1962) all have distributions than can be expressed in terms of the gamma function and distribution. For details, see Ref. 21.

3.3.4.8 *Mechanical wearout and statistical wearout*

"Wearout" is used in two senses in reliability engineering. Mechanical wearout is the physical phenomenon of loss of material during sliding, rolling, or other motion of materials against one another. Statistical wearout is the mathematical property of increasing hazard rate of a life distribution when the hazard rate does not decrease after the period of increase being described. The second interpretation arose because of the first: a population of devices subject to (physical) wearout will exhibit a life distribution with an increasing hazard rate in later life. The following example may help illustrate this phenomenon.

TABLE 3.2 Gamma Distribution Hazard Rate

If k is	Then the gamma hazard rate is
>1	Increasing
=1	Constant
<1	Decreasing

Example: A population of 5/8″ ball bearings is operated under nominal conditions under which their diameter decreases by X ten-thousandths of an inch per hour, where X is a random variable having a uniform distribution on [1, 4] (see Section 3.3.4.2). A ball bearing is declared failed when its diameter has decreased by 0.010″. What is the distribution of lifetimes L in this population of ball bearings? For a ball bearing that we label ω, the rate of decrease of its diameter is $X(\omega)$, and the amount of time (in hours) it takes for that ball bearing to decrease by 0.010″, which is 100 ten-thousandths, is $100/X(\omega)$ hours. Our task, then, is to find the distribution of $100/X$ when X has the stated uniform distribution. We know that

$$P\{X \le x\} = \begin{cases} 0, & x \le 1 \\ (x-1)/3, & 1 \le x \le 4. \\ 1, & x \ge 4 \end{cases}$$

Then

$$P\{L \le y\} = P\{100/X \le y\} = P\{X \ge 100/y\} = \begin{cases} 1, & y \ge 100 \\ \dfrac{4}{3} - \dfrac{100}{3y}, & 25 \le y \le 100. \\ 0, & y \le 25 \end{cases}$$

The density of this distribution is $100/3y^2$ for $25 \le y \le 100$, and zero elsewhere. So the hazard rate of this distribution is $100/y(100-y)$ for $25 \le y \le 100$, and zero elsewhere. This is clearly seen to be an increasing function of y as $y \to 100^-$ (i.e., as y approaches 100 from the left, or through smaller values, which is what the superscripted minus sign is supposed to convey). This example, while not generic, does illustrate the connection between physical wearout and the mathematical interpretation of wearout as an increasing hazard rate with increasing time. See also Exercises 6, 20, and 21. Further discussion may be found in Ref. 24.

Another way to understand this phenomenon is to imagine that all the ball bearings wear at exactly the same (constant) rate, say 2.5 ten-thousandths of an inch per hour. Then every ball bearing fails at 40 hours exactly. Then a small variation in the rate of wear (i.e., 0.00025″/hour ± a little bit) will translate into some variation in the failure times (40 hours ± a little bit[14]). The failure time density will be zero until shortly before 40 hours (i.e., up until 40 – the little bit) and then it will increase rapidly to a maximum near 40 hours and then decrease again rapidly to zero (at 40 + the little bit). The survivor function of the lifetimes will be zero until shortly before 40 hours, and then will decrease rapidly to zero shortly after 40 hours. Think about the quotient of these two quantities

[14] Usually, not the same "little bit" as that in the wear rate variation.

(the hazard rate): from shortly before 40 hours until at least 40 hours, the numerator is increasing rapidly while the denominator is decreasing. The quotient is therefore increasing, at least until the density peaks. Deeper analysis would reveal that the hazard rate continues to increase until "shortly after 40 hours," but that is not the point of this illustration. The point is that under very general conditions, physical wearout, even at random rates, leads to an increasing hazard rate life distribution, which is the characteristic of wearout in the statistical (or mathematical) sense.

3.3.5 Quantitative Incorporation of Environmental Stresses

In Chapter 2, we emphasized that three things must be present in a proper reliability requirement: a specification of a limit on some reliability effectiveness criterion, a time during which the requirement is to apply, and conditions (environmental or other) under which the requirement is to apply. In the discussion of earlier life distributions, no mention is made of conditions. In this section, we will discuss some modifications that enable us to incorporate the role of prevailing conditions into a life distribution model.

3.3.5.1 Accelerated life models

Accelerated life models are among the simplest models for relating the life distribution of a population of objects operated under a given set of environmental conditions to the life distribution of that population operated under a different set of environmental conditions. We describe two accelerated life models in this book, the strong accelerated life model and the weak accelerated life model, and the proportional hazards model which in analogous terminology might be called the accelerated hazard model.

The strong accelerated life model postulates that there is a linear relationship between the individual lifetimes at the different conditions. If L_1 and L_2 are the lifetimes of an object when the conditions under which it is operated are C_1 and C_2, respectively, then the strong accelerated life model asserts that $L_2 = A(C_1, C_2)L_1$, where A is a constant depending on the two conditions C_1 and C_2.[15] If many conditions change from one application to another, it is possible that C_1 and C_2 may be vectors. If the conditions are dynamic (may change with time), then C_1 and C_2 may be functions of time.

We begin our study with the simplest case in which the two conditions are constant. For example, condition C_1 may be a constant temperature of 10°C, while condition C_2 may be a constant temperature of 40°C. Typically, one of these conditions, say C_1, represents a "nominal" operating condition, that is, a condition under which life distribution estimates for the population are known (or the conditions prevailing when the data for these estimates were collected), and the other condition C_2 represents a condition under which operation of the

[15] Note that when $C_1 = C_2 = C$, $A(C, C) = 1$.

system is anticipated in service with the customer. The types of environmental conditions that are typically encountered in engineering systems include

- temperature,
- humidity,
- vibration,
- shock, and
- mechanical load.

This list is far from all-inclusive. It includes only those conditions that are commonly encountered. Other more specialized conditions may include salt spray and immersion for marine environments, dust and oil spray for automotive environments, etc.

If a population follows the strong accelerated life model, the life distributions at the different environmental conditions differ only by a scale factor. In fact we have, for F_1 the life distribution of L_1 and F_2 that of L_2,

$$F_2(t) = P\{L_2 \le t\} = P\{A(C_1,C_2)L_1 \le t\} = P\{L_1 \le t/A(C_1,C_2)\} = F_1(t/A(C_1,C_2)),$$

showing that the scale factor is $1/A(C_1, C_2)$. For the densities, we have

$$f_2(t) = f_1(t/A(C_1,C_2))/A(C_1,C_2)$$

and for the hazard rates, we have

$$h_2(t) = A(C_1,C_2)\, h_1(t/A(C_1,C_2)).$$

Example: Suppose that, under nominal conditions, a population of devices has a Weibull life distribution with parameters $\alpha=20{,}000$ and $\beta=1.4$ (see Section 3.3.4.3). Under the strong accelerated life model, what are the new parameters of the life distribution when the population is operated at conditions for which the acceleration factor is 8? Denote by the subscript 1 the nominal conditions and by the subscript 2 the operating conditions (for which the acceleration factor is 8). Then

$$F_2(t) = F_1(t/8) = 1 - \exp(-(t/8 \cdot 20{,}000)^{1.4}) = 1 - \exp(-(t/160{,}000)^{1.4}),$$

so the life distribution parameters under the operating conditions are $\alpha=160{,}000$ and $\beta=1.4$.

From the example, we may gather that if the life distribution at nominal conditions has a certain parametric form, then the life distribution at any altered conditions continues to have the same parametric form when the strong accelerated life model applies (see Exercise 8).

We summarize the strong accelerated life model in Table 3.3.

TABLE 3.3 Strong Accelerated Life Model

Description	Formula
Lifetime (or failure time)	$L_2 = A(C_1, C_2)L_1$
Life distribution	$F_2(t) = F_1(t/A(C_1, C_2))$
Density	$f_2(t) = f_1(t/A(C_1, C_2))/A(C_1, C_2)$
Hazard rate	$h_2(t) = h_1(t/A(C_1, C_2))/A(C_1, C_2)$

In the strong accelerated life model, the defining equation $L_2 = A(C_1, C_2)L_1$ shows that the individual lifetimes under the two conditions are proportional. In fact, the probabilist would write $L_2(\omega) = A(C_1, C_2)L_1(\omega)$ to emphasize that the proportional relationship holds for each individual lifetime (sample point ω in the sample space, or individual member of the population). This is a very strong assumption, but one that is in very common use. Weaker versions of the accelerated life model are available. One such is the weak accelerated life model that postulates the life distribution relationship $F_2(t) = F_1(t/A(C_1, C_2))$ without the assumption that the lifetimes are proportional as individuals. For this weaker model, all the relationships in Table 3.3 apply except for that in the first row. In practice, usually the weak accelerated life model is all that is needed to make sensible use of the accelerated life model ideas.

Requirements tip: In a reliability requirement, while you do specify the environmental conditions that will prevail during operation with the customer and under which the specified reliability is to be achieved, the model to be used when projecting reliability under the operating conditions when the base reliability estimates pertain under some other, "nominal," conditions is not normally part of the requirement. The choice of model to use when projecting potential system life distributions or when analyzing field reliability data would normally be made by a reliability engineer who is familiar with the system, its components, and the operating environment(s). Systems engineers, while not necessarily themselves carrying out the computations involved, need to be aware of the options available and be able to ascertain whether the reliability engineer's choice is suitable given all the conditions prevailing.

How do you tell whether an accelerated life model is appropriate? If you have lifetime data collected under two different operating conditions, then the strong accelerated life assumption is easily tested. From the defining equation of the strong accelerated life model, we have

$$\log L_2 = \log L_1 + \log A\,(C_1, C_2)$$

Therefore, if the strong accelerated life model applies, a quantile–quantile plot (Q–Q plot) [45] of the logarithms of the lifetimes should have slope 1 and vertical intercept log A. The Q–Q plot is a graphical aid for determining when the strong accelerated life model might be appropriate and provides a method for an initial guess at the value of A.

The foregoing development leaves open the structure of the function A. In practice, different functions A are associated with different types of stresses (temperature, voltage, vibration, etc.). One of the most commonly used in reliability modeling is the Arrhenius relationship (Svante August Arrhenius, 1859–1927) for temperature:

$$A(C,C_0) = e^{\frac{E}{k}\left(\frac{1}{C} - \frac{1}{C_0}\right)}$$

where C and C_0 represent the two temperatures in °K (Kelvins), E is an activation energy in electron-volts (eV) particular to the material, and k is Boltzmann's constant 8.62×10^{-5} eV/°K. This equation was first used to describe the speeding up of chemical reactions when heat is added and has been widely used in reliability engineering as an empirical acceleration factor, even for phenomena that do not involve heat.

Example: Suppose that, when operated at 10°C, a population of devices has a Weibull life distribution with parameters $\alpha = 20{,}000$ and $\beta = 1.4$ (see Section 3.3.4.3). Under the strong accelerated life model, what are the new parameters of the life distribution when the population is operated at 35°C? Assume the weak accelerated life model and that the Arrhenius relation holds for these devices with an activation energy of 1.2 eV.

Solution: The Kelvin temperatures corresponding to 10 and 35°C are 283.15 and 308.15, respectively. Then the acceleration factor is

$$\exp\left(\frac{120{,}000}{8.62}\left(\frac{1}{308.15} - \frac{1}{283.15}\right)\right) = 0.019$$

so the life distribution of the population at 35°C is

$$F_2(t) = F_1(t/0.019) = 1 - \exp\left(-(t/0.019 \cdot 20{,}000)^{1.4}\right) = 1 - \exp\left(-(t/380)^{1.4}\right)$$

Many other parametric acceleration functions are used for stress modeling. These include

- the Eyring equation $A(C_1,C_2) = C_1/C_2 \exp\left[\beta(1/C_1 - 1/C_2)\right]$, with the single parameter β, is used for temperature, humidity, and other stresses [16];
- the inverse power law model $A(C_1,C_2) = (C_1/C_2)^n$, with the single parameter n, is usually used for voltage [16]; and
- the Coffin–Manson equation [18], similar to the inverse power law model, used for modeling fatigue under thermal cycling and modeling solder joint reliability.

Environmental conditions in operation may also vary with time. In this case, C_1 and C_2 may be functions of time. Generalizations of the accelerated life model can be devised to cover this case. One such generalization is a *differential accelerated life model*. This model postulates that the differential change in the lifetime of a unit is proportional to the current value of stress on the unit. Begin with the equation for the strong accelerated life model:

$$L(C) = A(C_0, C)\, L(C_0)$$

and include the time dependence of C:

$$L(C(t)) = A(C_0, C(t))\, L(C_0).$$

If the life distribution in the population at condition C_0 is F_0, then the population life distribution after t time units has passed is

$$F_0\left(\int_0^t A(C_0, C(u))\, du\right)$$

(see Section 6.4.3 of [41]). This model is used in analysis of data from accelerated life tests that use time-varying stresses such as ramp stress in which the stress takes the form $C(t) = a + bt$.

Many other models exist for relating the life distribution of a population operated at some set of environmental conditions to the life distribution of the same population operated at a different set of environmental conditions. Perhaps the most flexible of these is the acceleration transform developed by LuValle et al. [42]. See also Refs. 20, 43.

3.3.5.2 The proportional hazards model

The proportional hazards model is similar to the accelerated life model in that it postulates that certain quantities are proportional: in this case, it is the hazard rates, not the lifetimes, that are proportional. That is, the model postulates that

$$h_2(t) = A(C_1, C_2)\, h_1(t),$$

where $h_1(t)$ (resp., $h_2(t)$) denotes the hazard rate of the population when the conditions under which the population is operated are described by C_1 (resp., C_2). Note the difference with the accelerated life model (Table 4.1). The proportional hazards model was first described by Cox [13] and is widely used in biomedical studies. $h_1(t)$ is referred to as the "baseline hazard rate" and is usually associated with a nominal set of conditions such as the conditions under which the data characterizing the population were collected (i.e., the same idea as seen in the accelerated life model, Section 3.3.5.1). See Exercise 22.

3.3.6 Quantitative Incorporation of Manufacturing Process Quality

A commonly accepted explanation for so-called early life failures is that the population contains items that have manufacturing defects (see Section 3.3.4.4). In other words, components or subassemblies used in the system are received from their manufacturer(s) with defects that are undetected and not remedied by the manufacturers' process controls. The model is that such a defect will activate, or "fire," at some later time and cause a failure at that time. Using this reasoning, one may seek to construct a model for the early-life reliability of a component or subsystem that contains some factor related to the manufacturer's process quality. This model may also be used for the whole system to describe the influence of manufacturing processes on its reliability although the generally larger number of manufacturing processes involved may make the model more complicated. One attempt at creating such a model is described in Ref. 56. A brief summary is given in this section.

We may think of manufacturing as an opportunity to add defects to a product in the sense that when a product (here interpreted broadly to include components, subassemblies, and entire systems) is designed, it has a certain reliability that is a consequence of the degree to which design for reliability (Chapter 6) for the product is successful. The reliability of any realization of this design in physical space can never be better than this because additional failure modes are introduced by this realization, some failure modes were not anticipated in the design for reliability process, etc. The approach taken in Ref. 56 to model this situation is to allow the life distribution for a product to depend also on a parameter that represents the quality of the manufacturing process for that product.

Suppose the lower and upper specification limits for the product's manufacturing process[16] are a_L and a_U, respectively, and $a_L < a_U$. The center of the process's specification window is $a^0 = (a_L + a_U)/2$, which we also assume to be the target of the process output. Finally, we postulate that the true process output is a random variable A that is normally distributed with mean μ and variance σ^2 (see Figure 3.5). The process meets "six sigma" goals [29] if there is an m, $4.5 \le m \le 7.5$, for which

$$a_L \le \mu - m\sigma = \mu + (12 - m)\,\sigma \le a_U.$$

If $m = 6$, the process is centered and the expected fraction of defective process outputs (those falling outside the specification window) is approximately 9.87×10^{-10}, or about one part per billion (PPB). The expected fraction of defective outputs is largest when the process center is as far away from a^0 as possible. The maximum deviation allowed by the six-sigma methodology is 1.5 standard deviations, corresponding to $m = 4.5$ (left of center) or $m = 7.5$ (right of center).

[16] There may be, of course, many manufacturing processes needed to realize the product. The extension of the method shown here to many processes is straightforward and is the subject of Exercise 3.

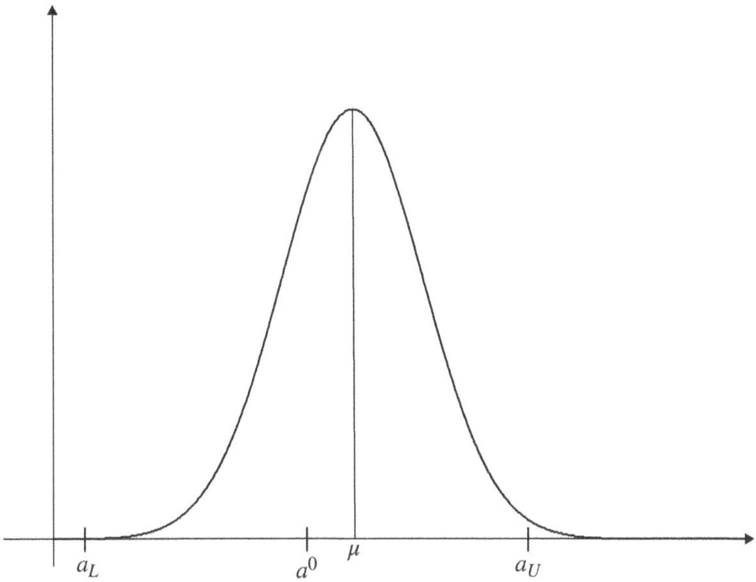

Figure 3.5 Specification limits and process output.

At the maximum deviation, the expected fraction of defective process outputs is approximately 3.4×10^{-6}, or about 3.4 parts per million (PPM).

To incorporate this understanding into a reliability model for the product, we postulate that when the process output is a, a defect may be introduced into the product that causes a failure at a later (random) time $X(a)$. Denote by $F(x, a)$ the life distribution of the failure mode attributable to this defect. That is,

$$F(x,a) = P\{X(a) \le x \mid A = a\}.$$

Then the distribution of the time at which the product fails (from this failure mode) is

$$\int_{-\infty}^{\infty} F(x,a)\, dP\{A \le a\}.$$

To make progress, we need to make some assumption about how $F(x, a)$ depends on a. For now, it is reasonable to assume that the further a process output is away from the process center, the more likely it is that the firing time of the associated defect is smaller. Formally, this is expressed as $F(x, a) \le F(x, a')$ whenever $|a - a^0| \le |a' - a^0|$, or $X(a)$ is stochastically larger [52] than $X(a')$ whenever $|a - a^0| \le |a' - a^0|$.

Suppose now that there are $M \ge 1$ downstream manufacturing and other product realization processes for this product, and that process j has lower and upper specification limits a_j^L and a_j^U, respectively, center a_j^0, and output A_j whose mean and standard deviation are μ_j^* and σ_j, respectively, for $j=1,\ldots, M$. As is customary in quality engineering studies, we assume all process outputs are normally distributed. The time at which an item chosen at random from this population of products fails is

$$Z = \min\{X(A_1),\ldots,X(A_M)\}$$

which, if the firing times are stochastically independent, has the survivor function

$$\bar{G}(z) = P\{Z > z\} = \prod_{j=1}^{M} \bar{G}_j(z) = \prod_{j=1}^{M} P\{X(A_j) > u\} = \prod_{j=1}^{M}\left[1 - \int_{-\infty}^{\infty} F(u,a)\varphi\left(\frac{a-\mu_j^*}{\sigma_j}\right)da\right]$$

where φ represents the standard normal density.

It follows that

$$\log[1 - G(z)] = \sum_{j=1}^{M} \log\left[1 - \int_{-\infty}^{\infty} F(u,a)\,\varphi\left(\frac{a-\mu_j^*}{\sigma_j}\right)da\right].$$

Using the result of Exercise 24 with $y = \int_{-\infty}^{\infty} F(u,a)\,\varphi((a-\mu_j^*)/\sigma_j)da$, you can show that

$$\exp\left\{-\frac{1}{2}\sum_{j=1}^{M}\left[\int_{-\infty}^{\infty} F(z,a)\varphi\left(\frac{a-\mu_j^*}{\sigma_j}\right)da\right]^2\right\} \le \frac{P\{Z > z\}}{\exp\left\{-\frac{1}{2}\sum_{j=1}^{M}\left[\int_{-\infty}^{\infty} F(z,a)\varphi\left(\frac{a-\mu_j^*}{\sigma_j}\right)da\right]\right\}} \le 1.$$

Multiplying everything by the denominator in the middle term gives lower and upper bounds for the survivor function of the product (considering these failure modes only).

3.3.7 Operational Time and Calendar Time

Throughout this section, the functions we use to describe reliability all use "time" as the argument. When so used, "time" almost always means operational time, or the amount of time the system is on and in use. Cumulative operational time does not increase when the system is off and not in use. When writing a reliability requirement, the "time" component of the requirement is intended to capture any increase in age of the system, where "age" refers to progression of any clock measuring time to failure of the system. Usually, this is operational time, so when you need to see how the requirement plays out in calendar time (e.g., warranties are usually written in terms of calendar time), it is necessary to understand the relationship between operational time and calendar time. This comes from how the customer uses the system. Some systems, like servers, broadcast transmitters, air traffic control radars, and the like, are intended to be used continuously and for such systems, operational time and calendar time are equal. Other systems, like refrigerators, do not run continuously and accrue age only when running (e.g., only when the compressor is running), so that operational time is less than calendar time for such systems. If there is a known relationship between operational time and calendar time, for example, as a function $\xi(t)$ giving elapsed calendar time ξ required to accrue an amount of operational time t, then the functions describing reliability can be transferred from one to the other using this relationship. This function is nondecreasing, satisfies $\xi(t) \geq t$, and may be deterministic or random. For example, if the refrigerator's compressor runs only 8 hours a day, then the relationship between operational time t and calendar time s for that refrigerator may be described by $s = \xi(t) = 3t$ when t is measured in hours.

The phrase "duty cycle" is also used to describe the fraction of (calendar) time that the system is in use. In the refrigerator example, the duty cycle is 1/3 or 33%.

If L is a lifetime random variable whose distribution in terms of operational time t is known, that is, $P\{L \leq t\} = F(t)$, say, then its distribution relative to calendar time s is given by $P\{L \leq s\} = P\{L \leq \xi(t)\} = F(\xi(t))$ where t is any value satisfying $\xi(t) = s$. For instance, if the refrigerator lifetime R has a distribution given by

$$P\{R \leq t\} = 1 - \exp\left[-(t/27{,}000)^{1.22}\right]$$

in terms of operational time t in hours, then the probability that the refrigerator fails before s calendar hours is given by

$$P\{R \leq s\} = 1 - \exp\left[-(s/3 \times 27{,}000)^{1.22}\right] = 1 - \exp\left[-(s/81{,}000)^{1.22}\right]$$

because, for this refrigerator, $s = 3t$.

Some systems may accrue age on some other clock besides the operational time or calendar time clocks. That is, progression to failure may be stimulated by some other mechanism in addition to time. A very common example is the automobile, in which age is measured not only by time but also by mileage. An electromechanical relay ages by number of operations in addition to time. Another way to look at this is to consider the system as having two failure mechanisms, one stimulated by the passage of time, and the other stimulated by a second "clock" like number of operations, mileage, etc.[17] Let L_1 and L_2 represent the lifetimes measured in terms of the first and second clocks, respectively, and let s and t denote the first and second clocks. Then the time at which the object fails is $\min\{L_1, L_2\}$.

[17] While this section is written in terms of two such failure mechanisms, similar considerations apply in situations where there are more than two failure mechanisms at play.

3.3.8 Summary

Section 3.3 has presented several diverse ways of describing the reliability of a nonmaintained system in quantitative terms: the lifetime, life distribution, density, and hazard rate. Most often, reliability engineers will use the hazard rate (force of mortality) as their preferred descriptor, and it will be called by them (almost universally, and inappropriately), "failure rate." It is vitally important to remember that when dealing with nonrepairable or nonmaintainable items, the use of the phrase "failure rate" should not lead you to think of the possibility of repeated failures of the same item (i.e., failures per unit time); only the most pernicious confusion will arise. It is best (although not yet common practice) to reserve "hazard rate" or "force of mortality" for this concept so that no confusion may arise.

The life distribution of a component may be altered if the component is exposed to environmental conditions other than those under which the data to estimate that life distribution were collected. Section 3.3.5 also discusses three forms of the accelerated life model that can be used to quantitatively describe such alterations. We also refer to the notion of acceleration transform that is a more general approach to this task.

When an explicit understanding of how manufacturing process(es) influence product reliability is needed, Section 3.3.6 provides a model connecting product reliability to the quality of the manufacturing processes for that product. This model provides a quantitative explanation of the phenomenon of "early-life" failures that are postulated to stem from latent defects introduced into the product by manufacturing process outputs that fall outside the process specification limits. More detailed models of this type can be constructed to capture the effects of more specific knowledge about downstream product realization processes.

Finally, we observe that the reliability descriptions that we introduce in this chapter are functions of operating time. When it is important to know how these are related to calendar time, we provide a means for moving easily from an operating time description to a calendar time description and back. We also discuss how this is related to the competing risk model (Section 3.4.4.2).

3.4 ENSEMBLES OF NONMAINTAINED COMPONENTS

3.4.1 System Functional Decomposition

3.4.1.1 System functional decomposition for tangible products and systems

Most often, nonmaintained items are not operated as individuals in isolation. There are exceptions, of course (the famous light bulb being a notable one), but real engineering systems almost always comprise many nonmaintained items and (possibly) maintained items and subassemblies operating together to perform the functions required of the system. So we would like to know how the lifetimes of such ensembles of nonmaintained items are related to the lifetimes of the individual items themselves. The system functional decomposition is a systematic description of how individual components and subassemblies operate together to enable the system to perform its required functions. There is a functional decomposition for every system requirement (of course it is possible that the same functional decomposition may apply to more than one requirement). The system functional decomposition is carried out to a level of detail necessary for which the life distributions of the components or subassemblies in the last layer of the decomposition are known or can be estimated. Before proceeding to the calculus of system reliability, that is, the methods for calculating the life distributions of higher level assemblies from their constituent components, we look at a few examples of system functional decompositions.

3.4.1.2 *Functional decomposition for services*

The services share of the world economy is large and growing. Our study of systems engineering for sustainability includes reliability, maintainability, and supportability of services. Services as understood here include not only the activities traditionally understood as "services" such as telecom services, auto repair service, fuel delivery service, and the like but also the emerging category of computer-based services such as personal computer and smartphone applications, cloud computing services, etc. All of these have the properties that they are intangible and do not exist outside the context of transactions taking place between users and service providers. The key to successful sustainability engineering for services is the realization that all such services are provided by elements of some tangible infrastructure of systems and humans that have to operate together in specified ways to deliver a transaction in the service.

Consequently, the functional decomposition required for reliability engineering for services requires peeling back an additional layer. To properly understand service reliability requires adopting the perspective of the user of the service, and the service functional decomposition consists of a detailed step-by-step description of how the service is provided. That is, a service functional decomposition acts as a bridge between the infrastructure the service provider uses to deliver the service and the user's understanding of the parts of a service transaction. Then with each step is associated the part(s) of the service delivery infrastructure that are involved in successful completion of that step. In this way, the reliability of the service (service accessibility, service continuity, and service release) [57] is expressed in terms of the reliability characteristics of those infrastructure components [58].

3.4.2 Some Examples of System and Service Functional Decompositions

3.4.2.1 *An automobile drivetrain*

The drivetrain in an automobile consists of those components that are required for the automobile to move forward. At a level of detail appropriate for this example, we may list these components as the engine, transmission, driveshaft, differential, and the four wheels[18] (a wheel comprises a rim and a tire). Each component listed is required for forward motion. If any component ceases to function, then the auto cannot be driven (for the purposes of this simple example, we are ignoring the possibility of partial failures such as loss of a single gear in the transmission) because one of the requirements of the auto is that be able to drive forward. Each component is a "single point of failure" in the sense that if it fails, then the system fails. Systems of this type are discussed in Section 3.4.4.

This example offers further instructional value. Most autos also carry a spare wheel so that if one of the active wheels fails (usually from a tire puncture), the wheel may be removed and replaced by the spare. This is an example of a system with "built-in redundancy." Redundancy means that there are additional components or subsystems built into the system that may be called on to restore the system to functioning condition when some component of the system fails (of course, the redundant component must be of the same type as the component that failed). In this example, the spare wheel assembly is a "cold standby" redundant unit. This terminology arises from the idea that the spare unit does not operate and accumulate age until it begins service. Systems containing redundant units are discussed in Section 3.4.5.

3.4.2.2 *A telecommunications circuit switch*

Automatic circuit switching has a long history in the telecommunications industry, starting with the panel office of the 1920s through the step-by-step, crossbar, and stored-program-control electronic systems that were the last generation of circuit switches. Many electronic switching systems were designed for high reliability by being "fully duplicated." That is, the system comprised two separate, identical call processing units that operated together. Every incoming call was handled by both processing units and sent on to its next destination. The idea was that should one of the call processing units suffer a failure, the other was operating right alongside it and would successfully route the call regardless. This is an example of a "hot standby" redundant system in which the standby or redundant unit(s) are operating (and aging) all the while the main or primary unit is providing service. Within each call processing unit, there are many line-replaceable units that are single points of failure for that individual call processing unit (but not for the system as a whole). This architecture, while costly, enabled extremely high availability: the availability objective for such systems in Bell System service was availability should be greater than 0.9999943, equivalent to an expected outage time of no more than 2 hours in 40 years of operation. See again Section 3.4.5 for discussion of redundant systems.

3.4.2.3 *Voice over IP service using a session initiation protocol server*

Here is an example of a functional decomposition for a service. Voice over Internet Protocol (VoIP) service is an example of a voice telecom service carried on a packet network. Session initiation protocol (SIP) is one of the ways

[18] Each of these elements may in turn be decomposed as an ensemble of other, simpler elements, but that is not needed for the purpose of this example.

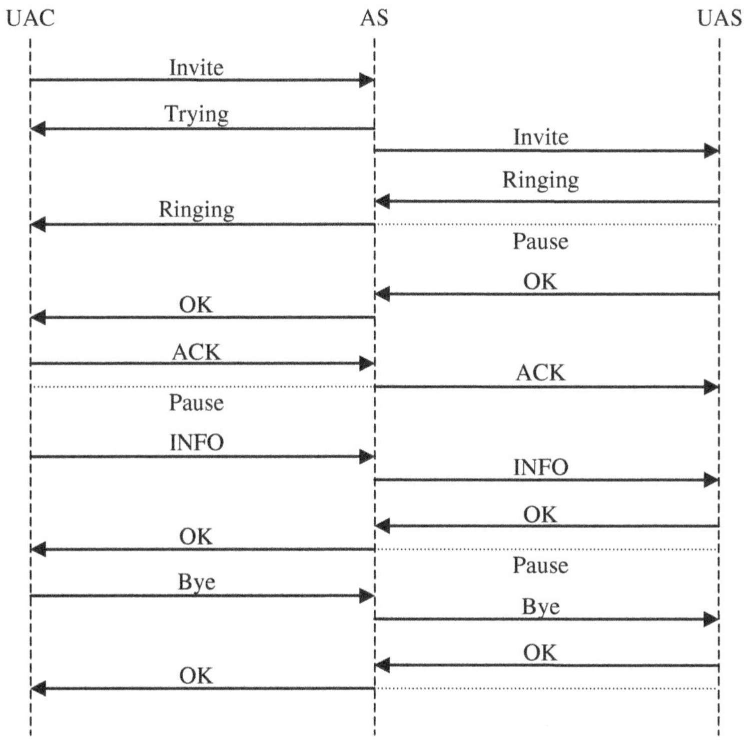

Figure 3.6 *UAC-UAS call flow.*

of setting up VoIP calls from one user to another. There are several different SIP implementations, but all SIP VoIP call setups involve interaction between the user (a "user agent client" (UAC) which is the user's local VoIP phone or computer) and a server ("user agent server" (UAS) which is part of the service provider's infrastructure). To successfully set up and carry an SIP VoIP call, certain messages must be exchanged between the UAC and the UAS; these messages are mediated by an application server (AS). Failure of the UAC, the UAS, or the application server at various times during the process of call setup and carriage will result in different kinds of user-perceived service failures. The service functional decomposition in this example consists of a chronological listing of those messages, called a "call flow," together with a description of how failures in the application server can disrupt the call flow. The listing of messages is facilitated by Figure 3.6.

In this diagram, time increases in the downward vertical direction. Failures of the UAC, AS, or UAS during the time from start to the first dotted horizontal line results in a call setup denial which is experienced by the user as a call attempt that did not succeed (a service accessibility failure). Failures at any time between the first and third dotted horizontal lines result in dropping the call which is already in progress, and is experienced by the user as a cutoff call (a service continuity failure). Failures after the third dotted horizontal line results in a call that is "hung" and the user perceives an inability to make his/her phone idle again (a service release failure). Additional discussion of service functional decomposition is found in Section 8.3.

3.4.3 Reliability Block Diagram

A *reliability block diagram* is a pictorial representation of the reliability logic of the system. The system functional decomposition is also a representation of the reliability logic of the system, so the reliability block diagram is simply a reliability-centered picture of the functional decomposition. It represents the manner in which failures of the components or subassemblies called out in the system functional decomposition lead to system failures. The reliability block diagram is drawn using boxes that represent the components and/or subassemblies, and lines connecting the boxes. See Figures 3.7 and 3.8 for two examples. A useful metaphor for reliability block diagrams is to imagine them as plumbing systems in which the lines are pipes and the boxes are valves that can be open (the unit represented by the box is working) or closed (the unit represented by the box is failed). The connecting lines are assumed to be irrelevant (always pass fluid). Then the system works if fluid can flow from one end of the diagram to the other. For the more mathematically inclined, it is also useful to think of the reliability block diagram as a graph in which the boxes are nodes (vertices) and the lines are links (edges). The presence of a node in the graph means the unit represented by that node is working. When that unit fails, the node is removed from the graph. In this metaphor, the system works if the graph is connected. Additional information about interpretation of reliability block diagrams can be found in Ref. 60.

For the remainder of this section, we will use the graph metaphor. In a reliability block diagram, a *cut* is any collection of nodes whose removal from the diagram (i.e., failure) disconnects the graph. For example, in Figure 3.7, every (nonempty) subset that can be formed from the five boxes in the diagram is a cut. There are $2^5 - 1 = 31$ cuts in this diagram. But you can see that there is a lot of redundant information in this formulation: it is enough that one of the boxes be removed to cause the graph to be disconnected. A *minimal cut* is a cut which is no longer a cut if one of its elements is removed from it. In the diagram of Figure 3.7, there are five minimal cuts, each consisting of one element. We will return to cut and path analysis in Sections 3.4.7 and 6.6.1.3 (see also Exercise 12).

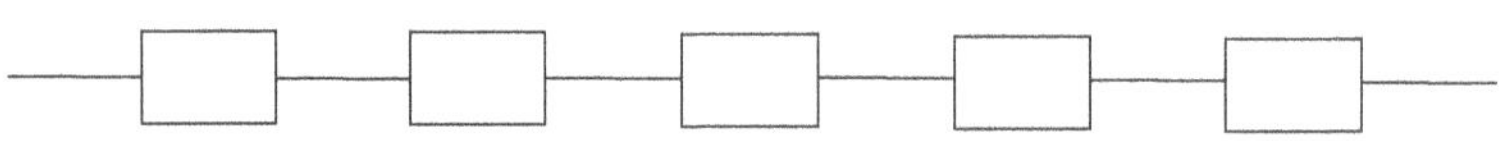

Figure 3.7 *An ensemble of five single-point-of-failure components.*

3.4.4 Ensembles of Single-Point-of-Failure Units: Series Systems

3.4.4.1 *Life distribution*

In many cases, the failure of a single item causes the system to fail. For example, consider the failure of a single diode in a four-diode-bridge balanced modulator in a single sideband transmitter. When the diode fails, the balanced modulator no longer functions as a mixer and the transmitter cannot emit properly formed single sideband signals. If emission of properly formed single sideband signals is a requirement for the transmitter, then the transmitter fails when the diode fails. In this situation, the diode is called a *single point of failure*, or a single-point-of-failure component, of the system. A single point of failure is a component of a system whose failure causes failure of the system (reminder: violation of one or more system requirements). A single-point-of-failure component may be a nonmaintained item, or it may be an ensemble comprising many items, and the ensemble may be maintainable or nonmaintainable (depending on the system maintenance concept).

The reliability block diagram for a series system is simply a picture of several (as many as there are single points of failure) elements in a linear configuration. An example with five single points of failure is shown in the Figure 3.7.

Reliability engineers call ensembles of single-point-of-failure components *series systems* because of the obvious nature of the reliability block diagram in Figure 3.7.

To introduce the method for quantitatively describing the life distribution of such ensembles, consider first an ensemble consisting of two (and only two) single points of failure. Letting L denote the lifetime of the ensemble and L_1 and L_2 denote the lifetimes of the first and second single points of failure, respectively, we can write

$$L = \min\{L_1, L_2\}$$

because the first lifetime to expire determines the lifetime of the ensemble. That is, the ensemble survives only as long as the shorter of the lifetimes of the two single points of failure comprising it ("a chain is only as strong as its weakest link"). It is then a straightforward matter to write

$$P\{L > t\} = P\{\min\{L_1, L_2\} > t\} = P\{L_1 > t, L_2 > t\}.$$

Provided we are willing to postulate that the two lifetimes L_1 and L_2 are stochastically independent, we may write

$$P\{L > t\} = P\{L_1 > t\}\, P\{L_2 > t\}$$

which brings us to the end of this story if we know the survivor functions of L_1 and L_2. For purposes of this exercise, we assume we do know these survivor functions, because what we were trying to do was write the life distribution of L in terms of the life distributions for L_1 and L_2, and so we have done (at least for the survivor functions). In terms of the life distributions, we have

$$P\{L \le t\} = 1 - \left(1 - P\{L_1 \le t\}\right)\left(1 - P\{L_2 \le t\}\right)$$

or

$$F_L(t) = 1 - (1 - F_{L_1}(t))\,(1 - F_{L_2}(t))$$

with the obvious notation. Absent independence, of course, we cannot go this far. All we can do is express the distribution of L in terms of the joint distribution of L_1 and L_2 as was shown earlier. Considerably more resources usually are needed to obtain the joint life distribution of L_1 and L_2 than are required to obtain the life distributions of L_1 and L_2 separately because a more complicated experimental design is required to collect suitable data. This is beyond the scope of this book. Interested readers may consult Ref. 45 for some ideas pertaining to this endeavor.

This argument generalizes to ensembles of many (more than two, say n) single points of failure. The formulas are

$$P\{L > t\} = \prod_{k=1}^{n} P\{L_k > t\}$$

and

$$F_L(t) = 1 - \prod_{k=1}^{n} (1 - F_{L_k}(t))$$

when it is possible to postulate that the individual lifetimes are stochastically independent. This principle takes its simplest form when written in terms of the survivor functions:

$$S_L(t) = \prod_{k=1}^{n} S_{L_k}(t).$$

Modeling tip: Almost all routine reliability modeling proceeds on the basis of stochastic independence (henceforth, simply: independence) of the constituent lifetimes. This is because the calculus of probabilities is simple for independent random variables or events, while accommodating random variables or events that are not independent requires dealing with joint distributions. As a rule, it is more difficult to ascertain the joint distribution of two or more random variables or events because the data collection and distribution estimation grows more complicated as the number of dimensions increases. We mention this here because it is often forgotten, and there are realistic reliability engineering situations in which independence cannot be assumed [44].

3.4.4.2 The competing risk model

It is not unusual that there may be more than one failure mechanism active in a single component. For instance, CMOS semiconductors are susceptible to failure by oxide breakdown, hot carrier damage, and electromigration. The series system model is readily adapted to use for this competing risk situation. The lifetime of the component is the minimum of the lifetimes of the competing failure mechanisms, that is, the component fails at the time the fastest failure mechanism has progressed to failure. On a macro level, every series system is a competing risk model: the system fails at the first time any of its elements fails.

3.4.4.3 Approximate life distribution for large series systems

Many practical engineered systems contain a large number of components. For example, printed wiring boards in defense and telecommunications systems typically contain thousands of components. Faced with such situations, the probabilist would inquire whether there might be some useful limit theorem that may simplify applications. In this case, Drenick's theorem [15] provides useful guidance. Drenick's theorem essentially says that in the limit as the number of components in a series system grows without bound, its life distribution tends to the exponential distribution, regardless what the life distributions of the individual components may be. But there are two important conditions: first is that the component lifetimes are stochastically independent; we have already discussed the use of this assumption in reliability modeling work (Section 3.4.4.1). The second is even more important: it requires that all the components have "similar" aging (aging in this context means "progression to failure") properties. That is, there is no one (or finitely many) component in the series system whose speed to failure dominates all the others. That is (and this condition is expressed in Drenick's work in technical terms which need not concern us now), there is no one component (or finitely many components) whose lifetime is so short that it is almost always responsible for the failure of the series system. This makes sense: if this one component almost always fails soonest, the life distribution of the ensemble is going to be very nearly the life distribution of that component.

Drenick's theorem is a kind of invariance principle. The limiting life distribution of the series ensemble is exponential, no matter what the individual life distributions may be (as long as the conditions of the theorem are satisfied). The mean of the limiting distribution is the harmonic mean of the mean lives of each of the constituent components:

$$\mu = \left(\frac{1}{\mu_1} + \cdots + \frac{1}{\mu_n} \right)^{-1}.$$

Referring to the formulas in Section 3.4.4.1, you can see that the life distribution of a series system of independent components (i.e., components whose lifetimes are stochastically independent) is not going to be exponential unless all the constituent life distributions are individually exponential. However, in many practical reliability modeling exercises, the life distribution of complex equipment, even equipment that is not an ensemble of single points of failure, is often postulated to be exponential. One reason for this is that the exponential distribution is particularly easy to work with in pencil-and-paper studies (although with the widespread availability of computer-based methods ([60] and many others), this is not a real attraction anymore; see also Section 4.6). Another reason is that enough data may be lacking to estimate more than one parameter, and the exponential, besides being simple, is a one-parameter family. These are rather weak justifications at best. But Drenick's theorem provides a more substantial justification for this. It is a sound theoretical basis for this choice, provided that the relevant conditions are satisfied. In particular, an exponential life distribution is often used for ensembles that are not series systems (the ease-of-use argument). Strictly speaking, this is not supported by Drenick's theorem. However, for subassemblies that are separately maintained (Section 4.4.4), the superposition theorem for point processes [25], [53] is employed to model the failure times of a complex repairable system as a Poisson process, which has exponentially distributed times between failures when the Poisson process is homogeneous [36]. In particular, the time to the first event (failure) has an exponential distribution under this model. Again, the superposition theorem is a kind of invariance principle in that it holds no matter what the point processes modeling the failure times of the individual subassemblies may be (again subject to a nondominance condition like that in Drenick's theorem). We will return to this discussion in Section 4.4.5.

Finally, the invariance principle represented by Drenick's theorem supports the following reasoning: if the limiting distribution of a series system is exponential, regardless of what the original component life distribution may be, and if the life distribution of a series system of exponentially distributed component lifetimes is also exponential (which it is), then you may as well assume the original component life distributions were exponential too because

- you get the same life distribution for the series system in either case and
- assuming the component life distributions are exponential will simplify any data collection and parameter estimation for the components.

This is not necessarily a bad approach as long as it does not hide unusual behavior in any of the components. The key concept in design for reliability is the anticipation and prevention of failures, and to do this effectively usually requires more, rather than less, detail. In particular, the response of a component's lifetime to various environmental stresses, and the stress–strength relationship for the component, may differ depending on the particular life distribution involved. We will see how similar reasoning is applied in repairable systems in Section 4.4.5.

The force of mortality for a series system

We know the life distribution and survivor function for a series system (Section 3.4.4.1). It is a simple matter to derive from this the hazard rate, or force of mortality (Section 3.3.3.4), of the life distribution of the series system. We will illustrate this for a series system of two components first, and ask for the full demonstration in Exercise 11.

Consider a series system of two components whose lifetimes are L_1 and L_2 with survivor functions S_1 and S_2 and hazard rates h_1 and h_2. Recall that the cumulative hazard function for unit i is $H_i(t) = -\log S_i(t)$, $i = 1, 2$. Then the cumulative hazard function for the series system is $H(t) = -\log S_1(t)S_2(t) = -\log S_1(t) - \log S_2(t) = H_1(t) + H_2(t)$. Consequently, when the hazard rates exist and the lifetimes are independent, the hazard rate of the series system is $h(t) = h_1(t) + h_2(t)$. This extends to any finite number of components; see Exercise 11.

This property is the basis for many reliability modeling software programs. When the components have an exponential life distribution with parameters λ_i, $i = 1, \ldots, n$, then the series system of those components has an exponential distribution whose force of mortality is $\lambda_1 + \cdots + \lambda_n$. In practice, the parameters λ_i are usually statistically estimated from some data or testing regime and so are not precisely known. Each estimate has some associated standard error, so the hazard rate of the series system comprising these components will also have some variability because it is a sum of the estimated hazard rates of the individual components. Some ideas for approximating this variability are given in the next section.

Confidence limits for the parameters of the life distribution of a series system

In practice, system subassemblies and line-replaceable units (LRUs) often are series systems of their constituent components. Reliability estimates for these components are derived either from life testing or from time-to-failure data collected during system operation. In either case, the component reliability estimates are *statistics*, or random variables, because they are a function of observational data. As such, they have distributions (Section 3.3.2). When they are combined using the formulas of Section 3.4.4.1, the result is another random variable (because the result is a function of the random variables that describe the component reliability). As such, it too has a distribution. In this section, we will describe a technique for obtaining information about this distribution when certain information about the component life distributions is available. This technique is based on the work of Baxter [6] which is in turn based on procedures developed by Grubbs [26], Myhre and Saunders [48], and others (see Ref. 6 for a review of the literature).

In this introductory material, we confine our discussion to the case in which the series system comprises components all of whose life distributions are exponential. Let the system contain n components and the parameters (hazard rates) of these components be $\lambda_1,\ldots,\lambda_n$. Suppose also that each parameter has been estimated from some data and has a 90% upper confidence limit (UCL) that for component i is denoted by u_i. To find an approximate 90% UCL (one-sided) for the hazard rate $\lambda=\lambda_1+\cdots+\lambda_n$ of the series system, first form the quantities $s_i=(u_i-\lambda_i)/1.282$, $S=s_1^2+\cdots+s_n^2$, and $\delta=2\lambda^2/S$. Then an approximate 90% UCL for λ is given by

$$\frac{S}{2\lambda}\chi^2_{\delta,0.9}=\frac{\lambda}{\delta}\chi^2_{\delta,0.9}$$

where $\chi^2_{\delta,0.9}$ is the 90th percentile of the chi-squared distribution on δ degrees of freedom. In most cases, δ will not be an integer, so interpolation between the nearest integers is used. If a UCL other than 90% is desired, change 1.282 to the appropriate confidence coefficient as found in Table 3.4.

We may also consider the use of two-sided confidence limits for λ. These would be useful when the quality of our knowledge about the λ_i is good as would be the case if u_i and λ_i are close together. The two-sided confidence 90% confidence interval for λ is then

$$\left[\frac{S}{2\lambda}\chi^2_{\delta,0.05},\ \frac{S}{2\lambda}\chi^2_{\delta,0.95}\right].$$

As with the one-sided case, if a confidence level other than 90% is desired, adjust the computation of the s_i according to Table 3.4. For details of these methods, consult Ref. 6.

Development of a more generally applicable method for confidence intervals for coherent systems with components whose life distributions are other than exponential has been attempted, but no satisfactory method yet exists that is fully applicable and easy to use. Use of the method given here will provide important insight into the quality of knowledge about the reliability prediction for a series system. In particular, this quality of knowledge information is important when using a reliability prediction of this kind to assess the likelihood that the design on which the prediction was made will meet its reliability requirements.

TABLE 3.4 Confidence Coefficients for UCL Computations

Confidence Level (%)	Confidence Coefficient
90	1.282
95	1.645
99	2.326

3.4.5 Ensembles Containing Redundant Elements: Parallel Systems

Many practical systems that require high reliability would be impossible to implement if they consisted only of single points of failure. Satellites, aircraft, undersea cable telecommunications systems, and many other systems we have come to accept as an ordinary part of daily life would be less effective without a reliability improvement strategy. Probably the reliability improvement strategy that most people think of first is the provision of spare units that will take over operation when another unit fails. This is called redundancy. Properly done, it can be very effective, but it is also costly and should not necessarily be the first thing the professional reliability engineer thinks of when reliability improvement is needed. This could be an introduction to the interesting and vital field of reliability economics, but that is beyond the scope of this chapter. This section discusses the reliability modeling issues pertaining to redundant systems.

Example: (Continuation of Section 3.4.2.1) An automobile requires four wheels to satisfy one of its most important requirements, namely that it be able to move forward under power provided by its engine. If a wheel fails (e.g., because of a tire puncture and consequent loss of air pressure), the vehicle fails because the requirement that it be able to move under power provided by its engine is violated. Thus each wheel constitutes a single point of failure for the vehicle. But most automobiles carry a spare wheel. When a wheel fails, it is possible to replace it with the spare. Thus, we may consider the spare wheel as a redundant unit that is provided so that the failure of a wheel may be remedied during a mission (driving trip). In the language we will introduce later, the wheel subsystem on the vehicle is a four-out-of-five cold standby redundant system. In the event that a wheel failure has occurred and the failed wheel was replaced by the spare, the vehicle is operating in a brink of failure state until a spare wheel is returned to the vehicle. This example is continued in Exercise 13.

Language tip: When a redundant ensemble operates with no spare units (e.g., all the spares may have been already used to cover failed primary units), we say the ensemble is operating in a brink-of-failure state. The terminology arises from the fact that in this scenario, the next unit failure to occur will cause the ensemble to fail. Some means of detecting when an ensemble is operating in a brink-of-failure state should be provided because if such operation is "silent," or undetected, failure of the ensemble may occur as a surprise. Provision of a brink-of-failure operation detector is an example of a predictive maintenance procedure; such procedures will be covered in greater detail in Chapter 11.

We examine three kinds of redundancy:

- Hot standby redundancy,
- Cold standby redundancy, and
- k-out-of-n redundancy.

Many more redundancy schemes exist; the reliability engineering literature concerning redundant systems is vast and untamed. In particular, there are many forms of "warm standby" redundancy in which the spare units are considered to be in various intermediate states between operation and complete inactivity. In addition, the reliability of the switching mechanism that implements the redundancy scheme is of great importance, but it is not included in any of the basic models discussed. We will present one example worked out in detail of a two-unit parallel (hot standby) system that includes the reliability of the switching mechanism, but the variety of switching mechanisms is too great to cover all of them completely. We hope that by following the ideas shown in the following example, you will be able to construct suitable models to include switching mechanism reliability when the need arises. Nor will we consider any warm standby models in this book; again, the fundamentals you will learn here will help you use the literature effectively and to model and work with other redundancy schemes when it becomes necessary.

The reliability block diagram for the three redundancy schemes studied here is drawn as a parallel ensemble of units. Figure 3.8 gives an example with four units.

There really is no way to distinguish one scheme from another on the basis of the diagram alone. That is, there is no universally accepted scheme to draw reliability block diagrams for parallel systems that distinguish, on the basis of

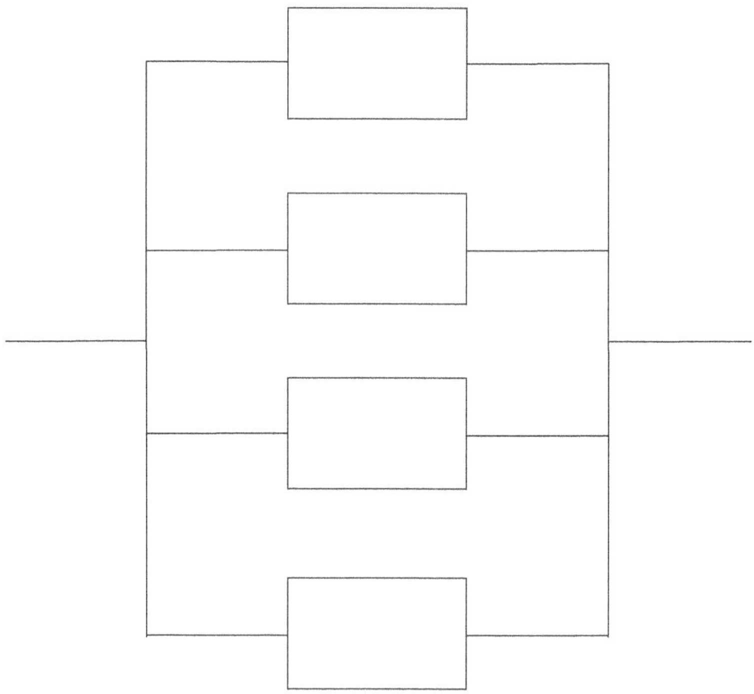

Figure 3.8 A parallel redundant system of four units.

the drawing alone, the different types of redundancy. The drawing in Figure 3.8 could represent any of the three types of redundancy listed earlier, or even another type (i.e., a warm standby scheme). Labels or color-coding may help when there is ambiguity that needs to be banished, but a universally accepted scheme has not yet been implemented.

3.4.5.1 Hot standby redundancy

The simplest redundancy scheme is hot standby redundancy in which a single unit (the "primary unit") is supported by one or more "redundant units" (or "backup units"), all of which are powered on and aging along with the primary unit. When the primary unit fails, some switching mechanism operates to bring the failed unit off line and one of the redundant units on line to take over the operation that was being performed by the primary unit before it failed (assuming the switching operation does not fail). Thus, the ensemble fails only when all units, the primary unit and all the redundant units, fail. In the case of hot standby, or *active*, redundancy, we have for the lifetime L of the ensemble in terms of the lifetimes $L_1,\ldots,$ L_n of its constituent units, again assuming the switching mechanisms does not fail,

$$L=\max\{L_1,\ldots,L_n\}.$$

From here, it is a routine matter to obtain the distribution of L:

$$P\{L\le x\}=P\{\max\{L_1,\ldots,L_k\}\le x\}=P\left(\bigcap_{k=1}^{n}\{L_k\le x\}\right)=\prod_{k=1}^{n}P\{L_k\le x\},$$

where the last equality is valid if the individual lifetimes are stochastically independent. We may also write

$$F_L(x)=\prod_{k=1}^{n}F_{L_k}(x).$$

In terms of the survivor functions, we have

$$S_L(x)=1-\prod_{k=1}^{n}(1-S_{L_k}(x)).$$

Note the duality between the series system discussed in Section 3.4.4 and the hot-standby parallel system discussed here: in the series system, the expression for the life distribution looks like the expression for the survivor function of a hot-standby parallel system, and vice versa.

Example: A Two-Unit Hot Standby Redundancy Arrangement with an Unreliable Switch

Consider the hot standby arrangement depicted in Figure 3.9.

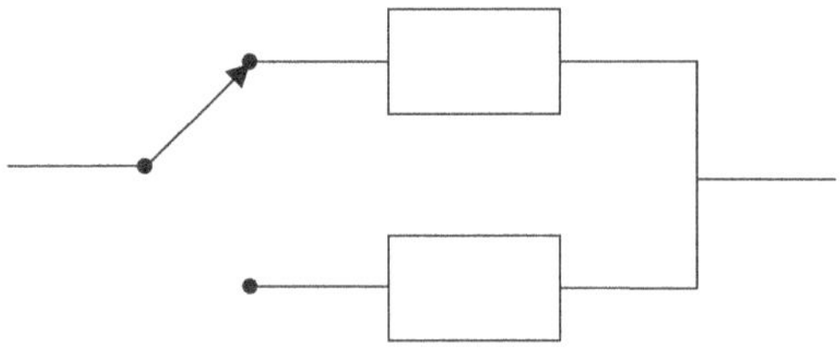

Figure 3.9 Two-unit hot standby ensemble with switch.

Denote by L_1 and L_2 the lifetimes of the first (topmost in the figure) and second units, respectively, let the corresponding survivor functions be denoted by S_1 and S_2, and denote by L the lifetime of the entire ensemble. Let $W(t)$ denote the indicator of the event that the switch operates correctly when it is called upon to do so at time t, let $p(t)=P\{W(t)=1\}$, and Z_t denote the lifetime of the switch given that it operates correctly at time t (if the switch fails to operate correctly when called for, then $L=L_1$ and the value of Z_t is irrelevant). Furthermore, we assume there is no "rejuvenation" of the switch if it fails when called for: once this failure occurs, the ensemble is failed. Let G_t denote the life distribution of Z_t. If there were no switch involved (usually not feasible in an engineering sense), then L would be equal to $\max\{L_1, L_2\}$, and the life distribution at time t of the ensemble would be $F_1(t)F_2(t)$. If the switch operation were perfectly reliable (i.e., $p(t)\equiv 1$ for all t), then L would be equal to the maximum of L_1, L_2, and $\min\{L_1, L_2\}+\min\{Z_T, \max\{L_1, L_2\}\}$ because the switch will be called for at time $T=\min\{L_1, L_2\}$. In this case, the survivor function of L would be

$$\begin{aligned}
P\{L>x\} &= P\left\{\max\left\{L_1,L_2,\min\left\{L_1,L_2\right\}+\min\left\{Z_T,\max\left\{L_1,L_2\right\}\right\}\right\}>x\right\}\\
&= P\left\{\max\left\{L_1,L_2,L_1+\min\left\{Z_{L_1},L_2\right\}\right\}>x \mid \min\left\{L_1,L_2\right\}=L_1\right\}\\
&\quad P\left\{\min\left\{L_1,L_2\right\}=L_1\right\}\\
&\quad + P\left\{\max\left\{L_1,L_2,L_2+\min\left\{Z_{L_2},L_1\right\}\right\}>x \mid \min\left\{L_1,L_2\right\}=L_2\right\}\\
&\quad P\left\{\min\left\{L_1,L_2\right\}=L_2\right\}\\
&= P\left\{\max\left\{L_2,L_1+\min\left\{Z_{L_1},L_2\right\}\right\}>x \mid L_1\le L_2\right\}P\left\{L_1\le L_2\right\}\\
&\quad + P\left\{\max\left\{L_1,L_2+\min\left\{Z_{L_2},L_1\right\}\right\}>x \mid L_1>L_2\right\}P\left\{L_1>L_2\right\}.
\end{aligned}$$

Now $P\{L_1\le L_2\}=\int_0^\infty F_1(y)dF_2(y)$ assuming that L_1 and L_2 are independent, and the distribution of Z_{L_i} is equal to $P\left\{Z_{L_i}\le t\right\}=\int_0^\infty G_y(t)dF_i(y)$ for $i=1, 2$, assuming that L_1 and L_2 are conditionally independent of Z_t for all t. These may now be substituted into the expression above for $P\{L>x\}$ to complete the development (see Exercise 16). In case the switching action may be unreliable (i.e., $p(t)<1$ for at least one t), then the survivor function of L is given by

$$P\{L>x\} = P\{\max\{L_1,L_2,\min\{L_1,L_2\}+\min\{Z_T,\max\{L_1,L_2\}\}\}>x\}$$
$$P\{W(T)=1\}+P\{L_1>x\}P\{W(T)=0\},$$

again assuming the necessary independence (in this case, that of $W(T)$ from everything else in sight).

Requirements tip: It is clear from this example that reliability models that include imperfect switching for redundant systems may be considerably more complicated than those that ignore the effect of potentially unreliable switching. Systems engineers need to be aware that switching mechanism unreliability can be a significant contributor to overall system unreliability in cases where redundancy is being used to increase system reliability overall. Be sure that reliability engineers on the project provide realistic reliability projections in these cases because the high cost of redundancy can be rendered for naught by a relatively low-cost switching mechanism that may be unreliable.

3.4.5.2 *Cold standby redundancy*

Cold standby, or *passive*, redundancy differs from hot standby redundancy in that the redundant units are not active while the primary unit is in service. This model postulates that the standby units do not age while they are not operating, that is, the "lifetime clock" does not start for these units until they are put into service. When the primary unit fails, the switching operation activates the first redundant unit and takes the failed unit off line, and puts the now-active redundant unit on line so that the ensemble continues to perform its function. In this case, the lifetime of the ensemble is the sum of the lifetimes of all its constituent units (again assuming that all the switching operations are perfect). That is,

$$L = L_1 + \cdots + L_n.$$

To find the distribution of L when the lifetimes $L_1,\ldots, L_n$ are independent, we introduce the notion of *convolution* of distribution functions. Suppose X and Y are independent lifetimes having distributions F and G, respectively. Then the distribution of $X+Y$ is given by

$$P\{X+Y\le t\} = \int_0^\infty P\{X+Y\le t \mid Y=y\}\, dP\{Y\le y\}$$
$$= \int_0^\infty P\{X\le t-y\}\, dG(y) = \int_0^t F(t-y)\,dG(y).$$

The last integral is called the *convolution* of F and G and is denoted $F*G(t)$. Thus, if $F_1,\ldots, F_n$ represent the life distributions of $L_1,\ldots, L_n$, respectively, then the life distribution of the cold standby ensemble is given by $F_1* \cdots *F_n$. The family of gamma distributions is closed under convolution. As was noted in Section 3.3.4.7, the sum of two independent random variables having gamma distributions with parameters (α_1, ν) and (α_2, ν) again has a gamma distribution with parameters $(\alpha_1+\alpha_2, \nu)$. So a cold standby redundant system whose first unit's lifetime has a gamma distribution with parameters (α_1, ν) and whose second unit's life distribution has a gamma distribution with parameters (α_2, ν) has a life distribution with parameters $(\alpha_1+\alpha_2, \nu)$. In most other cases, it is not possible to evaluate convolution integrals in closed form. Various numerical methods have been developed to enable computation of system reliability when cold standby redundancy is present. Among the simplest are the Newton–Cotes-like rules found in Ref. 54.

The number and variety of reliability models for imperfect switching incorporated into cold standby redundancy schemes is at least as great as the large number of such schemes. We will consider only a simple example to illustrate some possibilities. Suppose that in an n-unit cold standby redundancy scheme there is a switching mechanism whose duty is to switch in the next unit when the unit currently in service fails, and suppose that the indicator event that this switch performs its duty correctly when called upon at time t is $W(t)$, independent of everything else, $P\{W(t)=1\}=p(t)$, and if $W(t_0)=0$, then $W(t)=0$ for all $t\geq t_0$. Further suppose that the lifetime of the switch is infinite (i.e., the switch does not fail once the switching operation has completed successfully—the only possible failures of the switch are at the moments of switching). Then the life distribution of the ensemble is

$$P\left\{L_1+W\left(L_1\right)L_2+W\left(L_1+L_2\right)L_3+\cdots+W\left(L_1+\cdots+L_{n-1}\right)L_n\leq t\right\}.$$

See Exercise 17 to complete this example.

3.4.5.3 k-out-of-n redundancy

The final type of redundancy we study in this chapter is the k-out-of-n scheme. In this scheme, there are n units. The ensemble operates if and only if k of these n units are in an operating condition. One may think of this as a system that requires k units to operate properly and that has in addition $n-k$ spare units on site. This scheme may be implemented as hot standby or cold standby (and other types of warm standby which will not be covered here). In a hot standby arrangement, the lifetime of the ensemble is the kth shortest of the n unit lifetimes. This is an example of an *order statistic*. The life distribution of the hot standby k-out-of-n ensemble is the cumulative distribution of this order statistic which may be found in Ref. 14.

In a cold standby k-out-of-n scheme, the lifetime is determined by counting the total number of unit failures that occur by a given time. The first time this reaches $n-k+1$ is the time of ensemble failure. Let $N(t)$ represent the number of unit failures in the time interval $[0, t]$, and let L represent the system lifetime. Then the "counting argument" is that the time to system failure occurs after time t if and only if there have been no more than $n-k$ unit failures up to and including time t,

$$\{L>t\}=\{N(t)\leq n-k\}=\bigcup_{i=0}^{n-k}\{N(t)=i\}.$$

Our task will be to obtain the distribution of L as

$$P\{L\leq t\}=1-P\{L>t\}=1-\sum_{i=0}^{n-k}P\{N(t)=i\}=\sum_{i=n-k+1}^{n}P\{N(t)=n-i\}.$$

We will now assume that all primary units have the same reliability characteristics, all standby units have the same reliability characteristics (although these may be different from the primary units), and all units are mutually stochastically independent. We begin by defining the concept of "position." Consider the primary units that are started operating at time zero. The location or "slot" that each of these occupies is called a "position." At the time a primary unit fails, a spare unit is immediately placed in service in that position (the pool of spares initially contains $n-k$ units; if this is the $(n-k+1)^{st}$ failure in a primary slot, then the ensemble fails at this time and there are no spare units remaining in the pool). Thus, we record the failures in each "position" separately, or, in other words, each "position" is thought of as having a failure process of its own. To illustrate the idea, we will first work through the derivation in the simple case $n=2, k=1$. In this case, denoting by $N_1(t)$ the number of failures in position 1 that occur during $[0, t]$, we have $P\{L>t\}=P\{N_1(t)\leq 1\}$, because the system fails when the second failure in position 1 occurs. Define $T=\inf\{t: N(t)=1\}$ and $T+S=\inf\{t: N(t)=2\}$. Thus, $P\{N(t)\leq 1\}=P\{T+S>t\}=1-F*G(t)$. Obviously, in this case we have $L=T+S$, so there was really no need to go through the counting argument, but it is valuable to see how it works in this simple case first. In the general case, let $N_i(t)$ denote the number replacements in position i by spare units, $i=1,\ldots, k$. Then we have $N(t)=\sum_{i=1}^{k}N_i(t)$ because the spare units only operate (and fail) in the primary (first k) positions. This gives us the opening we need to get the distribution of $N(t)$. Define $W_0(t)=0$ and $W_i(t)=\sum_{r=1}^{i}N_r(t)$. Then $W_1(t)=N_1(t)$ and $W_k(t)=N(t)$. Using the relation $W_i(t)=W_{i-1}(t)+N_i(t)$, $i=1,\ldots, k$, and the mutual independence of $N_1(t),\ldots, N_k(t)$, we can get the distribution of each $W_i(t)$ by successive discrete convolutions:

$$P\{W_i(t)=j\}=\sum_{r=0}^{j}P\{W_{i-1}(t)=r\}P\{N_i(t)=j-r\},\quad i=1,\ldots,k,\ j=0,\ldots,n-k.$$

To model the failure counting processes $N_i(t)$, $i=1,\ldots, k$, in the k primary positions, we assume that the primary units are identical, and all have life distribution F, say, and the spare units are identical, and all have life distribution G, say. Then we have $P\{N_i(t)=0\}=1-F(t)$, $i=1,\ldots, k$, and

$$P\{N_i(t)=j\}=P\{N_1(t)=j\}=(G_{j-1}-G_j)*F(t),\quad i=1,\ldots,k;\ j=1,\ldots,n-k.$$

Here, G_{j-1} represents the convolution of G with itself $j-1$ times, and G_0 is the unit step function at the origin. Working backward to the equation for $P\{L\le t\}$ completes the derivation.

3.4.6 Structure Functions

In Section 3.4.3, we saw how to express pictorially the "reliability logic" of a system by using the reliability block diagram. We may also summarize the reliability logic of a system using a concept called the *structure function*.[19] The reliability logic of a system is a catalog of how system failure results from component failures. For instance, in the ensemble of single points of failure (the series system), the system fails whenever a component fails. The structure function is a mathematical representation of this logic. It is a Boolean function (its arguments and values come from $\{0, 1\}$ only); in this formalism, 1 is taken to mean the component or system is in an operating state and 0 is taken to mean the system is in a failed state. If C_i is the indicator that component i is working (i.e., $C_i=1$ if component i is working and is 0 otherwise), then the system structure function is $\varphi_R(C_1,\ldots, C_n)$ where $C_1,\ldots, C_n$ is the list of (indicator functions of) the components of the system. φ_R is the indicator that the system is working; the functional form in terms of the $C_1,\ldots, C_n$ tells whether the system works when its constituent components are working or failed. Sometimes, the vector $(C_1,\ldots, C_n)$ is called the *vector of component states* or the *state vector*.

For example, the structure function of an ensemble of single points of failure (a series system) is $\varphi_R(C_1,\ldots, C_n)=C_1 \cdots C_n$ because $\varphi_R(C_1,\ldots, C_n)$ is 1 if and only if all the C_i are 1. As soon as one of the C_i is zero, the structure function is zero. This is the logic of the series system (the system fails if and only if at least one of its components fails) expressed in mathematical terms. The structure function of a parallel (hot standby) system is given by $\varphi_R(C_1,\ldots, C_n)=1-[(1-C_1) \cdots (1-C_n)]$. The structure functions of ensembles comprising components in nested series and parallel configurations are readily expressible. See Exercise 18.

If we now allow the structure function and its arguments to take values in $[0, 1]$, we obtain a simple expression for the probability that the system operates

[19] When it is necessary to make the distinction, we will call this the "reliability structure function" because in 10 we will introduce a "maintainability structure function" to assist with maintainability modeling.

as a function of the probabilities of each component operating [9]. We can use this idea to write an expression for the survivor function of the system in terms of the survivor functions of its constituent components. Let $S_1,\ldots, S_n$ denote the survivor functions of components $C_1,\ldots, C_n$, respectively, and let S denote the system's survivor function. Then for each time t,

$$S(t) = \varphi_R\left(S_1(t),\ldots,S_n(t)\right).$$

A disadvantage of the structure function approach is that it is not easily possible to incorporate warm standby and cold standby redundancy into the structure function scheme. On the other hand, for systems not having these methods implemented, the structure function approach provides a compact and mathematically convenient approach to working with complex structures. Additional properties of structure functions are explored in detail in Refs. 5, 19.

3.4.7 Path Set and Cut Set Methods

The graph metaphor for reliability block diagrams was introduced in Section 3.4.3. Methods for determining the reliability of ensembles of single points of failure and ensembles with redundancy were reviewed in Sections 3.4.4 and 3.4.5. These methods rely heavily on the simple nature of the reliability block diagram graph when it can be represented in "series" or "parallel" form. Here we discuss some methods for determining the reliability of the diagram as a function of the reliabilities of its constituent components when the reliability block diagram graph is more complicated (not in "series" or "parallel" form). Most of this material originally appeared in Ref. 59 and is reprinted here with permission. This section reviews the concepts of connectedness, paths, cuts, path sets, and cut sets in the context of analyzing a system reliability block diagram described as a labeled random graph. The methods discussed also lend themselves to the development of bounds for system reliability. More general interpretations of this material can also be used to define and determine reliability for capacitated networks [50].

A *graph* is an ordered pair of sets $(\mathcal{N}, \mathcal{L}) := \mathcal{G}$ with $\mathcal{L} \subset \mathcal{N} \times \mathcal{N}$. $\mathcal{N}$ is called the set of *nodes* of the graph and $\mathcal{L}$ is called the set of *links* of the graph. Typically, a graph is pictured as a drawing in which the nodes are represented as points in the plane and the links are represented as lines drawn to join two points. In other terminology in common use, the nodes may be called vertices and the links may be called arcs or edges. A *labeled graph* is a graph in which the nodes and/or links have names. That is, there is a one-to-one correspondence between the nodes of the graph and a set of $|\mathcal{N}|$ objects (the node labels) and/or between the links of the graph and a set of $|\mathcal{L}|$ objects (the link labels). A *directed graph* is a graph in which each link is assigned an orientation or direction. In a directed graph, the links (i, j) and (j, i) are different, whereas in an ordinary (undirected) graph, they are identical. The concept of "a link from i to j" makes sense in a directed graph; in an undirected graph, it would be proper to say, rather, "a link between i and j."

Two nodes i and j are *adjacent* if $(i, j) \in \mathcal{L}$. A *path* in a graph is a sequence of adjacent nodes and the links joining them, beginning and ending with a node. Two nodes i and j are said to be *connected* if there is a path having i as its initial node and j as its terminal node. That is, the path takes the form $\{i, (i, \nu_1), \nu_1, (\nu_1, \nu_2), \ldots, \nu_k, (\nu_k, j), j\}$ for some $\nu_1, \ldots, \nu_k \in \mathcal{N}$ and $(i, \nu_1), (\nu_1, \nu_2), \ldots, (\nu_k, j) \in \mathcal{L}$. There is no loss in abbreviating this to $\{i, \nu_1, \nu_2, \ldots, \nu_k, j\}$. When it is necessary or desirable to explicitly indicate the nodes being connected, the path will be called an (i, j)-path. Clearly, adjacent nodes are connected, but connected nodes need not be adjacent. If the graph is directed, the links in the path must be considered with the proper orientation. A path connecting two given nodes is called *minimal* if it contains no proper subset that is also a path connecting the two nodes.

A *cut* for two given nodes is a set of nodes and/or links whose removal from the graph disconnects the two nodes. To explicitly indicate the nodes being disconnected, the cut may be called an (i, j)-cut. A cut for two given nodes is called *minimal* if it contains no proper subset that is also a cut disconnecting those nodes.

The Washington, DC, Metro subway system [32] may be modeled as a graph with the stations as the nodes. In this graph, the Pentagon and College Park–University of Maryland stations are connected but not adjacent. DuPont Circle and Farragut North are both connected and adjacent. The (Takoma, Union Station) path is a cut for the Silver Spring and Judiciary Square nodes. It is not a minimal cut because its subset (Brookland–CUA, Rhode Island Avenue) is also a cut for the Silver Spring and Judiciary Square nodes. See Exercise 26.

A *random graph* is a labeled graph in which the labels are stochastic indicator variables. When the variable is zero, it indicates that that node or link is not present in the graph. When it is one, it indicates that that node or link is present in the graph. Each choice of values for these indicator variables, by whatever random mechanism is at play, produces a different graph (the choice is not completely unrestricted; if the indicator of a node is zero, the indicators of all the links emanating from that node must be zero also). In the reliability modeling application, the indicator variable for a link or node describes the functioning or nonfunctioning of the link or node. The usual convention is that the indicator variable is 1 when the link or node functions and 0 when it does not function.

The system reliability block diagram is a labeled random graph whose nodes represent the components or subsystems whose reliability description is known or provided. The links are merely connectors and may be disregarded for these purposes.[20] The system reliability block diagram expresses the reliability logic of a system in the sense that it shows how the system fails when constituent components and subsystems fail. It is a pictorial representation of the system structure function. Two special nodes are called out: a source, or origin, node, and a terminal, or destination, node. The system

[20] In a reliability block diagram, the links are assumed not to fail. However, link failures are an essential part of graph models for network reliability.

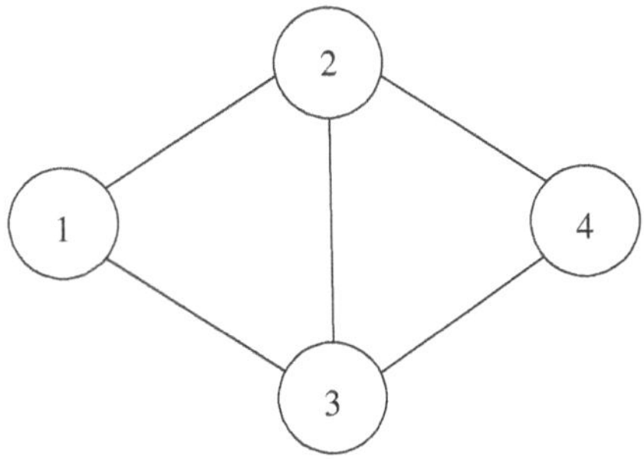

Figure 3.10 *Bridge network.*

functions if and only if in the random graph there is a path connecting the source node and the terminal node.

In many cases, the system reliability block diagram is a *series-parallel* structure. In such cases, the probability that the system functions is easily concluded from nesting of the standard formulas for the reliability of series systems and parallel systems (see Sections 3.4.4 and 3.4.5). Other structures, such as the k-out-of-n hot standby and k-out-of-n cold standby structures, are also amenable to similar probabilistic analysis as seen earlier. Some other structures, such as the bridge structure shown in Figure 3.10, lend themselves less readily to this type of analysis. In such cases, it may be convenient to use the path set or the cut set methods described here.

In the bridge network, the source is node 1 and the terminal is node 4. The (1, 4) paths are $\{1,2,4\}$, $\{1,3,4\}$, $\{1,2,3,4\}$, and $\{1,3,2,4\}$ and the minimal paths are $\{1,2,4\}$ and $\{1,3,4\}$. The (1, 4) cuts are $\{1\}$, $\{4\}$, $\{1,2\}$, $\{1,3\}$, $\{2,4\}$, $\{3,4\}$, $\{2,3\}$, $\{1,2,3\}$, $\{2,3,4\}$, and $\{1,2,3,4\}$. The minimal cuts are $\{1\}$, $\{4\}$, and $\{2,3\}$.

Given two nodes in a graph, the *path set* (sometimes called *tie set*) for those two nodes is the set of all paths connecting those two nodes. The *cut set* for two nodes is the set of all cuts for those two nodes. The *minimal path set* for a pair of nodes is the set of all minimal paths for that pair of nodes. The *minimal cut set* for a pair of nodes is the set of all minimal cuts for that pair of nodes. The key concepts for reliability modeling are

- The system functions if and only if there is at least one minimal path whose components are all in a functioning condition and
- The system does not function if and only if there is as least one minimal cut whose components are all in a failed (nonfunctioning) condition.

The random graph model provides a framework for computing probabilities of system functioning and failure (nonfunctioning) based on these concepts.

Example: Consider the reliability block diagram shown in Figure 3.11. The nodes representing subsystems that can fail individually in this system have been labeled by the letters A, B, C, and D. Note that, like the bridge structure

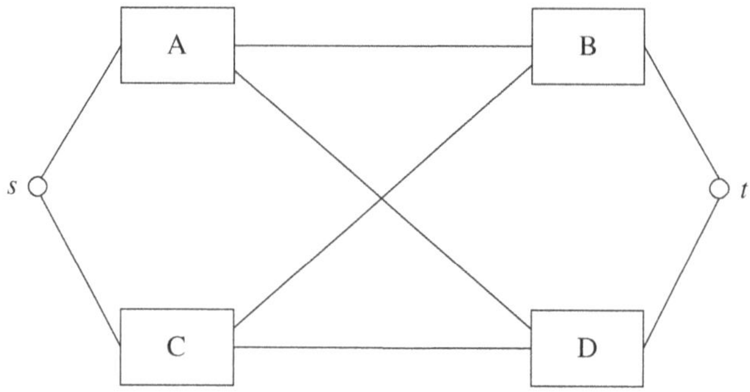

Figure 3.11 *Example of a system reliability block diagram.*

of Figure 3.10, this is not a series-parallel graph, so the methods of Section 3.4.3 do not apply. In this model, the system functions if the node s at the left-hand edge of the diagram and the node t at the right-hand edge of the diagram are connected. This representation indicates that the system functions if any one of the sets {A, B}, {A, D}, {C, B}, {C, D}, {A, B, C}, {A, C, D}, {B, C, D}, {A, B, D}, or {A, B, C, D} consists entirely of functioning units. Each of these is an (s, t)-path. The union of these nine paths constitutes the path set for the node pair (s, t). Note that not all these paths are minimal; for example, C can be removed from {A, B, C} and the result {A, B} is still an (s, t)-path. The minimal paths are {A, B}, {A, D}, {C, B}, and {C, D}, so {{A, B}, {A, D}, {C, B}, {C, D}} is the minimal path set. Similarly, the system fails to function if any of the sets {A, C}, {B, D}, {A, C, B}, {A, B, D}, {A, C, D}, or {A, B, C, D} consists entirely of nonfunctioning, or failed, units. Any one of these is an (s, t)-cut. The minimal (s, t)-cuts are {A, C} and {B, D}. The minimal cut set for (s, t) is {{A, C}, {B, D}}.

Because the path set contains all paths connecting s to t, for the system to function, it suffices that at least one path be made up entirely of functioning units. Therefore, the probability that the system functions is given by the probability of the path set in the labeled random graph representing the system reliability block diagram. However, only minimal paths need be considered because if a path is not a minimal path, then it has a proper subset that is still a path and is a member of the minimal path set. In other words, the union of all (s, t)-paths is equal to the union of all (s, t)-minimal-paths. Consequently, *the probability that the system functions is given by the probability of the system's minimal path set.* In general, the minimal paths will not be disjoint, so some version of the inclusion–exclusion formula [12] will have to be used to compute this probability.

Example (cont'd): Consider again the system shown in Figure 3.11. Letting $p_A = P\{A=1\}$ (where we have abused notation slightly by identifying the indicator random variable's letter with the unit's label) and similarly for B, C, and D, the probability that the system functions is given by

$$P(\{A=1, B=1\} \cup \{A=1, D=1\} \cup \{C=1, B=1\} \cup \{C=1, D=1\}).$$

This equation illustrates the strength and weaknesses of the path set method. Its strength is that it is completely straightforward and mechanical to write the expression for the probability that the system functions once the minimal path sets are known. Its weaknesses are that (1) enumerating the paths connecting s and t is tedious for all but the simplest of graphs, and (2) the expression that results is the probability of a large union of events that are not, in general, disjoint. However, these weaknesses pertain mainly to manual execution; the algorithmic nature of the procedure means that software for path set reliability analysis is within reach, and indeed has been available for some time [23, 40, 63].

For the general representation of system reliability using the minimal path set, let $\boldsymbol{x}=(x_1,\ldots,x_c)$ denote the vector of indicators of the functioning of the c components of the system. Enumerate the minimal paths of the system; suppose there are m of them called $\pi_1,\ldots,\pi_m$. Assuming independence of the units, the probability that the series system represented by the minimal path π_k is working is given by

$$\varphi_k(\boldsymbol{x})=\prod_{i\in\pi_k}x_i$$

for $k=1,\ldots,m$. The system functions if and only if at least one of the minimal paths consists entirely of functioning units, so it follows that the system reliability may be written as

$$\varphi(\boldsymbol{x})=1-\prod_{k=1}^{m}\left[1-\varphi_k(\boldsymbol{x})\right]=1-\prod_{k=1}^{m}\left[1-\prod_{i\in\pi_k}x_i\right]$$

This equation shows how the system structure function may be represented in terms of the structure functions of the system's minimal paths.

Because the cut set contains all (s,t)-cuts, for the system to fail it is necessary and sufficient that at least one cut be made up entirely of nonfunctioning units. However, only minimal cuts need be considered because if a cut is not a minimal cut, then it has a proper subset that is still a cut and is a member of the minimal cut set. In other words, the union of all (s,t)-cuts is equal to the union of all (s, t)-minimal-cuts. Therefore, the probability that the system fails to function is given by the probability of the minimal cut set in the labeled random graph representing the system reliability block diagram. Consequently, the probability that the system fails is given by the probability of the system's minimal cut set.

Example (cont'd): Consider again the system shown in Figure 3.11. The probability that the system fails to function is given by

$$P(\{A=0,C=0\}\cup\{B=0,D=0\}).$$

While this expression simplifies quickly because the two events in the union are disjoint, in general the expression that results from minimal cut set analysis will contain events that are not disjoint so that computation can become cumbersome. An algorithm for system reliability evaluation using cut sets may be found in Ref. 19.

For the general representation of system reliability via minimal cut set analysis, enumerate the minimal cuts $\chi_1,\ldots, \chi_n$ of the system. The probability that the series system represented by the minimal path χ_k consists entirely of nonfunctioning units is given by

$$\psi_k(x) = \prod_{i\in\chi_k}(1-x_i)$$

for $k=1,\ldots, n$. The system fails if and only if at least one of the minimal cuts consists entirely of nonfunctioning units, so it follows that the probability that the system fails may be written as

$$\psi(\mathbf{x}) = 1-\prod_{k=1}^{n}\left[1-\psi_k(\mathbf{x})\right] = 1-\prod_{k=1}^{n}\left[1-\prod_{i\in\chi_k}(1-x_i)\right]$$

Additional information on the use of path sets and cuts sets for system reliability modeling and computation may be found in Refs. 3, 49.

The minimal path set and minimal cut set representations for the system reliability lend themselves readily to the development for bounds on the system reliability. The first such bounds were developed by Esary and Proschan [19]. Letting C (resp., W) denote the minimal cut (resp., minimal path) set for the system, that is, for the nodes (s, t), Esary and Proschan's bounds for the system reliability R are

$$\prod_{\pi\in C}\left[1-\prod_{i\in\pi}(1-x_i)\right] \le R \le 1-\prod_{\chi\in W}\left(1-\prod_{i\in\chi}x_i\right).$$

The lower bound gives good approximations for highly reliable systems, while the upper bound works better for systems whose components have low reliability. Numerous improvements have been developed (see Refs. 22, 39 for further developments).

3.4.8 Reliability Importance

When designing for reliability, it is useful to expend resources on the parts of the system whose improvement causes the greatest improvement in system reliability. The concept of reliability importance formalizes this notion. The earliest definition of reliability importance was given by Birnbaum [8]. The reliability importance of component i in a system whose structure function is $\varphi_R(x_1,\ldots,x_n)$ is the partial derivative of φ_R with respect to x_i, evaluated at $x_1,\ldots, x_n$. For example, for a series system containing n components (single points of failure) having reliabilities $x_1,\ldots,x_n$, the reliability importance of component i is

$$\prod_{\substack{j=1 \\ j \neq i}}^{n} x_j = \frac{1}{x_i} \prod_{j=1}^{n} x_j$$

from which it can be seen that the least reliable component is the most important, that is, the one whose improvement would result in the greatest improvement of the system reliability.

Many other definitions of reliability importance, adapted to other applications, have been proposed [11]. Deeper discussion of reliability importance is beyond the scope of this chapter. Readers interested in pursuing this topic further would be well served by beginning with Ref. 11.

3.4.9 Non-Service-Affecting Parts

It may happen that some components of a system are irrelevant to system failure, that is, failure of one of these components has no effect on the operation of the system. Failure of the component is not only invisible when it happens but also it does not increase the load on other components. Obviously, this is an unusual situation, limited to such items as decorative trim, serial number labels, and so on. It is often obvious that components of this kind are not part of the system functional description and do not belong as part of the system reliability block diagram or the system structure function. In the mathematical theory of reliability [4], systems that contain irrelevant parts are called noncoherent.[21] Such components are also called *non-service-affecting*. The reliability importance (Section 3.4.8) of such parts is zero.

Unless there is a requirement for continued operation of decorative trim! This is not entirely facetious. For example, many electronic systems contain power supply bypass capacitors. Should one of these fail open (i.e., in such a way that the failed capacitor looks like an open circuit), usually no noticeable difference in operation can be discerned, and a single bypass capacitor may be considered a non-service-affecting part. The failure may cause increased noise on the power bus, and if enough bypass capacitors fail open, then the level of noise may increase to a point where a system bit error rate requirement may be violated, for example. Careful analysis may be required to determine the number of such failures tolerable before noise becomes an issue for other requirements. When this number is determined, the bypass capacitors may be incorporated into a system reliability block diagram as a k-out-of-n ensemble in series with the rest of the diagram, where n is the total number of such bypass capacitors in the system and $n-k$ is the number of (open) failures that need to occur before they are noticeable. Of course, there is another failure mode for capacitors, and that is to fail short (i.e., in such a way that the failed capacitor looks like a short circuit). A short failure should lead to a blown fuse,

[21] Other odd behaviors may make a system noncoherent. One of these is that the failure of a component of a system makes the system more reliable. Such systems are rarely encountered in realistic products or services.

or, if the power supply is not properly fused, a short failure can lead to failure of other power supply components or even cause the power supply to catch fire.

3.5 RELIABILITY MODELING BEST PRACTICES FOR SYSTEMS ENGINEERS

We defer this discussion until the end of Chapter 4 when we have covered reliability modeling for maintained systems also.

3.6 CHAPTER SUMMARY

This chapter has provided background material on reliability modeling systems engineers need in order to be good customers and suppliers in the development process. It is possible to use this chapter as a framework for advanced study of reliability modeling, but its primary intent is to equip systems engineers to be effective in dealing with the reliability engineering aspects of product and service development.

The chapter covers reliability effectiveness criteria and reliability figures of merit used for nonrepairable systems. Those in most common use are mission survivability and lifetime for nonrepairable systems. We caution extra care around "failure rate." The phrase is used for several different concepts, some of which require special conditions, so you need to be aware of which meaning is intended in any particular case.

3.7 EXERCISES

1. Suppose the strength of a population of devices is characterized by a random variable S having density f_V. Suppose the environment presents this population of devices with stresses Σ_i occurring at times T_i, $i=1, 2,\ldots$, where $\{T_1, T_2,\ldots\}$ is a homogeneous Poisson process with rate τ and Σ_i are independent and identically distributed random variables having density g_V. Determine the distribution of time to failure for a device chosen at random from this population subjected to these stresses.
2. Let $F(x)=1-\exp((-x/\alpha)^{\beta})$ for $x\geq 0$ and $F(x)=0$ for $x<0$, where α and β are positive constants. Show that F has the four properties of a life distribution listed in Section 3.3.2.3.
3. Show that the gamma distribution has a decreasing hazard rate when $0<\nu<1$.
4. Discuss the distribution $F(x)=1-\exp(-\alpha x^{\beta})$ where $x\geq 0$ and α and β are positive constants. Compare with Exercise 2.

5. Consider an ensemble of three single-point-of-failure components. The lifetime of the first component has an exponential distribution with parameter 0.001 failures per hour. The lifetime of the second component has a Weibull distribution with parameters 0.001 and 1.7, and the lifetime of the third component has a lognormal distribution with parameters 0.005 and 2.3. If the components are considered stochastically independent, what is the probability that the ensemble survives at least 40,000 hours? What information would be required to solve this problem if the components are not considered stochastically independent?
6. Derive the force of mortality for the uniform distribution on $[a, b]$ and show that it becomes infinite as $x \to a^+$ and $x \to b^-$. For a life distribution that has finite support $[a, b]$ with $0 \le a < b < \infty$ and that has a hazard rate, determine sufficient conditions for the hazard rate to become infinite as $x \to b^-$.
7. Write expressions for the life distribution and density corresponding to the Holcomb and North hazard rate model described in Section 3.3.4.4.
8. Show that, under the strong or weak accelerated life models, if the life distribution at nominal conditions has a certain parametric form, then the life distribution at any altered conditions continues to have the same parametric form.
9. Suppose a power supply choke inductor has a life distribution given by $F(t) = 1 - \exp(-(t/18{,}000)^{0.9})$, where t is measured in hours, when the ambient temperature is 15°C. Use the differential accelerated life model to determine the life distribution of the inductor when the operational temperature environment is 20°C with a diurnal variation of ±6°C. (Hint: represent the temperature environment as $T(t) = 20 + 6\sin(\pi t/12)$).
10. Suppose the system shown in Figure 3.9 is a cold standby system. Find the life distribution of the system in case
    (a) the switch is perfect and
    (b) the switch many fail.
11. Show that the force of mortality for the life distribution of a series system of an arbitrary (finite) number of components is the sum of the individual forces of mortality for the life distribution of each component. Is independence necessary? Is identical distribution necessary?
12. Find the cut sets and minimal cut sets for the reliability block diagram in Figure 3.10.
13. Develop further the example given in Section 3.4.1. Is it appropriate to consider the spare unit as a cold standby unit? What role does the replacement time (i.e., the time it takes to replace the failed wheel with the spare) play in the scenario? What are the consequences of the spare's being improperly inflated?
14. Consider a two-unit hot standby redundant system. Write an expression for the lifetime of this system in terms of its constituent component lifetimes when the switching mechanism may be imperfect (i.e., may fail when called for).

15. Find the life distribution of a two-out-of-three hot standby ensemble. Do the same for a two-out-of-three cold standby ensemble. Compare your results with the two-out-of-four and the three-out-of-four cases.
16. Complete the derivation of the survivor function shown in the example of a two-unit hot standby redundancy arrangement with an unreliable switch given in Section 3.4.5.1.
17. Complete the development of the life distribution of the cold standby ensemble with imperfect switching example given in Section 3.4.5.2.
18. Write the structure function for an ensemble consisting of a component in series with a parallel system of three components.
19. Regenerators for fiber-optic telecommunications systems are frequently located in remote, difficult-to-access areas. Consequently, a spare regenerator that is switched in automatically is provided for each active regenerator so that should the active regenerator fail, it is not necessary to incur the expense of sending a technician out to repair or replace the failed regenerator. The switching mechanism comprises a detection circuit (to determine that the active regenerator has failed), a switching mechanism to substitute the spare regenerator for the failed regenerator, and a communication mechanism that alerts (remote) staff to the success or failure of the switch when the active regenerator fails.
    (a) Should this be a hot standby or cold standby scheme? Discuss the advantages and disadvantages of each. How would you make this decision?
    (b) Make a reliability model for the switching mechanism.
    (c) Make a reliability model for the ensemble of the two regenerators and switch that is consistent with your solution to part (a).
    (d) Make a sensitivity study of the reliability of the ensemble as a function of your assumptions about the reliabilities of the components of the switching mechanism.
    (e) Write requirements for the reliability of the major components of the ensemble (the regenerators and switch components). Is there a reasonable way to do this if you do not yet have an overall system reliability requirement (i.e., a reliability requirement for the entire fiber-optic route of which this ensemble is a part)? How would your solution to part (d) contribute to the necessary understanding of the reliability economics of this ensemble and to the negotiation of reliability requirements with the systems engineer for the entire route?
20. Suppose a population of devices has a life distribution that is Weibull with parameters $\alpha=10{,}000$ and $\beta=2$. Find the expected number of device failures in the time intervals $[500k, 5000(k+1)]$ for $k=1, 2, \ldots, 10$. What is the probability that a device fails in each interval, given that it is alive at the beginning of the interval? What is the expected number of device failures in each interval among the devices that are still alive at the beginning of the respective intervals?
21. In the Example of Section 3.3.4.8, suppose the loss of material follows a normal distribution with mean 2.5 and standard deviation 1.5 instead of

the uniform distribution illustrated in the example. Repeat the steps in the example to show that the resulting life distribution of the population of ball bearings has an increasing hazard rate. Does the normal distribution assumption make sense here? Discuss.

22. Determine the relationships between the densities at the nominal and the operating conditions, and between the distributions at the nominal and the operating conditions, for the Cox proportional hazards model (Section 3.3.5.2).
23. Develop the model given in Section 3.3.6 to a product requiring two manufacturing processes. How would your solution generalize to more than two processes?
24. Use the Maclaurin series for $\log(1-y)$ to show that $-y^2/2 \leq y + \log(1-y) \leq 0$ for $0 \leq y \leq 1$.
25. Consider a lot of 200 circuit packs, each containing 10,000 solder attachments, manufactured by a wave soldering process whose lower and upper specification limits are a^L and a^U, respectively. Suppose that $F(x, a) = 0$ for $a^L \leq a \leq a^U$ and $F(x, a) = 1 - \exp(-\lambda x)$, independent of a, whenever $a \notin [a^L, a^U]$. Suppose further that the wave-soldering process just meets the six-sigma criteria (i.e., $m = 4.5$ or 7.5 in Section 3.3.6). Provide a lower bound and an upper bound on the survivor function for the solder attachments in this population of 200 circuit packs. How might you create a more realistic mathematical model of this process? Would it make a great deal of difference to the results?
26. Identify the cut sets and minimal cut sets for the Metro Central and Fort Totten nodes in the Washington, DC, Metro subway system.

REFERENCES

1. Abramowitz M, Stegun IA. *A Handbook of Mathematical Functions with Formulas, Graphs, and Mathematical Tables*. Washington: National Bureau of Standards; 1964.
2. Ascher H, Feingold H. *Repairable Systems Reliability: Modeling, Inference, Misconceptions, and their Causes*. New York: Marcel Dekker; 1984.
3. Barlow RE, Proschan F. *Statistical Theory of Reliability and Life Testing: Probability Models*. New York: Holt, Rinehart, and Winston; 1975.
4. Barlow RE, Proschan F. *Mathematical Theory of Reliability*. Philadelphia: SIAM Press; 1996.
5. Baxter LA. Availability measures for a two-state system. J Appl Prob 1981;18:227–235.
6. Baxter LA. Towards a theory of confidence intervals for system reliability. Stat Probab Lett 1993;16 (1):29–38.
7. Baxter LA, Tortorella M. Dealing with real field reliability data: circumventing incompleteness by modeling and iteration. Proceedings of the 1994 Reliability and Maintainability Symposium; 1994. p 255–262.
8. Birnbaum ZW. On the importance of different components in a multicomponent system. Technical report TR-54, Washington University (Seattle) Laboratory of Statistical Research; 1968.

9. Birnbaum ZW, Esary JD, Saunders SC. Multi-component systems and structures and their reliabilities. Technometrics 1961;3:55–77.
10. Bogdanoff JL, Kozin F. *Probabilistic Models of Cumulative Damage*. New York: John Wiley & Sons; 1985.
11. Boland PJ, El-Neweihi E. Measures of component importance in reliability theory. Comput Oper Res 1995;22 (4):455–463.
12. Chung K-L. *A First Course in Probability*. New York: Academic Press; 2001.
13. Cox DR. *Analysis of Survival Data*. London: Chapman and Hall; 1984.
14. David HA. *Order Statistics*. New York: John Wiley & Sons, Inc; 1970.
15. Drenick RF. The failure law of complex equipment. J Soc Indust Appl Math 1960;8 (4):680–690.
16. Elsayed EA. *Reliability Engineering*. 2nd ed. Hoboken: John Wiley & Sons, Inc; 2012.
17. Engel E. *A Road to Randomness in Physical Systems*. Volume 71, New York: Springer-Verlag; 1992.
18. Engelmaier W. 2008. Solder joints in electronics: design for reliability. Available at https://www.analysistech.com%2Fdownloads%2FSolderJointDesignForReliability.PDF. Accessed November 12, 2014.
19. Esary JD, Proschan F. Coherent structures of non-identical components. Technometrics 1963;5:191–209.
20. Escobar LA, Meeker WQ. A review of accelerated test models. Stat Sci 2006;21 (4):552–577.
21. Feller W. *An Introduction to Probability Theory and its Applications*. 2nd ed. Volume II, New York: John Wiley & Sons, Inc; 1971.
22. Fu JC, Koutras MV. Reliability bounds for coherent structures with independent components. Stat Probab Lett 1995;22:137–148.
23. Gebre BA, Ramirez-Marquez J. Element substitution algorithm for general two-terminal network reliability analyses. IIE Trans 2007;39:265–275.
24. Gertsbakh I, Kordonskiy K. *Models of Failure*. Berlin: Springer-Verlag; 1969.
25. Grigelionis B. On the convergence of sums of random step processes to a Poisson process. Theory Probab Appl 1963;8 (2):177–182.
26. Grubbs FE. Approximate fiducial bounds for the failure rate of a series system. Technometrics 1971;13:865–871.
27. Gumbel EJ. *Statistics of Extremes*. Mineola: Dover Books; 2004.
28. Gupta R, Goel R. The truncated normal lifetime model. Microelectron Reliab 1994;34 (5):935–937.
29. Harry MJ. *The Nature of Six-Sigma Quality*. Rolling Meadows: Motorola University Press; 1988.
30. Holcomb D, North JR. An infant mortality and long-term failure rate model for electronic equipment. AT&T Tech J 1985;64 (1):15–38.
31. http://www.cpii.com/division.cfm/11. Accessed November 12, 2014.
32. http://www.wmata.com/rail/maps/map.cfm.. Accessed November 12, 2014.
33. Jeong K-Y, Phillips DT. Operational efficiency and effectiveness measurement. Int J Oper Prod. Manag 2001;21 (11):1404–1416.
34. Johnson NL, Kotz S, Balakrishnan N. *Continuous Univariate Distributions*. Volume 1, New York: John Wiley & Sons, Inc; 1994.
35. Johnson NL, Kemp AW, Kotz S. *Univariate Discrete Distributions*. Hoboken: John Wiley & Sons, Inc; 2005.

36. Karlin S, Taylor HM. *A First Course in Stochastic Processes*. 2nd ed. New York: Academic Press; 1975.
37. Kline MB. Suitability of the lognormal distribution for corrective maintenance repair times. Reliab Eng 1984;9:65–80.
38. Kotz S, Lumelskii Y, Pensky M. *The Stress-Strength Model and its Generalizations: Theory and Applications*. Singapore: World Scientific; 2003.
39. Koutras MV, Papastavridis SG. Application of the Stein-Chen method for bounds and limit theorems in the reliability of coherent structures. Naval Res Logist 1993;40:617–631.
40. Kuo S, Lu S, Yeh F. Determining terminal pair reliability based on edge expansion diagrams using OBDD. IEEE Trans Reliab 1999;48 (3):234–246.
41. Lawless JF. *Statistical Models and Methods for Lifetime Data*. New York: John Wiley & Sons, Inc; 1982.
42. LuValle MJ, Welsher T, Svoboda K. Acceleration transforms and statistical kinetic models. J Stat Phys 1988;52 (1–2):311–330.
43. LuValle MJ, LeFevre BG, Kannan S. *Design and Analysis of Accelerated Tests for Mission-Critical Reliability*. Boca Raton: Chapman and Hall/CRC Press; 2004.
44. Marshall AW, Olkin I. A multivariate exponential distribution. J Am Stat Assoc 1967;62 (317):30–44.
45. Meeker WQ, Escobar LA. *Statistical Models and Methods for Lifetime Data*. New York: John Wiley & Sons, Inc; 1998.
46. Musa JD. Validity of execution-time theory of software reliability. IEEE Trans Reliab 1979;R-28 (3):181–191.
47. Musa JD, Okumoto K. A logarithmic Poisson execution time model for software reliability measurement. ICSE84, Proceedings of the 7th International Conference on Software Engineering. Piscataway, NJ: IEEE Press; 1984. p 230–238.
48. Myhre JM, Saunders SC. Comparison of two methods of obtaining approximate confidence intervals for system reliability. Technometrics 1968;10 (1):37–49.
49. Pages A, Gondran M. *System Reliability Evaluation and Prediction in Engineering*. New York: Springer-Verlag; 1986.
50. Ramirez-Marquez JE, Coit D, Tortorella M. A generalized multistate based path vector approach for multistate two-terminal reliability. IIE Trans 2007;38 (6):477–488.
51. Rice RE. Maintainability specifications and the unique properties of the lognormal distribution. Phalanx 2004;37 (3):14ff.
52. Shaked M, Shanthikumar JG. *Stochastic Orders*. New York: Springer; 2007.
53. Snyder DL. *Random Point Processes*. New York: John Wiley & Sons, Inc; 1975.
54. Tortorella M. Closed Newton-Cotes quadrature rules for Stieltjes integrals and numerical convolution of life distributions. SIAM J Sci Comput 1990;11 (4):732–748.
55. Tortorella M. Life estimation from pooled discrete renewal counts. In: Jewell NP *et al.*, editors. *Lifetime Data: Models in Reliability and Survival Analysis*. Dordrecht: Kluwer Academic Publishers; 1996. p 331–338.
56. Tortorella M. A simple model for the effect of manufacturing process quality on product reliability. In: Rahim MA, Ben-Daya M, editors. *Integrated Models in Production Planning, Inventory, Quality, and Maintenance*. Dordrecht: Kluwer Academic Publishers; 2001. p 277–288.
57. Tortorella M. Service reliability theory and engineering, I: foundations. Qual Technol Quant Manag 2005;2 (1):1–16.

58. Tortorella M. Service reliability theory and engineering, II: models and examples. Qual Technol Quant Manag 2005;2 (1):17–37.
59. Tortorella M. System reliability modeling using cut sets and path sets. In: Ruggeri F, editor. *Encyclopedia of Statistics in Quality and Reliability*. Hoboken: John Wiley & Sons, Inc; 2008.
60. Tortorella M, Frakes WB. A computer implementation of the separate maintenance model for complex-system reliability. Qual Reliab Eng Int 2006;22 (7):757–770.
61. Tsokos CP, Padgett WJ. *Random Integral Equations with Applications in Life Sciences and Engineering*. Volume 108, New York: Academic Press; 1974. Mathematics in Science and Engineering.
62. Viertl R. *Statistical Methods in Accelerated Life Testing*. Göttingen: Vandenhoeck and Rupprecht; 1988.
63. Willie RR. *Computer-Aided Fault Tree Analysis*. Defense Technical Information Center AD-A066567: Ft. Belvoir; 1978.

4

Reliability Modeling for Systems Engineers

Maintained Systems

4.1 WHAT TO EXPECT FROM THIS CHAPTER

This chapter continues the reliability modeling exposition that was begun in Chapter 3 for nonmaintained systems. Reliability models for maintained systems are built up out of reliability models for their replaceable units. Reliability models for the replaceable units are in turn built up out of reliability models for their constituent components that are now nonmaintained. This chapter covers details of how this is done.

The key point about reliability modeling for a maintainable system is that it may experience repeated failures. That is, a maintainable system may operate, fail, be repaired, operate again, fail again, be repaired again, etc. Reliability models for this behavior focus on describing the stochastic process of operating times, failure occurrence instants, and outage times. This chapter introduces basic ideas relating to this description and some of the specific models that are in common use. It will help you become familiar with terms commonly used in reliability requirements for maintained systems, such as times between failures, failure rate, outage duration, operating times, etc. The presentation emphasizes the separate maintenance model because it accords well with the maintenance concept of replace-and-repair which is very common in the defense, telecommunication, and other industries, and state diagram reliability model for maintained system is more than adequately covered in other books and papers.

Reliability, Maintainability, and Supportability: Best Practices for Systems Engineers, First Edition. Michael Tortorella.

The chapter begins with a discussion of reliability effectiveness criteria and figures of merit for maintained systems and proceeds to describe the two most frequently used maintained system reliability models, the renewal process and the revival process. Further developments include a brief introduction to state diagram reliability modeling for maintained systems and a discussion of why data collected from a large number of systems appear to follow a Poisson process.

4.2 INTRODUCTION

Reliability engineers use lifetimes and associated random variables to describe the reliability of nonmaintained components and for mission time studies of systems that may be maintainable but not while they are in use (Section 4.3.4). A key concept in the operation of maintained systems is the possibility of repeated failures (and repairs) as time passes. Reliability engineers use the formalism of stochastic processes to capture this phenomenon. A stochastic process is nothing more than a collection of random variables indexed by some parameter set. In this application, that parameter set is time. A stochastic process used to describe reliability of a maintained system has one or more random variables of interest attached to each time point. Reliability modeling for maintained systems amounts to creating descriptions of this process that are useful for learning about the quantitative properties of the operating times, failure occurrence instants, and outage times of the system so that appropriate reliability requirements can be constructed and verified by collection of data during operation. The stochastic process used for this purpose is the system reliability process (Section 4.3.2). It summarizes all the information about operating times, failure occurrence instants, and outage times that we need to describe the system's reliability as time proceeds. Most system reliability modeling for maintainable systems is directed at building a description of this process from what is known about the reliability of the components and replaceable units of the system and the way in which the system is operated.

4.3 RELIABILITY EFFECTIVENESS CRITERIA AND FIGURES OF MERIT FOR MAINTAINED SYSTEMS

4.3.1 Introduction

A maintainable, or repairable, system may suffer many failures in possibly various failure modes. That is, when a maintained system fails, it is repaired and restored to service, and this may happen repeatedly. Contrast this with the situation for a nonrepairable object which may suffer at most one failure. When a nonrepairable object fails, it is discarded. A new one may be installed in its place, and we shall consider this situation in Section 4.4.2. But all maintainable systems are characterized by a sequence of alternating on- and off-periods: the operating times and the downtimes (or outage times). Understanding the reliability of a maintained system amounts to coming to grips with the stochastic process describing the alternating time periods of proper operation and outage. The models discussed in this section all attempt to describe the properties of this sequence in some form. The literature contains many such descriptions; we shall confine ourselves to the two models that are most commonly used in practice, the renewal model (Section 4.4.2) and the revival model (Section 4.4.3). A brief introduction to other possibilities is given in Section 4.4.4, mainly as a way of reinforcing the detection of when neither the renewal nor the revival model is appropriate and of offering some other possibilities that can be further explored in the References. Typically, the decision to choose another model is best made by a reliability engineer who has experience with the several types of repairable system models.

Language tip: So far, we have used the words "maintainable system" and "maintained system" as though they were synonymous. However, there is good reason to make a distinction between them. Some applications of a maintainable system preclude its being repaired during use, so from here on we reserve "maintainable system" to include any type of system that could experience repeated failures and repairs, while "maintained system" will be used to exclude those for which repair during a mission is not possible (see Section 4.3.4).

4.3.2 System Reliability Process

We begin with the simplest characterization of the reliability of a repairable system. No assumptions are made here other than that the system alternates between periods of operation and outage. The generic diagram representing this situation will be called a "system history diagram," and we will make extensive use of system history diagrams in this chapter and elsewhere in the book. Figure 4.1 is an example of a system history diagram.

In this diagram, the horizontal lines at the level "1" represent time intervals during which the system is operating properly, that is, no requirements violations are in progress. The horizontal lines at the level "0" represent time intervals during which the system is not operating properly, that is, one or more requirements

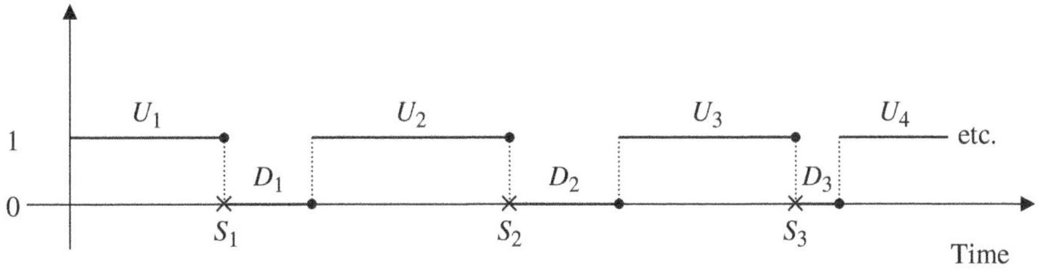

Figure 4.1 System history diagram.

violations are in progress during these time intervals. The diagram illustrates the alternation of the system between these two states. More complicated system behaviors (e.g., with multiple intermediate states between full operation and full outage) can also be accommodated using similar diagrams, although we will not study those in this chapter.

Figure 4.1 also introduces some notation that will be used throughout the remainder of this chapter. The lengths of the operating intervals are denoted by $U_1, U_2, U_3,\ldots$, and the lengths of the outage intervals are denoted by $D_1, D_2, D_3,\ldots$. A failure occurs whenever there is a $1 \rightarrow 0$ transition in the diagram; these are denoted by the × symbols on the time axis and are labeled as $S_1, S_2, S_3,\ldots$. Evidently, $S_1 = U_1$, $S_2 = S_1 + D_1 + U_2$, and, in general, $S_k = S_{k-1} + D_{k-1} + U_k$ for $k = 2, 3,\ldots$. In the models we study in this book, the operating times and the outage times are described as random variables, so the system history is a stochastic process that we may call the *system reliability process*. In this case, Figure 4.1 represents a sample path from that process.

4.3.3 Reliability Effectiveness Criteria and Figures of Merit Connected with the System Reliability Process

The reliability effectiveness criteria used for maintainable systems are different from those studied so far for nonmaintainable systems. Fortunately, many of them can be readily related to the constructs developed for the system history diagram (see Figure 4.1). For each reliability effectiveness criterion, we may

- construct a reliability model that enables making projections about important figures of merit connected with the effectiveness criterion (means, variances, distributions, etc.),
- given data (observations from systems in service), compute metrics that estimate figures of merit connected with the effectiveness criterion using data (observations from systems in service),
- compare the realized performance with the requirements for the effectiveness criterion or related figures of merit (to determine whether promises to customers are being kept) and with the projected values of the effectiveness criterion or figures of merit (so that the modeling process may be improved),
- compare different architectures to forecast their likely reliability, and
- compare the realized performance with the results of reliability modeling so that the reliability modeling process may be improved.

The first activity is undertaken during design and development, and supports the creation of effective and appropriate reliability requirements. The latter activities take place after systems are deployed and is used as feedback to determine how well the reliability requirements are being met. This feedback is also a key learning opportunity for improving not only the product or service but also the processes of creating reliability requirements and modeling the system reliability (see Chapter 5).

4.3.3.1 Number of failures per unit time, failure rate, and failure intensity

Let $N(t)$ denote the number of failures that occur between time 0 and time t, that is, in the time interval $[0, t]$. Then $N(t)=\max\{n : S_n \le t\}$, that is, the number of failures that occur before or at t is the index of the largest failure time that takes place before or at t. $N(t)$ is a random variable depending on t because the system history is a stochastic process, the system reliability process. So $N(t)$ is a stochastic process also. It is an example of a *point process*, a stochastic process having a continuum parameter space (in reliability applications this is usually the positive half-line $[0, \infty)$ representing time) and a discrete state space (in this case, the nonnegative integers) [7]. Figure 4.1 shows a basic fact about all point processes, namely that for all positive integers k,

$$\{N(t) > k\} = \{S_k < t\}.$$

That is, at least k events (in our model, system failures) take place before time t if and only if the kth event takes place before t. This is a basic tool that relates the count description of a point process (the left-hand side of the equation above) with the time description of the process (the right-hand side of the equation above).

To be consistent with the usual engineering interpretation of "rate," we would like to define failure rate[1] for a point process $\{N(t)\}$ to be something like failures per unit time. Let $0 \le t_1 < t_2$ denote two points on the time axis. Then the failure rate for the process over the interval $[t_1, t_2]$ is

$$\frac{N(t_2) - N(t_1)}{t_2 - t_1}.$$

Note that this is a random variable because $\{N(t)\}$ is a stochastic process. So *failure rate* is a reliability effectiveness criterion. For each point process model discussed in the following text as a model for the reliability of a repairable system, we will list the properties of the failure rate as defined here.

While the equation above is the most general definition of failure rate, it is a random quantity, and many applications are better served by a figure of merit than by an effectiveness criterion. Several reasonable possibilities exist:

- Number of failures per unit time: Whatever the units measuring time may be, the interval $[t, t+1]$ is a time interval of unit length. The number of failures occurring in this interval is given by $N(t+1) - N(t)$, and this, too, is

[1] The material here is generic to point processes, and we could use simply "rate" instead of "failure rate." However, the application here is always to reliability modeling, so we use "failure rate" as the preferred term.

a stochastic process (the same as setting $t_2=t_1+1$ in the equation above). $N(t+1)-N(t)$ is the number of failures per unit time at time t; note that this need not be a constant as it may vary with t. We may also consider the expected number of failures per unit time, $E[N(t+1)-N(t)]=EN(t+1)-EN(t)$.

- Overall failure rate: If $t_1=0$, then this version of failure rate is $N(t)/t$, also a random quantity. This is the quotient of the total number of failures during the entire time of system operation divided by that amount of time. When used in this fashion, $N(t)/t$ will be called the overall failure rate. We may also consider the expected value of this measure, $EN(t)/t$. Formally, this should be called the *expected overall failure rate.*

Requirements tip: Systems engineers and reliability engineers frequently talk of *failure rate.* Apart from the confusing use of "failure rate" when referring to the hazard rate of the life distribution of a population of nonrepairable items (see the "Language tip" in Section 3.3.3.5), one would expect that failure rate would have something to do with the frequency at which failures occur, or accumulate, per unit time. When used in this way, the phrase "failure rate" seems to imply that whatever it is referring to is a constant, because no reference to (absolute) time is made in this usage. Because of the stochastic nature of $N(t)$, we cannot expect $N(t+1)-N(t)$ to be constant, so for many purposes it's useful to have a definition for "failure rate" that at least has a chance of being constant in some circumstances. To do this, we first consider the expected value of the number of failures in $[0, t]$, namely $EN(t)$. In all the models we study in this book, this function of t is smooth and well-behaved (i.e., differentiable). There are now two reasonable possibilities for a definition of failure rate:

1. Define $r(t)=(d/dt)[EN(t)]$, that is, the time derivative of the function $E[N(t)]$. If this derivative is constant (does not depend on t), call r the failure rate of the system reliability process. In some, but not all, of the commonly used models for the system reliability process, $(d/dt)[EN(t)]$ is constant, and the phrase "system failure rate" makes sense without further explanation. Note that as $N(t)$ is nondecreasing, so is $EN(t)$, and so $r(t)\geq 0$ (even if it is not constant).
2. Define $r_\infty=\lim_{t\to\infty}(d/dt)[EN(t)]$ if the limit exists (and in all the models we study in this book, the limit does exist), and, as r_∞ is just a number, we call it the asymptotic failure rate of the system reliability process. Again, $r_\infty\geq 0$.

For requirements pertaining to "failure rate," be sure that all stakeholders interpret the phrase the same way. In particular, reliability engineers should advise the system engineer as to constancy of the "failure rate" in whatever model they use for the system reliability process.

Language tip: For a point process, "failure rate" as defined in item 1 is also referred to as "failure intensity." This is particularly prevalent in the software reliability field. Consistently using "failure intensity" for the concept defined in item 1, above, avoids the use of "failure rate" altogether in any facet of reliability engineering or modeling. Some may consider this desirable, but it by no means universally accepted as a common practice even though a strong case could be made for it as a best practice because it avoids any confusion about the varying interpretations of "failure rate" that plague this field.

Modeling tip: Much of the reliability engineering community uses "failure rate" for the concept we have labeled "hazard rate" or "force of mortality" in Section 3.3.3.4. Note in fact that we are now dealing with a completely different situation, namely, the system reliability process model for a repairable system, rather than the single failure that can occur for a non-repairable system, as dealt with in Chapter 3. In a field that has numerous opportunities for confusion because of reuse of terms in different contexts, this is perhaps the most vexing. Ascher and Feingold [1] finesse this issue by avoiding "failure rate" altogether; for nonmaintained systems, they use the terminology "hazard rate" and "force of mortality" as we do, and for maintainable systems they use "rate of occurrence of failures (ROCOFs)" for "failure intensity" as defined in this section. Possibly, the best way to keep all this straight is to recall that to engineers, "rate" usually means "count per unit time" and that is what the definitions in the list given earlier encapsulate (use of the derivative means we are really looking at an "instantaneous" rate, as explained in elementary calculus, but it is an honest engineering interpretation of rate as usually understood). Be alert for this potential confusion whenever you encounter "failure rate." In this book, whenever we use "failure rate," it will always pertain to a maintainable system, and be defined as in item 1 in the above list. Item 2 will always be referred to as "asymptotic failure rate."

Using the notation of Figure 4.1, the system failure rate refers to how rapidly the S_k are accruing on average. If the failure rate is small, the failure times tend to occur rather infrequently; while if the failure rate is large, the failure times tend to be denser on the time axis.

There are many system models in which the failure rate, as defined in item 1 of the above list, is not constant. One such, in very common use, is the revival model described in Section 4.4.3. In the revival model, the system failure rate (system failure intensity) may be decreasing, increasing, or variable. If the system failure rate is decreasing, then on average, failures are occurring less frequently and the times between outages tend to increase. If the system failure rate is increasing, then on average, failures are occurring more often and the times between outages tend to decrease. Ascher and Feingold [1] refer to the former as "happy systems" and the latter as "sad systems." It is possible that the system failure rate may fluctuate, that is, be neither increasing nor decreasing throughout; models exhibiting this behavior are unusual in practice, but there is no theoretical reason why they cannot be used.

4.3.3.2 Operating times, times between outages, and times between failures

In Figure 4.1, the operating times are the intervals U_1, U_2,... ; during these time intervals, the system is operating without violation of any requirements (i.e., without any failures). The sequence U_1, U_2,... of operating time random variables constitutes a stochastic process.

> **Language tip:** The operating time random variables are sometimes (usually!) called *times between failures*. This would be appropriate if we were to consider "failure" to refer to an entire outage interval instead of the instant at which the violation of a system requirement happens. This is the interpretation that commonly prevails even though it is an inconsistent use of terminology because a "failure" is something that takes place at a particular instant (the S_n in the system history diagram) whereas the "outage" is the period of time following a failure during which the failure condition (requirements violation(s)) persists. That the times between failures are in fact $S_2 - S_1$, $S_3 - S_2$,... has the perverse consequence that one may increase the times between failures by increasing the outage times while leaving the operating times alone. When you need to talk about the time periods during which the system operates properly, it is best to use the U_1, U_2,... as "operating times" or "times between outages." See Exercise 1.

As they are random variables, we may consider the sequence of their expected values EU_1, EU_2,... . In some reliability models (see especially Section 4.4.2), these expected values are all the same. In such cases, it makes sense to define the *mean time between outages* as this common value. This is often (usually!) called the "mean time between failures." Refer to the previous "Language tip" to clear up any ambiguity. In addition, specification of a single number for "mean time between outages" implies that there is a single mean time between outages, namely, that it is a constant. Be aware that many studies of repairable systems do not take pains to assure that this mean time between outages (MTBO) is a constant even though only a single number may be quoted. This is of some importance for systems engineers because a good grasp of potential system behaviors is required for effective design for reliability.

4.3.3.3 Outage times

In the system history diagram (Figure 4.1), the outage times are the intervals D_1, D_2,... ; during these time intervals, one or more system requirements are being violated (i.e., one or more system failures have taken place and have not yet been remediated). The sequence D_1, D_2,... of outage time random variables constitutes a stochastic process. As they are random variables, we may consider the sequence of their expected values ED_1, ED_2,... . In some reliability models, these expected values are all the same. In such cases, it makes sense to define the *mean outage time* as this common value.

4.3.3.4 Availability

Availability, denoted $A(t)$, is an important concept and is widely used as a reliability figure of merit in requirements. $A(t)$ is defined as the probability that the system reliability process is in the up state at time t. If we let $Z(t)=1$ when the system is in the up state and $Z(t)=0$ when the system is in the down state, then the stochastic process $\{Z(t): t \geq 0\}$ is the system reliability process. At each t, the value $Z(t)$ is a reliability effectiveness criterion and its expected value $A(t)=\mathrm{E}Z(t)$ is called the *system availability*, or simply *availability*. When the system reliability process starts in the up state, $A(0)=1$ and $A(t)$ decreases from there. As a rule, in most systems of practical interest to systems engineers, availability will vary up and down for a while but eventually settle down to a limiting value ($t \to \infty$) (further discussed in Sections 4.4.2 and 4.4.3); see also Refs. 12, 22. Methods for computing availability are discussed in Sections 4.4.2 and 4.4.3. In considering maintainability and supportability, special kinds of "availability" are defined by eliminating from consideration certain of the outage times that may stem from preventive maintenance, logistics delays, and other actions not having to do with correction of a failure. See Section 10.6.4.

It can be shown that the total time during the interval $[0, t]$ that the system spends in the "up" state is given by

$$U(t) = \int_0^t Z(u)\, du$$

and the total time during $[0, t]$ that the system spends in the "down" state is

$$D(t) = \int_0^t [1 - Z(u)]\, du = t - U(t),$$

from which it follows that the expected system downtime in $[0, t]$ is given by

$$\mathrm{E}D(t) = \int_0^t [1 - A(u)]\, du.$$

This vital connection was used in the example in Section 3.4.2.2.

Common practice distinguishes three different availability figures of merit, inherent, operational, and achieved availability. These differ in what components of outage times are counted in the downtime for each type. See Section 10.6.4 for the details (we have not yet discussed the corrective and preventive maintenance and supportability times that enter into these definitions).

4.3.4 When is a Maintainable System Not a Maintained System?

Up to now, we have used the words "maintained," "maintainable," and "repairable" interchangeably. There is one important case where we need to make a distinction, namely, where a maintainable system may have failure modes that cannot be corrected while it is in use. Consider a military aircraft. Such aircraft are used for "missions" of some duration, and commanders are vitally interested in the probability of successful mission completion (from many points of view, but these at least include no aircraft system failures during the mission). The important point is that there is no opportunity to repair some kinds of failures (especially hardware failures) during the mission, and while the aircraft is maintainable while on the ground, some kinds of corrective maintenance cannot be carried out while in flight. This does not preclude the possibility that some failures may be recoverable in flight. For example, a reboot of the affected subsystem may correct a software failure, and it may be possible to do this during flight if this capability has been provided in the system maintenance concept.

For such systems, a key reliability figure of merit is the probability that the system continues to operate properly for the entire duration of the mission. In this sense, techniques applicable to nonrepairable systems are used to say something about the probability of mission completion without failures because as far as the mission is concerned, the system is not repairable during the mission. In most cases, of course, the systems involved in the mission are not new at the start of the mission but may have already accumulated some age and can be described as some number of hours old at the start of the mission. The key variable of interest, then, is not the time to first failure of a brand new system, but rather the time to the next failure of a system (that may have already experienced many failures and repairs) that is a stated number of hours old at the start of the mission. This is the *forward recurrence time* from a stated age and will be discussed further in Sections 4.4.2 and 4.4.3.

4.4 MAINTAINED SYSTEM RELIABILITY MODELS

The most commonly used maintained system reliability models are the renewal model and the revival model. The words describe what kinds of repairs are performed to restore the system to functioning condition. These are discussed in detail in this section.

Many other maintained system reliability models are available in the literature. These are useful when more detailed modeling of special conditions is required. For example, the renewal model and the revival model do not account for the possibility that a repair person may introduce additional faults during the remediation of a failure (this is called the "clumsy repairman problem"). An example of a general model of this kind is given in Ref. 4. If you know that special conditions exist that are not part of the standard renewal or revival protocols, it is best to consult a reliability engineering specialist.

4.4.1 Types of Repair and Service Restoration Models

Whenever it is necessary to repair a system that has failed, some time is consumed. More descriptive system reliability models account specifically for these times, which is to say that the D_1, D_2,... intervals seen in the system history diagram are positive. This bears mention because some approximate system reliability models are used in which the outage times are assumed to be zero. In effect, this is an approximation that says that the outage times are so short compared to the operating times (times between outages) that making a model in which these are zero can provide a reasonable approximation for certain purposes (see Sections 4.4.2.1 and 4.4.3.1).

> **Modeling tip:** In a model in which all the outage times are zero, the system availability is always 1 (because $Z(t)=1$ except on a finite set of t-values). So such models are not appropriate where studies of availability and downtime are needed.

4.4.2 Systems with Renewal Repair

When a system is repaired according to a "renewal" protocol, it is returned to "good as new" condition after it fails and operation continues [2]. Most often, this means that the system is entirely replaced by a new one. This can be a costly strategy for large systems, but it is not unheard of. For example, weapons destroyed in battle may be replaced with new ones. Also, line-replaceable units (LRUs) in large, modular systems are often replaced with new ones when they fail (see Section 4.4.5.1 for coverage of this case).

4.4.2.1 Renewal models with instantaneous repair

In models in which the replacement by a new unit is assumed to consume no time, all the outage times in the system history diagram are zero. The resulting system reliability process $\{U_1, U_2, \ldots\}$ comprises a sequence of (necessarily nonnegative) stochastically independent, identically distributed operating times. The independence comes from an understanding that the length of time that a system has operated since its most recent failure has no influence on the length of time a replacement might operate: even though there may appear to be some correlation, we can't see a cause-and-effect relationship (of course, if this is not true, we need to adopt another modeling condition). The identical distribution comes from the similarity of the replacement to the original unit: the failed unit is replaced by one of the same type. These assumptions produce a system reliability process known as a *renewal process*: a sequence of nonnegative (because they are lifetimes), independent, and identically distributed random variables [11]. Denote the common distribution of the U_i by F and let $\mu<\infty$ denote its mean.[2] The renewal process is a simple, well-understood process for which many results of interest to systems engineers are known.[3] Some of these are

- The expected number of renewals (failures, in the reliability model) in the time interval $[t, t+h]$ is asymptotic to h/μ as $t\rightarrow\infty$.

[2] The theory does allow a renewal process with infinite mean ($\mu=\infty$), but these are not usually encountered in reliability modeling.

[3] This may to some degree account for its popularity as a reliability model.

- The expected number of renewals in $[0, t]$, divided by t, approaches $1/\mu$ as $t \to \infty$. In reliability language, the asymptotic failure rate (Section 4.3.3.1) for the renewal process is $1/\mu$.
- The asymptotic distribution of the forward recurrence time from time x is given by

$$\frac{1}{\mu}\int_0^t [1 - F(u)]\, du$$

as $x \to \infty$. If we imagine x as the current time, the asymptotic distribution of the time to the next failure after x describes the forward recurrence time from a given time point (x) for a system that has been in operation for a long time. In practical terms, a "long time" generally can be safely interpreted as about 5μ or greater.
- Let $N(t) = \max\{n : S_n \le t\}$ denote the number of renewals in $[0, t]$. The expected number of renewals in $[0, t]$, denoted by $EN(t) = M(t)$, is given by the solution of the integral equation:

$$M(t) = F(t) + \int_0^t M(t-u)\, dF(u).$$

When F is exponential, $F(t) = 1 - \exp(-t/\mu)$, then $M(t) = t/\mu$ for all t. Unfortunately, this equation has a closed-form solution for only a few other life distributions, but it is relatively easy to solve numerically [22]. The renewal process whose inter-renewal times are exponentially distributed is a homogeneous Poisson process with rate $1/\mu$ [11].
- The failure rate over the time interval $[t, t+h]$, as defined in Section 4.3.3.1, for the renewal process is

$$\frac{N(t+h) - N(t)}{h}.$$

Its expected value is $h^{-1}[M(t+h) - M(t)]$ which, for a renewal process whose inter-renewal time distribution has a finite mean μ, converges to $1/\mu$ as $t \to \infty$. By a slight abuse of terminology, it is often said that "the failure rate of a renewal process is $1/\mu$." What is correct is that the expected asymptotic failure rate of a renewal process is $1/\mu$ (the equality $h^{-1}[M(t+h) - M(t)] = 1/\mu$ holds for the homogeneous Poisson process as well as for a stationary renewal process [11] but does not hold for any other renewal process when $t < \infty$). Consequently, you may need to verify the way "failure rate" may be used in this context.

4.4.2.2 Renewal models with time-consuming repair

A more realistic model is obtained if we allow the repair (or outage) times to be nonzero. In this case, the $\{D_i\}$ are a significant part of the model. Following the renewal paradigm, we assume that $D_1, D_2, \ldots$ are independent and identically distributed with common distribution G which has mean $\nu<\infty$. The independence comes from a belief that there is no, or we can find no reason to suspect the existence of, mechanism that would cause an outage time to be influenced in any way by the outage times preceding it. The identical distribution comes from the idea that it is the same system that fails each time, so it is fair to describe the outage times for it as coming from a single population. Of course, these are idealizations and often ignore some facts that we may know or suspect to be relevant, so the model provides an approximation to reality that is more or less acceptable depending on the strength of the ignored facts. See Section 4.5 for some additional information.

The sequence $\{U_1, D_1, U_2, D_2, \ldots\}$ is called an *alternating renewal process* [11] because it consists of two renewal processes interleaved [12]. The system reliability process is the 0–1 process $Z(t)=I\{t \text{ falls into a } U\text{-interval}\}$. The reliability model for a system with time-consuming renewal repair consists of information that can be derived from these assumptions.

- The *system availability* (Section 4.3.3.4) is $A(t)=P\{Z(t)=1\}$, the probability that the system is operating at time t. System availability is a reliability figure of merit. In the alternating renewal process model, assuming the system begins operation at time 0 in the operating state, we may write an integral equation for system availability as follows:

$$A(t)=1-F(t)+\int_0^t[1-F(t-u)]dH(u),$$

where $H=F*G$ is the convolution of F and G (Section 3.4.5.2). The solution of this equation is given by

$$A(t)=1-F(t)+\int_0^t[1-F(t-u)]dM_H(u)$$

where M_H is the renewal function (Section 4.4.2.1) for H. Again, closed-form solutions are rare, but numerical evaluation is straightforward [22]. The asymptotic availability is given by

$$\lim_{t\to\infty} A(t)=\frac{\mu}{\mu+\nu}$$

which is the mean time between outages (mean operating time) divided by the mean cycle time (time from one failure to the next).
- The number of system failures in $[0, t]$ is given by $N(t)=\max\{n : S_n+U_{n+1}\le t\}$ with the convention that $S_0=0$ and again assuming the system begins operation at time 0 in the operating state. The expected number of system failures in $[0, t]$ when the system starts in the up state at time 0 is $1-F(t)+M_H(t)$.

- The system failure rate over a time interval $[t, t+h]$ is given by $h^{-1}[N(t+h)-N(t)]$, which has expected value $h^{-1}[M_H(t+h)-M_H(t)]$. As $t\to\infty$, this converges to $1/(\mu+\nu)$, the reciprocal of the mean cycle time. See the discussion of failure rate for the renewal process (Section 4.4.2.1).

Example: A fluorescent tube has the life distribution $F(t) = 1-\exp[-(t/24{,}780)^{1.2}]$ where t is measured in hours. When the tube fails, it is replaced with a new one and the replacement time has a uniform distribution $U_{2,6}$ on [2, 6], again with time measured in hours. What is the expected number of tube replacements per socket in 20 years of operation?

Solution: The conditions of the problem make it appropriate to use an alternating renewal process model to describe the system dynamics (by "system" here we mean the socket containing the tube, and by "dynamics" we mean the pattern of tube replacements in that socket). In the alternating renewal process model, the expected number of replacements in 20 years is the renewal function for $F*U_{2,6}$ evaluated at 175,320 hours (20 years). Numerical computation of this renewal function [22, 25] shows that $F*U_{2,6}(175{,}320)=7.367$. Absent the ability to carry out this numerical computation, we may reason that the average replacement time (4 hours) is so short compared to the average tube lifetime (23,305.6 hours) that the approximation afforded by a renewal process model (zero-duration replacement times) should be good. In that model, the solution is the renewal function for F evaluated at 175,320 hours. Numerical computation yields $F(175{,}320)=7.368$. In this example, using the nonzero repair time model changes the result only in the third decimal place, a meaningless change in this scenario.[4] Without the numerical computation, we may reason that 20 years is more than 5 mean tube lifetimes so that the asymptotic approximation shown in Section 4.4.2.1 may be used. The expected number of tube replacements in 20 years is then approximately 175,320/23,305.6 = 7.52, a slight overestimate of the true value. The availability in this system at 20 years is 0.999828728. The numerical computation shows that this value is reached at 12 years and remains steady (to this many decimal places) thereafter.

4.4.3 Systems with Revival Repair

The other most commonly used reliability model for repairable systems is the bad-as-old (BAO) model [1]. This model postulates that repair of a system is accomplished by returning it to operating condition without otherwise changing its age. That is, immediately after a repair, the system is working, but its age is the same as it was when it failed. The name BAO is intended to contrast with renewal repair that makes a system "good as new." The BAO model provides a good approximation to the common scenario in which repair of a large, complex

[4] Perhaps this might matter in a high-consequence system. Each case should be judged on its own merits.

system is effected by replacing some part or subassembly that constitutes only a small part of the system. That small part of the system may be new (if the repair part is new) or of some other age (if the repair part comes from a spares pool of repaired items); but because it constitutes only a small part of the system, the overall age of the system (i.e., the accumulation of time against a clock measuring the time to the next failure) is not much changed. We abstract this (or approximate it) by saying that the change in age is, in the model, zero.

Example: Most automobile repairs are accomplished by replacing a faulty part with a new (e.g., voltage regulator) or rebuilt (e.g., alternator) one. In either case, the replaced part is only one of a great many failure-susceptible parts in the automobile. All the parts that were not replaced retain their age at the time of the failure of the replaced part, so the next failure is much more likely to come from one of the parts that were not replaced because there are so many more of them. So the time to the next failure is mostly unaffected by the replacement, and this is the postulate of the BAO or revival model. See Exercise 3.

To make further progress, we need to distinguish the case in which repairs are considered instantaneous from the case in which repairs are time-consuming.

4.4.3.1 Revival models with instantaneous repair

Mathematically, the BAO model says that the conditional distribution of the time to the next failure, given the occurrence time of the failure, is the same as the distribution of the time to the first failure. From the point of view of aging of the system, it is as if all the failures, up to and including the one taking place at S_n, had never happened. That is, applying this reasoning to the nth failure that takes place at time S_n, the model postulates that the conditional distribution of the time U_{n+1} to the next failure satisfies

$$P\{U_{n+1} \le t \mid S_n = s, S_{n-1} = s_{n-1}, \ldots, S_1 = s_1\} = \frac{F(t+s) - F(s)}{1 - F(s)}, \quad t \ge 0$$

for all $n = 1, 2, \ldots$, where F is the distribution of U_1. Thompson [21] shows that this property entails that the sequence of failure times $\{S_1, S_2, \ldots\}$ forms a Poisson process whose intensity function is given by the hazard rate of the distribution of U_1. That is, if $U_1 \sim F$ and the hazard rate of F is h, then the number of failures in $[0, t]$ has a Poisson distribution

$$P\{N(t) = k\} = \frac{H(t)^k}{k!} e^{-H(t)}, t \ge 0, k = 0, 1, 2, \ldots$$

where H is the cumulative hazard function $H(t) = \int_0^t h(u)\,du$. It is easy to show also that $H(t) = -\log[1 - F(t)]$ for $t \ge 0$. From this, we can see that if $F(t) = 1 - \exp(-\lambda t)$, then $H(t) = \lambda t$ and the process is homogeneous with constant intensity function λ. For every other life distribution F, the process is nonhomogeneous.

Language tip: (possibly the most important one in this book): The revival process with an exponentially distributed time to first failure is a homogeneous Poisson process because $h(t)=\lambda$ and $H(t)=\lambda t$. This is the source of a great deal of confusion in both the language and the modeling in reliability engineering. Foremost is that the failure rate (failure intensity) λ of the process is equal to the hazard rate (usually called "failure rate," see Section 3.3.3.5) of the distribution of the time to the first failure. The simplest system reliability model is a single unit that has an exponential life distribution and is replaced upon failure by another of the same type. The process of replacement times in this case is both a renewal process and a homogeneous Poisson process having a constant failure rate. It is easy to overapply this simple model to situations where it does not apply. Many engineers mistakenly believe that this model is the beginning and end of reliability modeling. Even a cursory review of this chapter will show that this is far from accurate. Yet the language and modeling errors brought on by this point of view persist: all failure rates are constant, all times between outages have the same mean, availability is always equal to its asymptotic value, and many other such beliefs are common. Fortunately, increasingly capable reliability modeling software is beginning to make more complex reliability modeling readily accessible, and the persistence of these mistaken ideas should diminish. As a systems engineer, be aware that errors and oversimplifications are common in reliability modeling, and be prepared to ask for clarification where required. See also Section 4.4.6.

Example: A sonar system contains 16 individually replaceable beam-former units. When one beam-former fails, the system is inoperative. The beam-former lifetime has a gamma distribution (Section 3.3.4.7) with location parameter 44,350 hour^{-1} and shape parameter 2. The remainder of the system comprises power supplies, displays, signal processing units, antennas, and other essential equipment. When a beam-former fails, it is replaced by a working unit from an inventory of spares. The replacement process takes about an hour. What is the expected number of beam-former failures in the first 5 years of operation of the system?

Solution: We may use a revival process model for the sequence of failure times of the beam formers because when we replace a failed beam former, we are only affecting one-sixteenth of the ensemble. The approximation afforded by assuming that the replacement time is negligible compared to the mean operating time is good because the mean operating time is more than four orders of magnitude larger than the replacement time. So we may use the revival model with instantaneous repair for this scenario. The time to the first failure of the system we are considering (the ensemble of 16 beam formers, each of which is a single point of failure) has the survivor function

$$\left[1+\frac{t}{\alpha}\right]^{16} e^{-16t/\alpha}.$$

The mean beam-former life is 88700 hours = 10.12 years. For this survivor function, the cumulative hazard function is $H(t) = (16t/\alpha) - 16\log(1 + (t/\alpha))$. Thus the sequence of beam-former failure times is approximately a (nonhomogeneous) Poisson process with intensity function $H'(t)$. When $t = 5$ years = 43,830 hours, $H(t) = 4.82$. This is the expected number of individual beam-former failures, out of the group of 16 beam formers, in 5 years. See Exercise 4.

4.4.3.2 Revival models with time-consuming repair

So far, we have discussed the revival model only with zero repair times. The same reasons that we needed to consider time-consuming repair in the renewal model apply in the revival case as well. This aspect of the theory is not yet completely developed, however. We will review here what is known. The basic model begins with a nonhomogeneous Poisson process describing the operating times $\{U_1, U_2, \ldots\}$ of the unit being studied, so that repair of the unit is done according to the revival protocol. Each time a failure occurs, a repair is undertaken, and the repair times $\{D_1, D_2, \ldots\}$ form a renewal process with $ED_1 = D$, $0 < D < \infty$. This model is used when there is no reason to believe that the repair times depend on the index number of the failure being corrected. This is usually a reasonable assumption. We assume that the sequences of operating times and repair times are independent to reflect a belief that the length of time it takes to complete a repair has nothing to do with how long the unit was operating before it failed.

Denote by F the distribution of U_1, the time to the first failure, and suppose F has a continuous density f and cumulative hazard function $H = -\log(1 - F)$. We further suppose that $\lim_{t\to\infty}(H(t)/t) = \eta$ with $0 < \eta < \infty$. With $S_n = U_1 + \cdots + U_n$ and $N_U(t) = \min\{n : S_n \le t\}$, it is shown in Ref. 24 that $\lim_{n\to\infty}(ES_n/n) = 1/\eta$ and $\lim_{t\to\infty}(EN_U(t)/t) = \eta$ so that the process $\{S_1, S_2, \ldots\}$ is tame [12].

Now let $N(t)$ denote the number of failures in the revival process with nonzero repair times $\{U_1, D_1, U_2, D_2, \ldots\}$. As a result of these preliminary considerations, the salient facts about the revival process with nonzero repairs times are

- The asymptotic overall failure rate is

$$\lim_{t\to\infty}\frac{N(t)}{t} = \frac{\eta}{1 + \eta D} \text{ almost surely}$$

- The asymptotic average availability is

$$\lim_{t\to\infty}\frac{1}{t}\int_0^t A(u)\,du = \frac{1}{1 + \eta D}.$$

Note that this expression is only for the asymptotic <u>average</u> availability. This is weaker than the corresponding result for alternating renewal processes. It is not yet known whether the stronger result can be established for the revival process with nonzero repair times.

It is also shown in Ref. 12 that the availability in the revival process with time-consuming repair can be computed by evaluating the integral

$$A(t)=1-\int_0^t[1-G(t-x)]d\bar{N}(x)$$

where $\bar{N}(x)=\mathrm{E}N(x)$ and G is the common distribution of the downtimes. Unfortunately, this is not yet of much practical help because a satisfactory expression for E$N(x)$ is not yet known.

Requirements tip: If the system has an availability requirement, and if the system can reasonably be modeled using the revival process with time-consuming repair, then demonstrating compliance with the availability requirement will be difficult until an expression for availability, rather than only the average availability over a long time interval, is developed. Usually, when an availability requirement like "The system availability shall not be less than 0.98..." is put in place, reliability modeling to asses compliance with the requirement uses the formula for asymptotic availability in an alternating renewal process (Section 4.4.2.2) to compare with the requirement. Even this does not completely cover the requirement because it is possible under not unusual conditions for the system availability to decrease below its asymptotic value, and even oscillate above and below. The asymptotic value alone does not reveal the transient behavior of availability. In case an alternating renewal model is appropriate, transient availability can be solved for by numerically solving the availability integral equation found in Section 4.4.2.2. This is less readily accomplished in the revival model with time-consuming repair: while the previous equation shows how to compute the transient availability in this model, the difficulty is transferred onto the computation of E$N(t)$ in this model, and this is not yet a routinely solved problem. See Exercise 5.

Example: Consider the sonar system example in Section 4.4.3.1, but now the repair times are independent and identically distributed with a mean of 2 weeks (336 hours). What are the asymptotic overall failure rate and the asymptotic average availability of the beam-former ensemble?

Solution: We continue to use the revival model for beam-former replacement, but now we treat the repair times as nonnegligible. When we assumed instantaneous replacement, the expected number of beam-former replacements in 5 years is 4.82. Taking into account the nonzero repair times, we find that $\eta=16/\alpha=0.000361$, $D=336$, and the asymptotic overall failure rate is 0.000322. The asymptotic average availability is 0.89.

4.4.3.3 Approximations

The most prominent approximation used in reliability modeling for maintained systems is the one we have already discussed in Sections 4.4.2.1 and 4.4.3.1, which is that in most cases, typical operating times are so long compared to typical repair times that the repair times may be taken to be zero at least for the purpose of computing the failure rate. If an estimate of availability is required, this will not do, and a model that explicitly incorporates nonzero repair times will have to be used.

Other approximations have been proposed, including the method of phase-type distributions [13, 15] which approximates an arbitrary life distribution by a special distribution that is the distribution of the time until absorption in a Markov process with one absorbing state. This approximation permits the use of matrix methods and well-known linear algebra techniques to obtain numerical results for several useful reliability models [14]. Advancing capabilities in applied probability numerical computations have rendered these methods of more theoretical than practical interest, and this trend is likely to continue.

4.4.4 More-General Repair Models

In reality, the repair models we have described so far rarely capture perfectly everything we might know to be pertinent in a maintenance situation. The BAO or revival model when used to describe the replacement of a single failed component or subassembly in an ensemble of many components or subassemblies that are not replaced is not quite exact, yet it provides decent results in most cases in practice. But beyond the matter of model stability, some repair situations require more detailed treatment. One example is the case of the clumsy repairman. While repairing an item, even if the target failure is remedied, other faults that may later cause failures may be introduced inadvertently. This phenomenon is common enough in software projects that it has been studied extensively in the software engineering community [8, 9]. A general framework has been proposed [4] in which other important phenomenon like imperfect or incomplete repair may be introduced. Of course, the proper approach to this issue is to train staff so that erroneous repairs are avoided. The model is there to help with the inevitable occasional error that may occur even with highly trained staff.

As usual, the decision to employ more complicated models like these is guided by the amount of precision needed in the study. In all but the most critical cases, the approximations afforded by the commonly used models described earlier give acceptable results. For high-value and/or high-consequence systems, the need for additional precision may warrant the extra prevention cost that would be incurred to acquire or develop more specialized models. Use of higher precision models carries with it the obligation to use more precise information as input (life distribution estimates, etc.). Most routine reliability modeling is well-served by the approximations in common use because the quality of the input information usually available does not justify the use of extremely precise models. Indeed, the understanding of how component reliability estimate errors combine to produce errors in the reliability estimates ("predictions") for higher level units and assemblies does not yet exist except for ensembles of single points of failure (series systems); see Section on "Confidence limits for the parameters of the life distribution of a series system."

4.4.5 The Separate Maintenance Model

It is very common in defense systems, telecommunications systems, and other highly complex technological systems to find that the system maintenance concept involves correcting failures by replacing a failed subassembly with a working one drawn from a pool of spares. This is discussed further in Chapter 10. The separate maintenance model is convenient for creating a reliability model consistent with this maintenance concept.

> **Language tip:** Beware of possible confusion in terminology: the separate maintenance model is a reliability model, intended to describe the reliability of a system that is maintained using the plan described earlier. In Chapter 11, we will consider a maintainability model based on the same maintenance plan. The goal of the maintainability model is to describe the number of maintenance actions taking place during some stated time interval.

To implement the separate maintenance model, arrange the system's reliability block diagram so that there is a one-to-one correspondence between units (subassembly or subsystem) designated as replaceable in the system maintenance concept (Chapter 10) and blocks in the diagram. Then, write the structure function (Section 3.4.6) $\varphi_R(X_1,\ldots, X_n)$ associated with this reliability block diagram. Denote by $Z_1(t),\ldots, Z_n(t)$ the reliability processes (Section 4.3.3.4) of the replaceable units or subassemblies[5] of the system. The separate maintenance model is the reliability process $Z(t)=\varphi_R(Z_1(t),\ldots, Z_n(t))$ of the ensemble of the replaceable units. Different separate maintenance models, all following this same general layout, are obtained when different descriptions of the individual LRU reliability processes $Z_1(t),\ldots, Z_n(t)$ are imposed.

> **Example:** Consider a single server rack in a server farm. The rack contains 12 servers, a two-element hot-standby redundant power supply, a cooling fan assembly, a cabling harness, and a backplane. The 12 servers, each power supply, the cooling fan assembly, and the backplane are individually replaceable or serviceable. Backplane failures can occur because they often contain passive components, and connectors wear from insertion and reinsertion of circuit cards. All other elements of the rack (the rack frame and the cable harness) are permanent installations and their failure would require reconstruction of the rack; we will not include these failures in this model. Rack failure is defined as failure of any server, failure of the redundant power supply pair, failure of the cooling fan assembly, or failure of the backplane. In the hot-standby power supply arrangement, each power supply is individually replaceable. In order for the power supply pair to fail, one power supply would need to fail and the other would need to fail during the time the first supply is being replaced. Failure of one power supply puts the rack in a brink-of-failure state. Hot standby systems with individually replaceable units fail infrequently under normal circumstances. If it

[5] In fact, this is a description of the operating and outage times of the socket containing the replaceable unit.

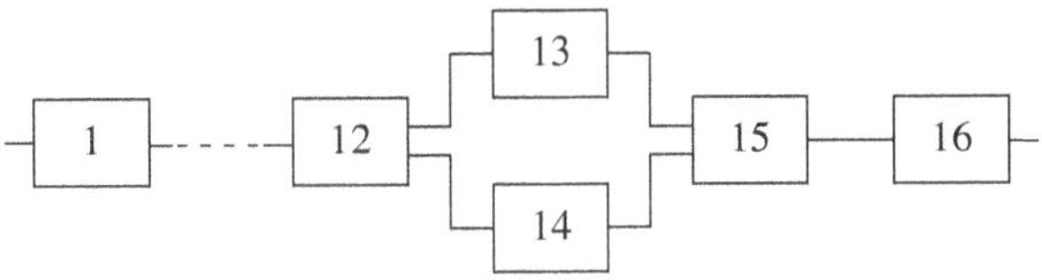

Figure 4.2 *Server rack example reliability block diagram.*

takes a positive amount of time to switch from a failed power supply to its redundant unit, that time must be accounted for in total rack outage time. If the power supplies are configured in a hot standby load-sharing arrangement, there is no switchover time. Backplanes are not generally replaceable but are repaired in place by removing and replacing individual components. This is generally a time-consuming activity: Telcordia GR-418 [20] specifies a mean time to repair of 48 hours for backplanes.

Label the servers 1–12, the two power supplies 13 and 14, the fan assembly 15, and the backplane 16. Then a reliability block diagram for this rack is shown in Figure 4.2.

The structure function associated with this diagram is

$$\varphi_R(X_1,\ldots,X_{16}) = [1-(1-X_{13})(1-X_{14})]X_{15}X_{16}\prod_{i=1}^{12} X_i.$$

Label the reliability process for the servers as $R(t)$, that for the power supplies $P(t)$, that for the fan assembly $F(t)$, and that for the backplane $B(t)$.

Then the reliability process for the rack is

$$R(t)^{12}F(t)B(t)\left[1-(1-P(t))^2\right].$$

Here, we have assumed that all the servers have the same reliability characteristics and both power supplies have the same reliability characteristics. If this is not true, the labels will have to be redefined accordingly. Modeling continues by selecting a repairable system model for each of R, F, B, and P from those described in Sections 4.4.2 and 4.4.3, or some other source. See Sections 4.4.5.1 and 4.4.5.2.

For availability in the separate maintenance model, let $A_1(t),\ldots,A_n(t)$ denote the availabilities for each of the reliability processes $Z_1(t),\ldots, Z_n(t)$. Then the system availability is given by $A(t)=\varphi_R(Z_1(t),\ldots, Z_n(t))$ [3]. Computing the number of failures or failure rate for the separate maintenance model is more complicated. The details may be found in Ref. 23.

4.4.5.1 Separate maintenance with renewal LRU replacement

The separate maintenance model can be used with any reliability process (Section 4.3.2) description for each of its components. Not all components need have the same reliability process description. In this section, we study the separate maintenance model when the individual component reliability processes are renewal or alternating renewal processes.

Consider a system whose structure function is $\varphi_R(x_1,\ldots, x_k)$. The individual component reliability processes are alternating renewal processes (Section 4.4.2.2) $Z_1(t),\ldots, Z_k(t)$ whose availabilities are denoted by $A_1(t),\ldots, A_k(t)$, respectively. Then the system availability is given by $\varphi_R(A_1(t),\ldots, A_k(t))$ [3]. If the mean operating time and mean outage time for component i are μ_i and ν_i, respectively, for $i=1,\ldots, k$, then the asymptotic system availability is given by

$$\varphi_R\left(\frac{\mu_1}{\mu_1+\nu_1},\ldots,\frac{\mu_k}{\mu_k+\nu_k}\right).$$

If the entire system consists only of single points of failure (i.e., is a series system), then the number of system failures is $N_1(t)+\cdots+N_k(t)$ where $N_i(t)$ is the number of failures in time t in alternating renewal process i, $i=1,\ldots, k$. The expected number of failures and availability for each individual component reliability process may be computed using the expressions given in Section 4.4.2.2. If the system is not a series system, the system availability is still given by $\varphi_R(A_1(t),\ldots, A_k(t))$, but computation of the number of system failures is more complicated. The method is outlined in Ref. 23. For a parallel (hot standby) system comprising m units whose structure function is given by $\varphi_R(x_1,\ldots,x_m)=1-(1-x_1)\cdots(1-x_m)$, the expected number of system failures in $[0, t]$ is given by

$$\sum_{i=1}^{m}\int_0^t A_i(u)\frac{1-A(u)}{1-A_i(u)}\,d(F_i*(\mathbf{1}+M_{H_i}))(u)$$

where $A(t)=1-(1-A_1(t))\cdots(1-A_m(t))$, F_i and G_i are the operating time and outage time distributions for unit i, $H_i=F_i*G_i$, and M_{H_i} is the renewal function (Section 4.4.2.1) for H_i, $i=1,\ldots,m$ (equation (4.10) of Ref. 23).

When the reliability process descriptions are renewal processes (zero outage times), the system availability is always 1. Counting the number of system failures is again accomplished by the method outlined in Ref. 23 but now $H_i=F_i$.

Example: Return to the server rack example shown in Figure 4.2. We will determine the mean and standard deviation of the time to the first rack failure and the expected number of maintenance actions, availability, and cumulative expected downtime for the rack under the separate maintenance model with renewal repair. To do this, we need to know the operating time and outage time distributions for each of the rack's components. These are as given in Table 4.1.

The mean time to the first rack failure is 2481.9 hours, and its standard deviation is 1345.1 hours. Over the first 40,000 hours of operation (~5 years), the cumulative expected number of rack failures is shown in Figure 4.3,

TABLE 4.1 Server Rack Example

Rack Component	Uptime Distribution	Downtime Distribution
Server	Gamma, mean 40,000 hours, standard deviation 15,000 hours	Lognormal, median 6 hours, shape factor 2
Power supply	Weibull, $\alpha = 2{,}000$, $\beta = 1.4$	Lognormal, median 6 hours, shape factor 2
Fan	Exponential, mean = 10^5 hours	Lognormal, median 6 hours, shape factor 2
Cable harness	Exponential, mean = 10^6 hours	Lognormal, median 6 hours, shape factor 2
Backplane	Exponential, mean = 5×10^5 hours	Lognormal, median 168 hours, shape factor 1.8

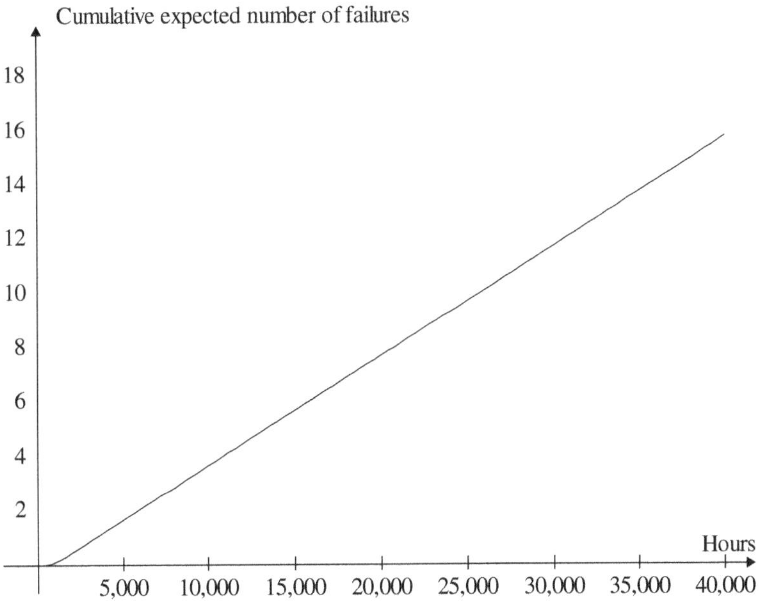

Figure 4.3 *Server rack example—number of failures.*

the availability is shown in Figure 4.4, and the cumulative expected downtime is shown in Figure 4.5.

At the end of the study period, 40,000 hours, the cumulative expected number of failures is 15.76, the availability is 0.97695, and cumulative expected downtime is 496.42 hours. It appears that the limiting value of availability has not yet been attained by the end of the study period. The minimum availability over the study period is 0.9766 (note the expanded vertical scale in Figure 4.4). To satisfy an availability requirement, it may be enough to show that the minimum value of availability is greater than the requirement. Numerical computations are again from Refs. 22, 25.

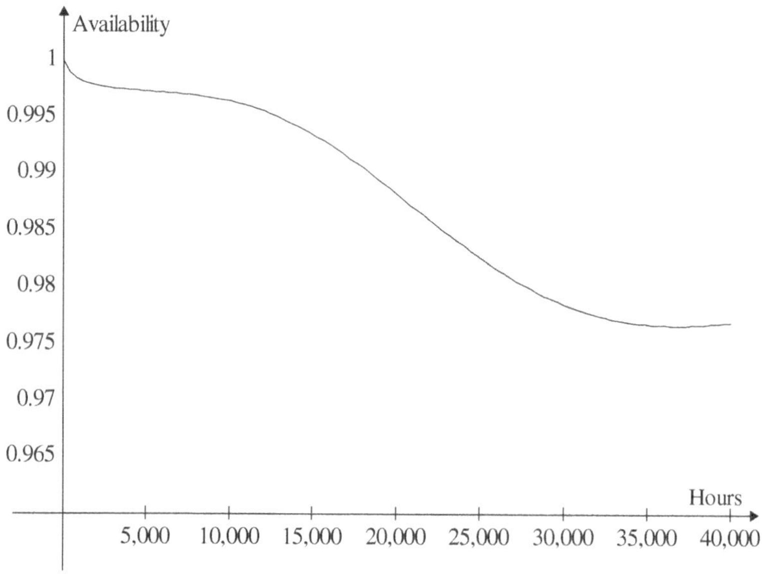

Figure 4.4 Server rack example—availability.

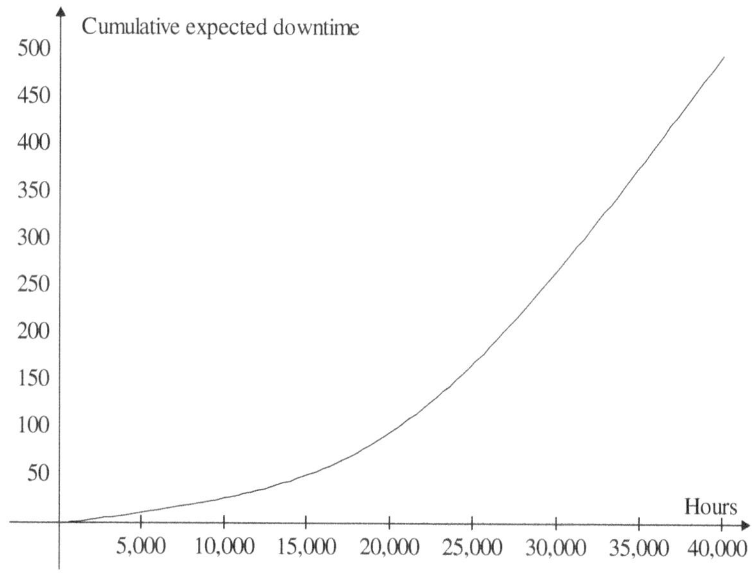

Figure 4.5 Server rack example—downtime.

4.4.5.2 Separate maintenance with revival LRU replacement

We now consider the case where $Z_1(t),\ldots, Z_k(t)$ are revival processes with zero repair times or revival processes with nonzero repair times. If repairs are instantaneous, the system availability is always 1 and the expected number of system failures for a series system is

$$-\sum_{i=1}^{m}\log(1-F_i(t))$$

where F_i is the distribution of the time to the first failure for component i, $i=1,\ldots,m$. An expression for the expected number of system failures for systems other than series systems is not known at this time.

When repair times are nonzero, little can yet be said because results relating the asymptotic average failure rate and asymptotic average availability for the system to the asymptotic average failure rates and asymptotic average availabilities for its components remain to be developed.

4.4.5.3 Separate maintenance with a spares pool of repaired LRUs

In a remove-and-replace maintenance concept with repair of failed LRUs, neither of the above two models is quite correct because replacement LRUs come from a spares pool that contains not only new LRUs but also LRUs that have previously failed and been repaired. If we assume repair of an LRU is accomplished by a renewal procedure, this is tantamount to assuming that all LRUs are new, and the model of Section 4.4.5.1 applies. However, realistic repair of an LRU is usually accomplished by replacing one or a few components on the LRU that had failed, and this is more appropriately described by a revival model (Section 4.4.3). After some period of operation, spares in the pool will have failed and been repaired perhaps several times, and a spare drawn at random from the pool will have a lifetime whose distribution is the same as the distribution of the length of the ith interval in a nonhomogeneous Poisson process for some (unknown) i. The number of times an LRU has been used previously should not decrease as time passes, but beyond that it is a complicated function of the maintenance concept's dynamics. Solution of this challenging research problem would be useful in developing greater understanding of system reliability when the remove-and-replace maintenance concept is used.

4.4.6 Superpositions of Point Processes and Systems with Many Single Points of Failure

Consider a system containing N replaceable units, each of which is a single point of failure. In this system, there are no redundant units (or this could be a maintainability block diagram (Section 11.3.1) by which maintenance actions are being counted). Associated with each individual replaceable unit is a sequence of times at which failures of that unit occur. This sequence is an example of a *point process*, a stochastic process that has a continuous parameter space (here, time) and a discrete state space (here, number of failures). Because each replaceable unit is a single point of failure in this system, every time one of these units fails, the system fails. The sequence of failure times of the system is then the *pool* or *superposition* of the N sequences of failures times of the individual replaceable units. We may picture a superposition of point processes like this:

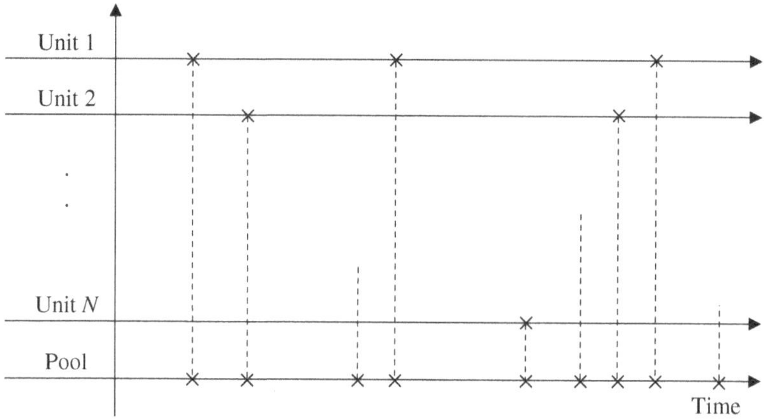

Figure 4.6 Generic superposition of point processes.

In Figure 4.6, the pool or superposition process is shown containing nine points over the time interval depicted. Three of these come from unit 1, two from unit 2, one from unit N, and three come from unspecified units (between 3 and $N-1$) not shown in the picture. Note that each time a point appears in one of the unit failure time point processes, that same time point appears in the pool. The superposition is also a point process, and its intensity (Section 4.3.3.1) is the sum of the intensities of the pool's constituent processes.

Superpositions are useful in reliability modeling because of two relevant applications and one invariance property. One way that a superposition can arise is as the set of times at which system failures occur in a system containing only single points of failure, that is, series systems. Each time one of the system's constituent units fails, the system fails, and the superposition of the individual unit failure time sequences is the system failure time sequence. Another way a superposition can arise is as the collection of times at which failures occur in a population of systems that is being tracked as part of data collection and analysis. Each time a failure occurs in one of the systems in the population, the count in the superposition process that describes the failure times of the members of the whole population increases by one. The invariance property is that under conditions that are usually satisfied in practice, the superposition of a large number of (independent and uniformly sparse) point processes becomes approximately a Poisson process as the number of constituent processes grows without bound. It is called an invariance principle because it doesn't matter what the characteristics of the constituent point processes are—as long as there is no one (or some finite number of) processes that are so fast that almost all the points in the superposition come from those processes (this is roughly what is meant by "uniformly sparse") and they are mutually stochastically independent.

The formal statement of this invariance property is called Grigelionis's theorem [10]. It is a generalization of the easy-to-demonstrate property that the superposition of a finite number of independent Poisson processes (homogeneous or not) is also a Poisson process [11] (see Exercise 6). So a superposition is easier to deal with because it can be treated as a Poisson process in most applications.[6] This fact may account for the common practice of treating the times between outages of a large system, or the times at which failures occur in a large population of systems being tracked, as though they had an exponential distribution. That is, the results of reliability modeling for some system may show that it has some life distribution. The invariance principle says that if the population of those systems being tracked is large enough, then failures will arrive in that population approximately according to a Poisson process regardless what the individual system life distribution was. Using this reasoning, it is possible to claim that you may as well make the system life distribution be as simple as possible, that is, exponential, but while this may not be inappropriate for collection of data from a large population of these systems, it may obscure needed information, such as availability, about the individual system.

4.4.7 State Diagram Reliability Models

We have so far discussed so-called *structural* reliability models based on the reliability block diagram and the life distributions of the diagram's elements. These models are particularly suitable for the kind of system maintenance concept, often encountered in defense, telecommunication, and other large-scale systems, in which certain subassemblies of the system are designated as replaceable and remediation of a system failure is accomplished by replacing one or more of these subassemblies (the "remove-and-replace" maintenance concept). The separate maintenance model matches this operation well. However, there are other types of system maintenance plans and system operations for which the separate maintenance model is less well adapted. For these systems, an alternative reliability modeling strategy based on *state diagrams* can be useful.

A state diagram is a graph in which the nodes represent system states and the links represent transitions from state to state. If we number the states from 1 to N and let $X(t)$ denote the state that the system is in at time t, then it is possible to posit conditions that make $\{X(t) : t \geq 0\}$ a Markov process. Among other things, this will mean that the sojourn time in each state (i.e., the time the system spends in each state) will have an exponential distribution, and the transitions from state to state are governed by a mechanism for which the probability of transition to another state depends only on the state the system is currently in and does not depend in any way on previously visited states. A Markov process of this type is an example of a continuous-time Markov chain (CTMC), a Markov process having a continuum parameter space (time) and a discrete state space.

[6] The conditions that would cause Grigelionis's theorem to fail are rarely encountered in common reliability engineering practice, but good practice would include checking them nonetheless.

Example: Let us describe the three-unit hot-standby redundant system with a state diagram. Label the units 1, 2, and 3. In Figure 4.7, a bubble containing some numbers indicates a system state in which the units numbered in the bubble are operating. A unit whose number does not appear in the bubble is failed (in that system state).

In the state represented by the empty bubble at the bottom of the diagram, no units are operating, and this is the state in which the system (the three-unit hot standby ensemble) is failed. In all other states, at least one unit is operating, and thus the system is operating when it is in any of those states. The same diagram can also be used to describe the reliability of a three-unit series system. The states and transitions are the same, but the only state in which the system is not failed is the one labeled "1, 2, 3" at the top of the diagram.

It would appear that this method for modeling the reliability of the three-unit hot standby arrangement is more complicated than the structural method discussed in Section 3.3.4.5, and it has the additional disadvantage that the sojourn time (time spent in a state) distributions must all be exponential, so this would not be a good choice for modeling the reliability of the three-unit hot standby ensemble. It is also apparent that the number of states required for a state-diagram model of any system of substantial size is extremely, perhaps unmanageably, large. However, for more complicated systems, such as system operations involving queueing for repair, queueing networks, etc., where the structural approach is too complicated to use effectively, the state diagram approach provides a better (if not the only) alternative. A more comprehensive examination of factors to consider when choosing whether to

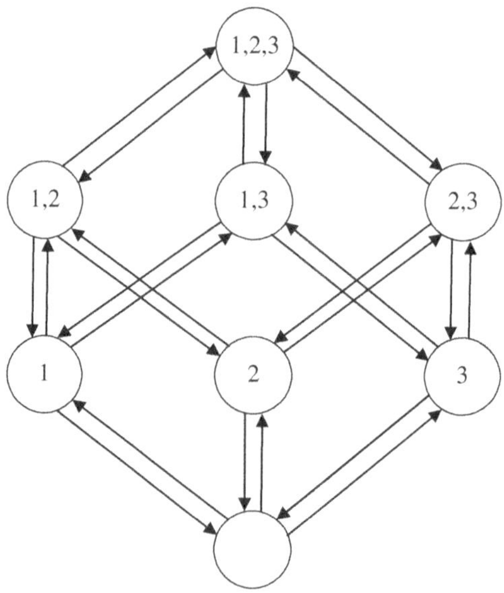

Figure 4.7 *State diagram for three-unit hot-standby redundant system.*

use a separate maintenance model or a state diagram model for reliability is found in Ref. 25. Many treatments of the use of the state diagram approach in reliability modeling are available (see Refs. 16, 26 and others).

Yet other reliability modeling approaches are available. While citing the stochastic Petri net model [19] as another approach that is particularly suitable when the sequence (chronological order) of system operations influences its reliability, we make no pretense here to a thorough review of all possible models. An older review that is nonetheless helpful is Ref. 17.

4.5 STABILITY OF RELIABILITY MODELS

George Box has said "All models are wrong, but some are useful" [6]. This idea speaks to the impossibility of incorporating into a model all the factors that are known to be at play. Generally, the modeler's judgment is the most important determinant of which factors will be included and which ignored. Consequently, every model of the type discussed in this chapter is at best an approximation and at worst mistaken, misleading, and dangerous. Systems engineers are perhaps the best suited by training and experience to sort out which is which in situations where time is short and information is sparse. However, there are mathematical results that give sufficient conditions for a model to have certain desirable properties, such as continuous dependence on the initial conditions of the problem. In high-consequence systems embodying mission criticality, economic make-or-break situations, or life-and-death safety situations, incorporating a formal study of reliability model stability may be worth the resources expended to assure that conclusions drawn from the models are supported with a degree of knowledge consistent with the seriousness of the scenario. A full treatment of these ideas is beyond the scope of this book, but one important point falling into this general area deserves mention. While the decision to adopt a more detailed reliability model depends to a great deal on the customer's need for precision, it also depends on the precision of the information available as input to the model. This includes reliability estimates of the components and subassemblies making up the system. As discussed in Section on "Confidence limits for the parameters of the life distribution of a series system," it is possible to aggregate precision information from components to form precision information for a series system, but how to do this for other structures remains an open problem. Nevertheless, it is fair to say that larger standard errors in the estimates of component reliabilities should lead to a larger standard error in the reliability estimate for any coherent structure containing those components.[7] So if component reliability estimates are quite uncertain, then it may not be justifiable to use a very detailed system reliability model.

[7] This assertion is a theorem in the mathematical theory of reliability that remains yet unproven, but it would be a perverse world indeed if it were not true.

While it does not deal with reliability models specifically, the authoritative treatment of stability in stochastic models is [18].

4.6 SOFTWARE RESOURCES

Time was when reliability models of the sort described in this chapter had to be built from scratch and by hand for every new application. More recently, much of this knowledge has become commoditized and several vendors offer system reliability modeling software for either the separate maintenance model or the state diagram model approach. It is not the purpose of this book to recommend any software product or vendor over any other. Rather, your choice of software should be guided by the learning found in this chapter. Some questions to consider when choosing software include

- Does the software offer a variety of life distribution options?
- Does the software contain appropriate adjustments for environmental stresses?
- Do you need to see confidence interval information for series system models?
- Do the assumptions of the separate maintenance model or the state diagram model fit better with your understanding of system operation?
- Does the software offer sensitivity analysis capability and/or reliability importance computations so that you can see how possible errors in component reliability specifications may be reflected in the system reliability model?
- Does the software offer any reliability optimization models that may be useful in reliability budgeting (Section 4.7.3)?

Software is also available for other reliability engineering tasks, including fault tree analysis and failure modes, effects, and criticality analysis (Chapter 6), analysis of reliability data (including data from life testing and data from operation of installed systems) (Chapter 5). In addition, simulation can be a very effective tool for reliability modeling in complicated situations.

4.7 RELIABILITY MODELING BEST PRACTICES FOR SYSTEMS ENGINEERS

So far, Chapters 3 and 4 have introduced several reliability modeling tools that are useful for systems engineers to understand as part of the process of setting and evaluating the suitability of reliability requirements. How should these tools be put to use in the systems engineering process? Or, how should systems engineers direct the reliability engineering function for maximum added value to the product or service? In this section, we provide four perspectives on these questions.

4.7.1 Develop and Use a Reliability Model

The major purposes of the reliability modeling that have been discussed at some length in this chapter are

- to assess the potential reliability of the design at any stage and to compare the reliability of alternative design proposals,
- to check whether design for reliability activities have been successful in the sense that they have resulted in a system that has the potential to meet its reliability requirements,
- to provide guidance for the data analysis needed for determining whether systems in operation do meet their reliability requirements.

The statistical analyses needed for these activities are described in Chapter 5. In all cases, we aim for the result of reliability modeling to be expressed in the same terms that will be used in the data analysis. That is, if the data to be collected are outage durations, times between outages (operating times), number of failures per stated time interval, etc., then a reliability model should be chosen that will produce as its output some information about outage durations, times between outages (operating times), number of failures per stated time interval, etc. These are reliability effectiveness criteria, and are as such random variables, so the output of the reliability model will necessarily be some abbreviation or summary of the reliability effectiveness criterion, that is, a reliability figure of merit. When comparing a reliability model with reliability requirements or with reliability data, arrange the data analysis so that the same reliability figure of merit appears on both sides of the comparison. For example, if a reliability requirement involves expected outage duration, then both the reliability model and the data analysis should address expected outage duration. Discussion of data analysis for both reliability effectiveness criteria and reliability figures of merit is found in Chapter 5 as well as in some examples in other chapters.

4.7.2 Develop the Reliability–Profitability Curve

The costs of reliability to the supplier include

- prevention costs,
- appraisal costs, and
- external failure costs.

Prevention costs include everything the supplier does to influence the reliability of the product or service. These include developing reliability requirements and reliability engineering activities including reliability modeling and design for reliability. Appraisal costs include the cost of any reliability testing that may be performed during development as well as the portion of failure reporting, analysis and corrective action system or FRACAS (Section 5.6) operating costs attributable to the product. External failure costs include a tangible component (the cost of servicing any warranty that may be offered) and an intangible component (loss of potential sales because customer perception of the product's reliability is poor). The cost of warranty servicing is direct, easily understood, may be offset by the sale of extended warranties, and immediately comprehensible to executives. The cost of loss of reputation is more difficult to pin down precisely, as it represents more abstract items like loss of business from potential customers who may be discouraged from purchasing from the supplier because the supplier's reputation may have suffered from previously selling products with poor reliability.

Formal methods may be used to determine an appropriate balance between cost of reliability and expected profitability, but this is rare. Contemporary quality engineering principles support putting more resources into prevention with the goal of driving down failure costs to a far greater degree (the "1–10–100 rule"). Even if the reliability—profitability balance is addressed only informally, and the results are not very precise, there is value in the exercise because answers may be quite different for different classes of products. For example, it is likely that prevention costs will represent a higher proportion of development costs for an airplane, with its long useful life and high reliability needs, than for a consumer entertainment product which may rapidly become obsolete.

4.7.2.1 Warranty cost modeling

A typical warranty offered by a supplier contains conditions like: the supplier will repair or replace an item that fails when in use by the customer, for some period of time after the initial purchase. Projecting the likely cost of various warranty schemes is an important part of the planning a supplier must do to determine the likely profitability of a product or service. Detailed examination of warranty cost models is outside the scope of this book. The reliability engineering literature contains many studies of this kind; Blischke and Murthy [5] provide a good introduction. Because warranties are usually specified in terms of calendar time, the relationship between operational and calendar time is important for warranty modeling. See Sections 3.3.7 and 9.3.3.

4.7.3 Budget for Reliability

A complex system contains many components, subassemblies, and subsystems that may "fail," that is, behave in such a way as to cause system failure (requirements violations). When reliability requirements are defined for a complex system, it is necessary to determine how the frequency and duration of component, subassembly, and subsystem failures and outages may be arranged so that the frequency and duration of system failures and outages satisfies the system reliability requirements. The process of parceling out the system reliability requirements to the components, subassemblies, and subsystems is called reliability budgeting. The goal of reliability budgeting is to assign to each component, subassembly, and/or subsystem a reliability characterization—probability of failure, life distribution, reliability process, etc., as appropriate—in such a way that

- the system reliability requirements are met,
- cost is minimized, and
- the component/subassembly/subsystem can be designed to meet that target.

The formalism of reliability budgeting is mathematical programming or optimization. We can ask for an assignment of reliability to components, subassemblies, and subsystems to meet the system reliability requirements; or, if a fixed budget is given, we can ask to maximize the system reliability subject to the cost constraint implied by the budget. Either of these is a type of system reliability budget. An introduction to these ideas is found in Section 2.8.4.

For systems describable by the formalism developed in Sections 2.8.4 or 4.3, the system reliability budget begins with a reliability block diagram and its associated structure function. Many formulations of the reliability budget problem are possible, depending on the system reliability requirements. Here are a few examples, all for systems having a structure function φ_R and components whose reliability indicator functions are $C_1,\ldots, C_n$ and whose costs are $a_1,\ldots, a_n$:

- A nonrepairable system with a mission time requirement: In this case, the system survivor function is required to be at least as large as a given value s at a specified time T. If the component/subassembly/subsystem survivor functions are $S_1(t),\ldots, S_n(t)$ having parameters $\lambda_1,\ldots, \lambda_n$, respectively (these may be vectors), then we may formulate the reliability budget problem for this system as

$$\text{Minimize} \sum_{i=1}^{n} a_i(\lambda_i) \text{ subject to } \varphi_R(S_1(T),\ldots,S_n(T)) \geq s.$$

- A repairable system with an availability requirement: In this case, the system availability is required to be at least as large as a given value α over a specified time interval $[0, T]$. If the component/subassembly/subsystem availability functions are $A_1(t),\ldots, A_n(t)$, then we may formulate the reliability budget problem for this system as

$$\text{Minimize} \sum_{i=1}^{n} a_i(A_i) \text{ subject to } \min_{0\leq s\leq T} \varphi_R(A_1(s),\ldots,A_n(s)) \geq \alpha.$$

Formulating the budgeting problem as an optimization problem is one thing, getting to a solution is quite another.[8] This is not because the mathematics is complicated or beyond current capabilities. Rather, it is because of the difficulty in acquiring reliable information about the component/subassembly/subsystem costs as functions of their reliability or availability. A related difficulty is that there is usually not a continuum of choices for the components/subassemblies/subsystems as a function of their reliability or availability. It is usually not possible to marginally increase the reliability of a component by spending an additional cent on it. There may be only one or two choices for a component. This does not adversely affect the solution procedure—it is easy to write the cost function of such a component as a step function—but the greater practical difficulty is that this information is hard to come by and is usually so imprecise that making the effort to solve the budget problem formally is not justified by the quality of the information being used to feed it. In many cases, budgets are devised informally, and often this is "good enough" because the input information is so diffuse. Regardless, a reliability budget is necessary to proceed with design for reliability and creation of a budget, by whatever means, is a best practice.

4.7.4 Design for Reliability

Once you know what the reliability requirement for each component, subassembly, and/or subsystem is, you can begin to apply the design for reliability ideas found in Chapter 6. Best practice indicates that design for reliability is preferable to other reliability management practices inasmuch as design for reliability attempts to be proactive, uses lessons learned from prior experience, and is consistent with contemporary quality engineering principles.

4.8 CHAPTER SUMMARY

This chapter has provided background material on reliability modeling for maintainable systems that systems engineers need in order to be good customers and suppliers in the development process. It is possible to use this chapter as a framework for advanced study of reliability modeling, but its primary intent is to equip systems engineers to be effective in dealing with the reliability engineering aspects of product and service development.

The chapter covers reliability effectiveness criteria and reliability figures of merit used for maintainable systems. Those in most common use are times between outages, failure intensity (failure rate), and availability. Several types of repair models are covered.

We again caution extra care around "failure rate." The phrase is used for several different concepts, some of which require special conditions, so you need to be aware of which meaning is intended in any particular case.

[8] For a more informal approach, see Section 6.6.1.4.

4.9 EXERCISES

1. A maintained system in operation yielded the following data: $U_1=28, D_1=3$, $U_2=50$, $D_2=8$, $U_3=17.6$, $D_3=2$, $U_4=33.9$, $D_4=4.5$, $U_5=71$, $D_5=2.7$, $U_6=108$, $D_6=11$, and at this point data collection ceased.
   a. What is the total time of operation covered by this data collection?
   b. Estimate the mean time between failures of this system over this period of operation.
   c. Estimate the mean time between outages of this system over this period of operation.
   d. Explain any differences you see.
2. Show that a system whose life distribution has an increasing hazard rate and which is repaired according to the revival protocol has an increasing failure rate. What can be said about this system if it is repaired according to the renewal protocol?
3. Suppose a system contains 1000 single-point-of-failure components, each of which has an exponential life distribution with parameters (mean lifetimes) $\lambda_1, \ldots, \lambda_{1,000}$.
   a. What is the life distribution of the system?
   b. What is the mean life of the system?
   c. Suppose component 1 fails at time T and is replaced instantaneously by a new one of the same type. What is the distribution of the time to the next failure after T? What is its mean?
   d. Work the same problem with the life distributions of the components being Weibull with parameters $(\lambda_1, \beta_1), \ldots, (\lambda_{1000}, \beta_{1,000})$.
4. Consider the example in Section 4.4.3.1. Suppose instead that each beam former is replaced by a new one when it fails and that the replacement time is negligible compared to the expected life of one beam former. Discuss the use of a superposition-of-renewal-processes model for this scenario. (Hint: you may also wish to consult Section 4.4.6.)
5. *From Section 4.4.3.2, find the expected value of $N(x)$, the number of failures in the time interval $[0, x]$. Develop an expression for $EN(x)$ so that the integral formula for availability may become useful. Hint: $N(x) = \max\{n : S_n + D_1 + \cdots + D_{n-1} \leq x\}$ and condition on the distribution of $D_1 + \cdots + D_{n-1}$.
6. Let $N_i(t)$ be a Poisson process with intensity function $\lambda_i(t)$, $i=1, 2$. Suppose that the two processes are mutually stochastically independent. Show that the superposition of $N_1(t)$ and $N_2(t)$ is a Poisson process with intensity function $\lambda_1(t) + \lambda_2(t)$. Hint: the superposition can be written as $N_1(t) + N_2(t)$.
7. Consider a system composed of three units A, B, and C. A and B are in a hot standby redundant arrangement, and C is in series with this ensemble.
   a. Draw a reliability block diagram for this system.
   b. Write a structure function for the diagram.
   c. Suppose the operating probabilities are p_A, p_B, and p_C, and that the cost of each unit is proportional to $(1/p) - 1$ $(p = p_A, p_B, p_C)$. Suppose further that the system reliability is to be at least 0.99. Determine the values of p_A, p_B, and p_C that meet the system reliability requirement at minimal cost.

d. Suppose the units have survivor functions $S_i(t) = 1 - \exp(-\lambda_i t)$, $i = $A, B, C, and that the system reliability requirement is that it survive for 10,000 hours with probability at least 0.99. Determine the values of λ_A, λ_B, and λ_C that meet the system reliability requirement at minimal cost.
e. *Suppose now that A, B, and C are repairable, all have repair time distributions that are uniform on $[1, 4]$, and that the system availability requirement is 0.99 or better. Determine the values of λ_A, λ_B, and λ_C that meet the system reliability requirement at minimal cost.
f. *Suppose that A and B are a cold standby arrangement instead of hot standby. How does your solution change?

8. Consider the server rack example in Section 4.4.5. How would the analysis change if the power supplies were in a cold standby configuration?

REFERENCES

1. Ascher H, Feingold H. *Repairable Systems Reliability: Modeling, Inference, Misconceptions, and their Causes*. New York: Marcel Dekker; 1984.
2. Barlow RE, Proschan F. *Statistical Theory of Reliability and Life Testing: Probability Models*. New York: Holt, Rinehart, and Winston; 1975.
3. Baxter LA. Availability measures for a two-state system. J Appl Probab 1981; 18:227–235.
4. Baxter LA, Kijima M, Tortorella M. A point process model for the reliability of a maintained system subject to general repair. Commun Stat Stoch Models 1996;12 (1):12–19.
5. Blischke WR, Murthy DNP. *Warranty Cost Analysis*. New York: John Wiley & Sons, Inc; 1994.
6. Box GEP, Draper NR. *Model-Building and Response Surfaces*. New York: John Wiley & Sons, Inc; 1987.
7. Cox DR, Isham V. *Point Processes*. Boca Raton: CRC Press; 1980.
8. Fakhre-Zakeri I, Slud E. Mixture models for reliability of software with imperfect debugging: identifiability of parameters. IEEE Trans Reliab 1995;44 (1):104–113.
9. Gokhale SS, Philip T, Marinos PN. A non-homogeneous Markov software reliability model with imperfect repair. Computer Performance and Dependability Symposium, 1996, Proceedings of IEEE International; 1996. p 262–270.
10. Grigelionis B. On the convergence of sums of random step processes to a Poisson process. Theory Probab Appl 1963;8 (2):177–182.
11. Karlin S, Taylor HM. *A First Course in Stochastic Processes*. 2nd ed. New York: Academic Press; 1975.
12. Marlow NA, Tortorella M. Some general characteristics of two-state reliability models for maintained systems. J Appl Probab 1995;32:805–820.
13. Neuts MF. Probability distributions of phase type. Liber Amicorum Prof Emeritus H. Florin 1975;173:206ff.
14. Neuts MF, Meier KS. On the use of phase type distributions in reliability modelling of systems with two components. Oper Res Spectr 1981;2 (4):227–234.
15. Neuts MF, Pérez-Ocón R, Torres-Castro I. Repairable models with operating and repair times governed by phase type distributions. Adv Appl Probab 2000;32: 468–479.

16. Osaki S. *Stochastic System Reliability Modeling*. Singapore: World Scientific; 1985.
17. Osaki S, Nakagawa T. Bibliography for reliability and availability of stochastic systems. IEEE Trans Reliab 1976;25 (4):284–287.
18. Rachev ST. *Probability Metrics and the Stability of Stochastic Models*. New York: John Wiley & Sons, Inc; 1991.
19. Sahner RA, Trivedi KS, Puliato A. *Performance and Reliability Analysis of Computer Systems*. Dordrecht: Kluwer Academic Publishers; 1996.
20. Telcordia Technologies. Generic reliability assurance requirements for fiber optic transport systems. GR-418 Core Issue 2. 1999. Piscataway, NJ: Telcordia Technologies.
21. Thompson WA. On the foundations of reliability. Technometrics 1981;23 (1):1–13.
22. Tortorella M. Numerical solutions of renewal-type integral equations. INFORMS J Comput 2005;17 (1):66–74.
23. Tortorella M. On cumulative jump random variables. Annal Oper Res 2013;206 (1):485–500.
24. Tortorella M. Strong and weak minimal repair and the revival process model for maintained system reliability. 2014.
25. Tortorella M, Frakes WB. A computer implementation of the separate maintenance model for complex-system reliability. Qual Reliab Eng Int 2006;22 (7):757–770.
26. Trivedi KS. *Probability and Statistics with Reliability, Queueing, and Computer Science Applications*. Hoboken: John Wiley & Sons, Inc; 2008.

5

Comparing Predicted and Realized Reliability with Requirements

5.1 WHAT TO EXPECT FROM THIS CHAPTER

Once reliability requirements have been established and a system has been developed and deployed, important information can be gained from determining whether the requirements are being met. This chapter will provide you with some ideas and tools you can use to do this. The presentation keys on the two types of reliability requirements introduced in Chapter 2: those based on reliability effectiveness criteria and those based on reliability figures of merit. We offer some statistical procedures appropriate to each case and discuss how to interpret the results they lead to. The chapter closes with some ideas for a failure reporting analysis and corrective action system (FRACAS) that systematizes the process of data collection, archiving, and analysis for more effective and efficient feedback about the reliability of deployed systems.

5.2 INTRODUCTION

Systems engineers can get vital feedback about the reliability experienced by customers of their products and services from appropriate collection and analysis of relevant reliability data. This feedback supports important learning about

Reliability, Maintainability, and Supportability: Best Practices for Systems Engineers,
First Edition. Michael Tortorella.

the degree to which realized reliability of the system conforms to its requirements. Any gaps identified are opportunities for improvement of:

- the process of creating reliability requirements,
- the design for reliability process, and
- the reliability modeling process.

The generally accepted way to do this is to collect reliability data from systems in the field and analyze these data with the aim of making some comparison with the reliability requirements. In this chapter, we discuss the types of data appropriate for this endeavor, several techniques for collecting and analyzing reliability data, and making comparisons among the results.

5.3 EFFECTIVENESS CRITERIA, FIGURES OF MERIT, METRICS, AND PREDICTIONS

5.3.1 Review

The material briefly reviewed here was first presented in Section 2.4. An *effectiveness criterion* is a quantitative descriptor of some phenomenon of engineering interest. In most engineering contexts, and in the sustainability disciplines in particular, an effectiveness criterion is conceptualized as a random variable for reasons described in Chapter 2. Some examples of reliability effectiveness criteria include time between outages of a repairable system, number of failures per unit time, time to first failure, etc. A *figure of merit* is a summary description of an effectiveness criterion that uses probability concepts like distribution, density, hazard rate, mean, variance, percentile, etc. Some examples of reliability figures of merit include the mean time between outages, hazard rate of the distribution of the time to the first failure, variance of the number of failures in the first year of operation, etc. A *metric* is the result of a computation on some data, or a function of some data. For instance, the sample mean of 35 outage durations recorded on a group of 10 systems is a metric. Statisticians call a metric a *statistic*, but the "metric" terminology is more common in engineering. It is important to note that in common systems engineering usage, in almost no case is the word "sample" included in any definition of a metric or statistic; whereas in the theory of statistics, "sample" is almost always included in such definitions. Consequently, the systems engineer needs to be aware that sampling is often present in data collection and be prepared to explain to other stakeholders the origin of any quantities displayed.

Language tip: "Effectiveness criterion," "figure of merit," and "metric" are not always used the same way throughout the sustainability engineering community. The definitions we have chosen serve to make it easier to develop a consistent framework for interpreting reliability requirements and analyzing the data needed to determine conformance to those requirements. In particular, "metric" is sometimes used for any of these concepts, so be particularly careful to discern the correct meaning and context when "metric" is used.

5.3.2 Example

Consider the following example of a reliability requirement for a servomechanism motor: "When operated in an environment characterized by Condition A, the mean time between unscheduled outages for the motor shall be no less than 11,800 hours." Condition A lists all the environmental parameters pertinent to operation of the servo motor, including but maybe not limited to, temperature, humidity, vibration, air quality, and lubrication interval, and specifies the range of values for these parameters under which it is anticipated that the servo motor should do its job. The definition of proper functioning of the servo motor may include the maximum time for the motor to move to a commanded position, the maximum overshoot permissible, etc., so that the definition of failure can be unambiguously determined, that is, all failure modes in the servo motor can be listed. The reliability effectiveness criterion covered by this requirement is the time between unscheduled outages, and the requirement is written in terms of a related reliability figure of merit, the mean time between unscheduled outages. Restricting only the mean time between outages may allow some unduly small individual time between outage values, and the systems engineer should make a determination (possibly by contracting this work to a reliability engineering specialist) that restricting the mean time between outages to at least 11,800 hours is consistent with acceptable reliability (customer satisfaction, profitability) of the larger system of which the servo motor is a part. This means that, among other things, the systems engineer has some understanding of the possible variability in the times between outages so that the system containing the servo motor experiences times between outages on the left tail of the distribution (i.e., small values) infrequently enough so that it still provides acceptable reliability to the customer.

Requirements tip: Imagine, for a moment, that we are dealing with a system whose time between outages is exactly 976 hours in every instance. It would be an easy matter to make a plan to keep the system operational at all times in the sense that it continues to provide the service it is used for without (unplanned) interruption: simply replace it with a new one every 975.9 hours. The difficulty with such plans in reality is that we rarely know the times between outages that precisely—they are always different across installations to a greater or lesser degree depending on how many noise factors may be present and how influential they may be. Reliability (or maintainability or supportability, as appropriate) requirements and modeling are tools we use to cope with this variability. When using (only) the mean of the distribution of a reliability effectiveness criterion as a requirement, consider that this makes most sense when

1. the variation in the values of the reliability effectiveness criterion across installations is anticipated to be small, or
2. the customer owns a large number of installations of the system and may be aggregating costs or other operational measures at a high enough level that a mean value is enough to provide the guidance the customer needs.

In the first instance, consider that requiring a mean of 1,000 hours, for example, is more useful when the range of values of the reliability effectiveness criterion is from 900 to 1,150 hours than when the range is from 100 to 25,000 hours. We understand the system whose range is 900–1,150 hours much better than we do the system whose range is 100–25,000 hours (sometimes we say that the quality of our knowledge about the former system is better than that of the latter; see Section 5.4.1), and the customer is able to make a better maintenance plan then too. Examples of the second instance are defense systems that may sell in the hundreds, thousands, or even millions. Here, the purchaser (say, a national defense agency) acquires many of the same system and is responsible for maintenance on all of them. Such a purchaser may aggregate all the relevant costs over the entire fleet of systems it owns and the mean may provide all the information it needs (because of the strong law of large numbers [5]). Even in this instance, though, the owner needs to make maintenance plans for individual systems in its fleet, and more individualized information would still be a benefit here.

In Parts II and III of this book, we will also examine effectiveness criteria, figures of merit, and metrics for maintainability and supportability, respectively. These are treated in a similar fashion as we treat them for reliability.

5.3.3 Reliability Predictions

When the systems engineer completes reliability requirements, developers should attempt to determine at intermediate stages of development whether the system is capable of meeting these reliability requirements. A determination of this capability should be scheduled as a regularly recurring part of every development program. Developers use a process of reliability modeling (Chapters 3 and 4) to assist in this determination. The quantities that result from the modeling work represent a prediction or forecast about the potential reliability of the system in operation. It is vital to realize that any reliability prediction of this kind should be used to identify parts of the system design where improvements can be made most cost-effectively when shortfalls are uncovered. Treating any reliability modeling effort as a formal "numerical exercise" with no feedback into the design process is a waste of time and resources. Furthermore, reliability modeling is most effective when it is used together with focused design for reliability techniques (Chapter 6) because complementary information is acquired that, when acted on appropriately, should lead to improved reliability of the system in operation.

Because much of the raw material used in reliability modeling comprises statistical estimates of component and subsystem reliability taken from test or operational reliability data, a reliability prediction is also a random variable (see Section "Confidence limits for the parameters of the life distribution of a series system" of Chapter 3). Systems engineers who need to make sense of the reliability predictions provided by the development team's reliability engineers need to use the statistical techniques contained in this chapter to make the necessary decisions: do we continue with the current course of development (we believe with high probability that the system will meet its reliability requirements if its development continues as currently envisioned), or are changes needed so that the system reliability should be improved (we believe with high probability that the system as currently envisioned will not meet its reliability requirements). The answers to these questions often appear as probabilities, or odds. The techniques discussed in this chapter are intended to help systems engineers make this decision.

5.4 STATISTICAL COMPARISON OVERVIEW

5.4.1 Quality of Knowledge

Before treating any of the technical issues, we emphasize that the comparisons discussed in this chapter are statistical in nature because the quantities involved are random variables. This means that any judgments about relative positions of the numbers involved can only be probabilistic. That is, we can only say in many instances that the operational reliability estimated from the current data set shows that this reliability is at least as good as the requirement with probability 0.9 (or some other number). Alternatively, we may find that the data support only a finding that the difference between the estimated operational reliability and the requirement is not statistically significant. These examples show that comparisons like these serve to support a risk-based view of the systems engineering process. It is rare that quantitative judgments of this nature can be made with certainty because most of the raw material for them comprises data, and the metrics drawn from data, namely statistics, are random variables. Systems engineers need to be conversant and comfortable with this approach to gaining knowledge about the systems they work with. It is worth remembering that every system development constitutes a bet by the company that it will be profitable. Among other things, this means that the direct and indirect costs of failures will not be so great that they impair profitability. One purpose of reliability modeling and design for reliability is to improve the odds on this bet, and the business benefits from an understanding of the threshold for the decision to redesign or go ahead with the current design. This reasoning leads directly to a risk-based understanding of system development. Full treatment of this is outside the scope of this book, and interested readers are referred to Refs. 1, 8 for more detail.

The statistics that should be monitored by systems engineering are those from which knowledge of some quantity that has systems engineering importance can be derived. For example, a reliability engineer may create an estimate of the mean number of replacements per month of a particular type of line replaceable unit (LRU) across a population of installed systems. This estimate is likely to be the sample mean of a data set containing the number of replacements per month of every LRU of this type in all the systems surveyed. This statistic is subject to sampling error [4, 11], a source of variability stemming from the fact that the sample chosen is only a random sample from the population, and not a census of the entire population. That is, another sample chosen from the same population may produce a different value for the statistic. With statistics of this kind, we are concerned with precision and accuracy. The accuracy of a statistic is measured by what statisticians call bias, the difference between the expected value of the statistic and the true value of the quantity in the population. Most statistics that are used in systems engineering are unbiased under the usual sorts of conditions that apply in systems engineering studies (of course, this statement needs to be made precise in any particular case). Perhaps, more significant is the issue of precision. Precision of a statistic is measured by how diffuse its cumulative distribution (called the *sampling distribution*) is. This, in turn, is indicated by the standard error of the statistic, which is related to its variance. If the standard error of a statistic is small, the knowledge imparted by the value of the statistic is more precise than if its standard error is large, and we say that the quality of our knowledge about the variable being described by the statistic is high. On the contrary, if the standard error is large, the knowledge imparted by the value of the statistic is of relatively poorer quality—because there is more uncertainty associated with the value when the standard error is larger. In this way, we can describe the quality of knowledge that may be imparted by a statistic so that judgments about the cost and worth of collecting data to improve the quality of knowledge about particular systems engineering variables can be made impartially. Sometimes, the standard error of a statistic is too large to allow confident assessments of downstream risk, and additional resources must be expended, or additional time must pass if the number of systems under observation is small, to improve the quality of that knowledge so that a sensible decision can be taken. Managers make decisions under uncertainty all the time; this intellectual framework provides a way to understand the cost and value of improving the quality of knowledge (decreasing the uncertainty) connected with a particular quantitative variable of importance to the decision-maker.

5.4.2 Three Comparisons

We consider now three important quantities relevant to system reliability. There are the requirements: the guiding quantities for developers and operations managers. There are reliability predictions that come about at various times in the development process. And finally there are estimates of system operational reliability that come from reliability data collected during system operation. There are good reasons for comparing each one of these to the others. We will discuss those reasons in this section, and cover techniques for making the comparisons in Section 5.5.

5.4.2.1 Comparing operational reliability with reliability requirements

It is very clear why systems engineers would want to compare an operational reliability estimate with the system's reliability requirements: knowing whether a system meets its reliability requirements gives important insight into how a customer may view the system and its supplier. Failure of a system to meet its reliability requirements provides an opportunity for improvement of:

- systems engineering understanding of how the customer uses the product,
- the process by which reliability requirements are created, and
- development actions taken to design for reliability (Chapter 6).

Any estimate of operational reliability created from reliability data collected during actual system operation is a statistic, and hence a random variable. Some statistical techniques that can be used to compare operational reliability with requirements are discussed in Section 5.5.

5.4.2.2 Comparing operational reliability with a reliability model

Reliability modeling should produce a picture of what to expect after a system is put into operation. That is, after a system is put into operation, staff responsible for determining the realized reliability of the system, or collection of systems, will collect and analyze various reliability-related data from the operation. Reliability modeling should give staff members an idea of what to expect. In particular, a reliability model should be arranged so that one of its outputs is in the same form as the data expected to be gathered when monitoring operational reliability. For example, if a requirement specifies a mean time between outages, modeling ought to include a statement about mean time between outages, and data collection should include times between outages that can be used to make inferences about their mean.

This comparison supports improvement of the reliability modeling process. Both the reliability model output and the estimated field reliability are random variables, and somewhat different statistical techniques are needed to make suitable comparisons. In practice, though, dispersion information about a reliability prediction is hard to come by except in limited circumstances (Section "Confidence limits for the parameters of the life distribution of a series system" of Chapter 3). So most reliability engineers treat a reliability prediction as a deterministic quantity, and when doing so, the same techniques for comparing operational reliability with requirements can be used.

The one exception we have studied is the series system with components all of whose life distributions are exponential. In that case, Section "Confidence

TABLE 5.1 Example Subassembly Component Reliability Parameters

Component No.	Hazard Rate	90% UCL
1	38	50
2	75	90
3	600	660
4	100	105
5	55	71
6	20	22
7	52	64

limits for the parameters of the life distribution of a series system" of Chapter 3 outlines a method for obtaining a confidence interval for the hazard rate of the system (i.e., the parameter of the exponential life distribution for the system). Using this upper confidence limit, we gain additional information about the probability that a requirement for the hazard rate of the series system may be met. The following example illustrates the general idea.

Example: Suppose the reliability requirement for a small subassembly containing one of each component listed in Table 5.1 is that the mean time to the first failure of the subassembly shall not be less than 1000 hours and that the components on the subassembly are reasonably well modeled by exponential life distributions with the properties listed in Table 5.1. The hazard rate is in units of failures per 10^6 hours. The requirement for the subassembly is equivalent to a hazard rate of 1000 (in the same units) or less. What is the probability that the subassembly will meet this requirement?

Solution: The subassembly hazard rate is 940 failures per 10^6 hours (the sum of the entries in the second column of the table). In the notation of Section "Confidence limits for the parameters of the life distribution of a series system" of Chapter 3, $S = 2676$ and $\delta = 660.4$. The 90% UCL for the subassembly hazard rate[1] is approximately $1006.3 > 1000$ so our confidence that the subassembly meets its hazard rate requirement is low (<90%). The sum of the hazard rates of the components on the subassembly is $940 < 1000$ so we would have reached a different conclusion if no dispersion information were available, but the additional dispersion analysis quantifies the degree of confidence we are justified in having about this conclusion.

5.4.2.3 *Comparing a reliability model with reliability requirements*

Reliability requirements are created to satisfy customer needs for cost-effective operation and performance with few interruptions. As such, they should be seen as a driving force in the system design. Designers should take actions to

[1] Note that the method given in Section "Confidence limits for the parameters of the life distribution of a series system" of Chapter 3 does not assert that the subassembly hazard rate has approximately a χ^2 distribution with 660.4 degrees of freedom. The χ^2 approximation described there is only for the purpose of computing the UCL.

create a system that will meet the reliability requirements; tools for doing this are discussed in Chapter 6. But designers also need a way to know whether the design as currently envisioned is capable of meeting its reliability requirements so that they can make corrections or improvements as necessary. Before a system is completed and operated, reliability modeling is used to assess the current state of the reliability of a design, and the result is called a reliability prediction (Section 4.7.1). The purpose of comparing reliability predictions to reliability requirements is to determine whether the design in its current state is likely to meet the reliability requirements. Because the reliability prediction is a random variable, this decision can never be made with certainty. We can only ask that the probability that the design will meet its reliability requirements be high enough that the remaining risk to profitability (that the design will not meet the reliability requirements and will consequently force additional resources to be spent on remedying failures in the field) and customer satisfaction be acceptably small.

However, as noted in Section 5.4.2.2, in almost all cases reliability engineers treat a reliability prediction as a deterministic quantity, so comparing a reliability prediction with a requirement usually reduces to a simple yes-or-no decision.

5.4.3 Count Data from Aggregates of Systems

When count data (e.g., number of replacements over a study interval) are collected from a large number of systems (of the same type), the aggregate count over all systems is a superposition of the counts from the individual systems. If they are operated independently, the conditions of Grigelionis's theorem [10] are satisfied and the superposition is approximately a Poisson process (Section 4.4.6) whose intensity is the sum of the intensities of the contributing system counts. If you know that a large number[2] of systems is to be put into service, then a simple system reliability model using an exponential distribution for all single point of failure components or subassemblies may be appropriate because the times at which failures occur in this model form a Poisson process and the superposition of independent Poisson processes (the counts from each individual system) is again a Poisson process. Analysis techniques for data generated by a Poisson process are numerous and well-developed [6], making analyses of these kinds of data sets convenient and informative.

5.4.4 Environmental Conditions

So far, we have said little in this chapter about a very important component of reliability requirements, namely, the conditions under which the requirement is supposed to hold. It is frequently the case that the data collected to carry out

[2] A practical rule of thumb is that "large" means the number of counts per system over the study interval, multiplied by the number of systems, should be at least 15, and no system should contribute fewer than five counts over the study interval.

the analyses discussed in this chapter are collected under a variety of conditions, mostly unknown, that may or may not match the conditions specified in the requirement. It is usually the case that environmental conditions have an effect on reliability (Section 3.3.5), so the data will contain these effects. However, most of the time data collection does not include a precise description of the conditions prevailing for the unit from which the data come, and those conditions usually vary in an unknown way from time to time as well. A precise statistical analysis would take account of these environmental conditions, but when they are unknown, analysis that ignores them is not fully informative.

A full treatment of this problem is beyond the scope of this book. Furthermore, the required analyses are specialized and resource-intensive, which means that they are going to be reserved for those infrequent cases in which very precise conclusions are warranted by the economics involved. One approach to this sort of analysis is presented in Ref. 3.

5.5 STATISTICAL COMPARISON TECHNIQUES

One of the major themes of this book is that is it important for systems engineers to monitor the achievement of reliability requirements through appropriate data collection and analysis. This section supports this theme by providing analysis procedures for data collected to carry out this task for the different kinds of reliability requirements discussed in the book. In the sustainability disciplines, the quantities of interest are usually durations of certain events (outages, maintenance tasks, etc.) and counts of certain events (failures, repair shop buffer overflows, etc.) during given time intervals. Table 5.2 guides you to the appropriate analysis procedure for the type of data you have. It may also be helpful to consult Section 12.6 for more background on data collection and analysis for verifying requirements that are written in terms of effectiveness criteria or figures of merit.

Many of the techniques described here in the context of reliability management also apply to comparisons concerning maintainability and supportability variables. Some examples will be discussed in Chapters 10 and 12.

5.5.1 Duration Requirements

Reliability, maintainability, and supportability engineering involves events whose duration and frequency of occurrence are of interest to system engineers and planners. Most requirements for these events are restrictions on their duration and/or frequency of occurrence. In this section, we study statistical procedures appropriate for determining compliance with requirements for durations. Section 5.5.2 covers requirements for counts, or frequency of occurrence.

TABLE 5.2 Data Analysis Decision Tree

Requirement for	Written as	Applying to	Data	Section
Duration	Effectiveness criterion	Entire population	Census	Duration effectiveness criterion applied to entire population, census data
			Sample	Duration effectiveness criterion applied to entire population, sampling data
		Proportion of the population	Census	Duration effectiveness criterion applied to part of the population, census data
			Sample	Duration effectiveness criterion applied to part of the population, sampling data
	Figure of merit	Entire population	Census	Duration figure of merit, applied to entire population, census data
			Sample	Duration figure of merit, applied to entire population, sampling data
Count	Effectiveness criterion	Entire population	Census	Count effectiveness criteria, applied to entire population, census data
			Sample	Count effectiveness criteria, applied to entire population, sampling data
		Proportion of the population	Census	Count effectiveness criteria, applied to part of the population, census data
			Sample	Count effectiveness criteria, applied to part of the population, sampling data
	Figure of merit	Mean	Census	Count figure of merit, mean, census data
			Sample	Count figure of merit, mean, sampling data
		Proportion	Census	Count figure of merit, proportion, census data
			Sample	Count figure of merit, proportion, sampling data

5.5.1.1 Duration effectiveness criterion requirements

Duration effectiveness criterion requirements are usually written to apply to an entire population (e.g., the time between outages shall not be less than 1880 hours) or to some proportion of a population (e.g., 98% of times between outages shall not be less than 1880 hours). In this section, we will discuss these two cases when the data collected represent a census of the population or a sample from the population.

To illustrate the analysis for an effectiveness requirement for a duration, we examine a reliability effectiveness criterion requirement for outage times of a helicopter rotor transmission assembly. When a requirement is written in terms of an effectiveness criterion, the requirement may be interpreted as applying to every member of the population of deployed products, systems, or services, or it may apply to some portion of the population (Section 2.6.3). For instance, a reliability requirement for a helicopter rotor transmission assembly may be written as "the times between unscheduled outages shall not be less than 100 hours," or it may be written as "95% of the times between unscheduled outages shall not be less than 100 hours." In the first case, the requirement applies to the entire population of helicopters in service. It is easy to tell whether the requirement is being met for each helicopter for which outage time data have been collected (see Sections "Duration effectiveness criterion applied to entire population, census data" and "Duration effectiveness criterion applied to entire population, sampling data"). In the latter case, two reasonable interpretations are possible. First, 95% of the times between outages for each individual helicopter should be no less than 100 hours; second, 95% of all times between outages, over all the helicopters in service, should be no less than 100 hours. It may be unreasonable to ask that this requirement hold for each individual helicopter; by the time enough outages have accumulated on one helicopter to be able to reasonably estimate the fifth percentile of the distribution of times between outages, that helicopter may be considered too old to fly any more. Be that as it may, appropriate analysis procedures for this case is given in Sections "Duration effectiveness criterion applied to part of the population, census data" and "Duration effectiveness criterion applied to part of the population, sampling Data."

Duration effectiveness criterion applied to entire population, census data

If a census of the entire population of helicopter rotor transmission assemblies in service is available (i.e., times between outages are recorded for every assembly in the population of helicopters in service), it is straightforward to tell which helicopters are meeting the requirement and which are not. Simply catalog the times between unscheduled outages for each helicopter's rotor transmission assembly (as, e.g., in Table 5.3) and see if any of them are less than 100 hours. Any helicopter for which all its times between unscheduled outages are greater than 100 hours meets the requirement. If a helicopter experiences a time between outages of less than 100 hours, it does not meet the requirement. Of the 15 helicopters in Table 5.3, numbers 2, 6, 9, and 11 do not satisfy the requirement while the remaining assemblies do.

TABLE 5.3 Sample Times between Unscheduled Outages Data

Assembly No.	Hours between Unscheduled Outages
1	151.4, 108.7, 211.5, 185.5, 110.1
2	98.6, 141.4, 173.2, 314.1, 100
3	318.4, 400.9
4	255.0, 221.6, 160.4, 284.7, 252.8
5	135.4, 223.8, 260.3, 256.2, 169.0
6	186.7, 192.8, 83.6, 130.4
7	202.0, 176.5, 122.2
8	172.9, 177.3, 138.2, 146.7, 101.3
9	126.1, 69.6, 109.8, 79.0, 65.4
10	107.4, 198.1, 118.1, 198.4, 126.5
11	116.0, 67.1, 181.4, 92.1, 150.7
12	181.1, 186.8, 111.4, 168.7
13	133.7, 118.3, 137.7, 111.8
14	131.1, 166.0, 110.1
15	155.0, 121.6, 260.4, 154.7, 131.8

Duration effectiveness criterion applied to entire population, sampling data
If data from only a sample (a subset) of the helicopters in service are available, we can still estimate the probability[3] that the requirement is being met in the population. This is a slightly different piece of information, though: in the census case, we can make a judgment about every individual helicopter that is clear-cut, yes-or-no. In the sampling case, we don't have access to data from all members of the deployed population. For the members of the sample, the clear, yes-or-no judgment is still possible; but for other members of the deployed population for which data are not available, this judgment is not possible. The procedure to estimate the probability that the requirement is being met in the population as a whole is straightforward: estimate the proportion of the population that meets the requirement (or, equivalently, the probability that the requirement is being met) by the proportion of the helicopters in the sample that meets the requirement (this is called the "sample proportion").

> **Example:** From a population of 120 deployed helicopters, data on the times between outages of 15 rotor transmission assemblies (one per helicopter) were collected as Table 5.3. The sample proportion of assemblies satisfying the requirement is 11/15 = 0.733, and this a point estimate of the proportion of the (entire) population of deployed helicopters that satisfies this requirement (or the probability that the population satisfies

[3] The probability referred to here is a result of the sampling nature of the procedure. There is no randomness in whether the population meets the requirement, either it does or it does not, we simply do not know which. The probability that the population meets the requirement is interpreted as follows: if the sampling is repeated a large number of times, the probability that the requirement is met is approximately equal to the proportion of those times the procedure results in a conclusion that the requirement is met.

the requirement). A 90% confidence interval for this proportion is approximately given by

$$\left[0.733-1.645\sqrt{\frac{0.733\times 0.267}{15}},\ 0.733+1.645\sqrt{\frac{0.733\times 0.267}{15}}\right]=[0.545,0.921];$$

see section 9.4 of Ref. 4. Based on the data in Table 5.3, then, we conclude that, with 90% confidence, somewhat more than half the population of deployed helicopters meets the requirement. A larger sample would produce a tighter confidence interval (other things being equal). The probability statement in the confidence interval (namely, that the probability that the 90% confidence interval covers the true population proportion satisfying the requirement) comes from the sampling scheme, not from any randomness in the population proportion (in which there is no randomness, each individual member of the population either satisfies the requirement or it doesn't). In other words, if we were to repeat the sampling experiment (drawing a sample of 15 from this population) a large number of times, then the 90% confidence interval (and there would be a different one each time because the data would be different) would include the true population proportion in about 90% of cases.

The confidence interval given above was constructed using a normal approximation. When the sample size is small, say 10 or fewer, use the t-distribution approximation instead. See again section 9.4 of Ref. 4.

Three other points about this analysis need to be noted:

1. First, the size of the entire population of helicopters is only 120, so it is worth examining whether a finite population correction [4] should be applied. The finite population correction is $(N-n)/(N-1)$ applied to the sample variance, where N is the population size and n is the sample size. In this case, the correction is $105/119=0.94$, which is close enough to 1 that not using it is not a major source of error.
2. Second, the standard Wald confidence interval formula (which is what we have been using throughout) is not quite accurate when the sample size is less than about 100–150. A correction is available [17]. Its use in this example yields a confidence interval of [0.50, 0.89], which is slightly more pessimistic (but more accurate) than the analysis given earlier.
3. Third, by pooling the observations, it may be possible to estimate the distribution of the outage times using a parametric or nonparametric model if it is reasonable to assume that all helicopters have the same distribution of times between unscheduled outages.[4] This topic is beyond the scope of this treatment and requires familiarity with statistical issues including censoring, truncation, maximum likelihood methods, and others that are covered thoroughly in Refs. 12, 14, 16.

[4] A χ^2 test [4] could be developed to test this assertion.

Duration effectiveness criterion applied to part of the population, census data

In this case, the requirement is different. It is still an effectiveness criterion requirement, but it is written so that it applies to a portion of the population, as, for example, "95% of the times between unscheduled outages shall not be less than 100 hours." This could be interpreted as applying to

1. every item in the population individually (in the helicopter example, this would mean that for each helicopter, no more than 5% of its times between unscheduled outages should be less than 100 hours) or
2. the population as a whole (cataloguing all the unscheduled outage times over all helicopters in service, no more than 5% of these should be less than 100 hours).

In either case, this requirement says the fifth percentile of the distribution of the times between unscheduled outages is 100 or greater (at most 5% of the durations are allowed to be less than 100 hours).

The first interpretation is less frequently encountered. To tell whether any particular helicopter in the sample meets the requirement, treat each helicopter's data as sample from its own distribution. Count the number of observations less than 100 and apply the usual binomial proportion estimation methods. What's new here is that some helicopters have not had an observation less than 100, that is, we have 0 "successes" out of those helicopters' four or five observations. Statistically, it has not been satisfactorily resolved what one should do in such cases. One approach is to do a simple Bayesian analysis using a flat prior on the proportion p. After seeing n observations with none less than 100, the updated distribution for p is Beta$(1, n+2)$. As you can readily imagine, one shouldn't expect much decisiveness in such cases.

Example: Assembly no. 8 in Table 5.3 satisfies the requirement so far because all of its times in Table 5.3 (i.e., times so far recorded) exceed 100 hours, so the fifth percentile of those times exceeds 100 hours. We may ask how confident should we be that assembly no. 8 will meet the requirement in the future? For this, treat the observations 172.9, 177.3, 138.2, 146.7, and 101.3 as a sample from an unknown distribution of (future) times between unscheduled outages and estimate the proportion of times less than 100 hours from the data. The sample mean of the times between unscheduled outages for helicopter no. 8 is 147.3 hours and the sample standard deviation is 30.64. Because there are only five data points, we use the $t_{(4)}$ distribution (t distribution with 4 degrees of freedom) to estimate the probability that future times between unscheduled outages will be greater than 100 hours. For helicopter 8, this is 0.987, so we are very confident that helicopter 8 will experience few times between unscheduled outages less than 100 hours. In practice, we could perform this analysis on all 15 helicopters and rank them to determine in which order helicopters should receive attention to correct unusually poor reliability.

The second interpretation is more common. There are two strategies for answering this question:

1. Estimate the fifth percentile and compare to 100.
2. Estimate the proportion less than 100 and compare with 5%.

Estimating percentiles can be difficult and the results usually have large uncertainties. Treating the problem as estimating a binomial proportion is simpler; we've done several examples already and the results can have reasonable uncertainties. Gather all the data from the census and use these to estimate the proportion of durations less than 100 hours using the same procedures as above. In Table 5.3, 7, or 10.8%, of the 65 observations are less than 100. This group of 15 assemblies does not meet the requirement up to the time the data were collected. To see how confident we should be that this group of 15 assemblies may meet the requirement in the future, treat the 65 observations in the table as a sample from a population of future observations of times between unscheduled outages and estimate the proportion of times less than 100 from these data. This is the content of Exercise 2.

If the data contain right-censored observations, simple binomial procedures do not apply. The Kaplan–Meyer method [12, 14] for estimating a survivor function (complementary distribution) can be used. This provides an indirect estimate of the fifth percentile of the durations (first the distribution is estimated and then the fifth percentile is derived from the distribution estimate), so there is diffusion of information and the resulting confidence intervals are likely to be larger than they would have been if it were possible to estimate the percentile directly. Data with right-censored observations force a tradeoff: if there are only a few right-censored observations,[5] it might be more efficient to ignore them and estimate the required proportion directly from only the noncensored observations, while if there are many right-censored observations, estimation of the survivor function first, using the right-censored observations, could produce better results.

Duration effectiveness criterion applied to part of the population, sampling data

Finally, we consider the case in which the data collected are a sample from the population of installed systems. This case is treated in the same way we treated the observations in Table 5.3 as a sample from an (unseen) stream of future observations.

5.5.1.2 Duration figure of merit requirements

Duration figure of merit requirements usually apply to an entire population (e.g., the mean time between outages shall not be less than 1880 hours). In this section, we will discuss duration figures of merit when the data collected represent a census of the population or a sample of the population.

[5] Rough rule of thumb: 5–10% of the total number of observations are censored.

Duration figure of merit, applied to entire population, census data
When a duration requirement is written in terms of a reliability figure of merit, it is almost always interpreted as applying to a population of deployed systems (Section 2.6.4). If a census of the deployed population is possible, then the value of the figure of merit is computed from the census data and compared with the value stated in the requirement. A clear-cut, yes-or-no judgment is possible.

Example: Suppose the helicopter rotor transmission assembly reliability requirement is now that the mean time between unscheduled outages shall not be less than 100 hours and that the data shown in Table 5.3 constitute a population census (only the 15 helicopters from which the data have been collected are in service). The mean of the 65 data points in Table 5.3 is 167.3 hours which is greater than 100 hours, so the requirement is met in this population.

Duration figure of merit, applied to entire population, sampling data
If a census of the population is not available, a definite statement about whether the requirement is being met is not possible, but data collected from a sample from the population can be used to estimate the probability that the requirement is being met.

Example: Suppose the helicopter rotor transmission assembly reliability requirement is now that the mean time between outages shall not be less than 100 hours and that the data shown in Table 5.3 constitute a sample from a larger population. We wish to estimate the probability that the requirement is being satisfied in this population. There are 65 times between outages recorded in Table 5.3. The sample mean of these outages is 167.3 hours and the sample standard deviation is $65.76/\sqrt{65}=8.16$ hours. The estimated mean time between outages in the population then has approximately a normal distribution with mean 167.3 and standard deviation 8.16, so the probability that the population mean is less than 100 hours is $\Phi_{(0,1)}((100-167.3)/8.16)=\Phi_{(0,1)}(-8.25)=0$. Thus the probability that the requirement is being met in this population is 1. From the same data, we may construct a two-sided confidence interval for the mean of the population. The resulting 95% confidence interval is

$$\left[167.3-1.96\frac{65.76}{\sqrt{65}},\ 167.3+1.96\frac{65.76}{\sqrt{65}}\right]=[167.3-15.99,\ 167.3+15.99]$$
$$=[151.3,\ 183.3].$$

For other confidence intervals based on the normal approximation, Table 5.4 of confidence coefficients below can be used. If the number of data points is small, say fewer than 10, use confidence coefficients based on the Student's t-distribution with $n-1$ degrees of freedom (n is the sample size). For example, when $n=6$, the two-sided confidence coefficients based on the $t_{(5)}$ distribution are 2.01, 2.57, and 4.03 for the 90, 95, and 99% confidence intervals, respectively.

TABLE 5.4 Normal Confidence Coefficients

	Confidence Coefficient	
Confidence Level (%)	One-Sided	Two-Sided
68	0.75	1.0
90	1.28	1.645
95	1.645	1.96
99	2.33	2.58

See also Section 2.7.5. Computation of the normal distribution, the t-distribution, and their inverses may be accomplished with Microsoft Excel™.

Example: Early in the deployment of a new fiber-optic transport system that is expected to sell in large numbers, 10 units are in service. A reliability requirement is that the mean time between unscheduled outages be at least 1900 hours. From these 10 units, the following times between unscheduled outages were recorded (time unit is hours, clock starts at 0, all systems operational at time 0, renewal repair (Section 4.4.2)): 1715, 2128.5, 1254.8, 1634, 2528, 1830.1, 1419, 2030.5, 857, 1328, 467, 1335.7, 3150, and 2530.8. What is the probability that the requirement is being met?

Solution: Renewal repair was specified, so we may treat the data set as a random sample from a single distribution of unscheduled times between outages. It is apparent that one or more systems in the sample have experienced at least two outages (there are 10 units and 14 observations). The sample mean of these data points is 1729.2 and the sample standard deviation is 710.4. The sample mean is the point estimate of the population mean, which we are taking as the value characterizing the population of systems yet to be built (assuming that they are the same as those in the sample) and deployed. If the point estimate were all we knew, we would say that the requirement was being met. However, this reasoning does not take into account the sampling variation that arises because the future stream of times between unscheduled outages, from members of the population yet to be installed and from the members in the sample too, is unknown and the 14 observations in the data set are only a sample from this stream. The estimated probability that the requirement is will not be met in the larger population is the probability that a normally distributed random variable having mean 1729.2 and standard deviation $710.4/\sqrt{14} = 189.9$ is less than 1900 which is approximately 0.82. We could also try to approach this problem by constructing a one-sided confidence interval for the population mean. A one-sided 90% confidence interval for the population mean, based on these data, is $[1729.2, 1729.2 + 1.28 \times 710.4/\sqrt{14}] = [1729.2, 1972.2]$. This interval contains the requirement value 1900, so it is possible that these data support a conclusion that the requirement is being met. However, <u>no definitive statement about the probability that the requirement is being met can be made using the confidence interval approach unless the requirement value is outside the</u>

constructed interval. In that case, it would be appropriate to assert that the requirement is not being met with probability no larger than the confidence level at which the interval was constructed. The largest one-sided confidence interval (for the population mean) that does not include the requirement value is [1729.2, 1729.2 + 170.8] = [1729.2, 1900] = [1729.2, 1729.2 + 0.90 × 710.4/$\sqrt{14}$]. The (one-sided) confidence level corresponding to the confidence coefficient 0.90 is 82%, consistent with the earlier analysis. It is very unlikely that this requirement will be met.

Finally, we need to consider whether the way in which the data are collected and recorded influences the analysis that is possible. If we did not know that these data were from a system with renewal repair, we would not be justified in analyzing it as a sample from a single distribution, and the foregoing analysis would not be valid. Absent some assumption of this kind, it would be difficult to draw any firm conclusions from these data, not least because we would not have a good description of the mechanism generating the data—the first prerequisite of any statistical analysis.

One additional point is germane here: The data and the renewal repair model support the use of the Kaplan–Meyer procedure [12, 14, 16] to construct an estimate of the survivor function (Section 3.3.2.2) of the times between outages. At the time the data were taken, it is likely that some, if not all, of the systems were in the operating state. The times since the end of the most recent outage to the time of data collection are right-censored observations [12] from the same distribution and contain valuable information which should not be discarded. With the data as given in this example, we don't know what these observations are (because we don't know the times at which the data were recorded and the states of the systems at those times), so we can't use them in the analysis, although it is worth some effort to gather such data because they contain information that should not be ignored.

5.5.2 Count Requirements

In this section, we study statistical procedures appropriate for determining compliance with requirements for counts, or frequency of occurrence. See Section 5.5.1 for procedures concerning requirements for durations.

5.5.2.1 Count effectiveness criterion requirements

Count effectiveness criterion requirements are usually written to apply to an entire population (e.g., the number of failures over 25 years of operation shall be no more than 3) or to some portion of a population (e.g., 98% of installed systems shall have a number of failures over 25 years of operation not exceeding 3). In this section, we will discuss these two cases when the data collected represent a census of the population or a sample of the population.

Count effectiveness criteria, applied to entire population, census data

If a count effectiveness criterion requirement is applied to an entire population and a census of that population is available, determining compliance with the requirement is a matter of comparing the data for each member of the population against the requirement. A yes-or-no judgment about compliance with the requirement is possible for each member of the population. Because these are census data, there is no need to consider sampling error.

Example: Suppose a reliability requirement for a marine diesel engine is "no more than 10 failures in 10 years of operation." Sixteen engines are monitored and the following data are collected:

TABLE 5.5 Marine Diesel Engine Failure Counts

Engine No.	Failures in 10 Years
1	1
2	9
3	11
4	6
5	4
6	6
7	2
8	0
9	0
10	13
11	3
12	5
13	3
14	2
15	0
16	1

Engines no. 3 and 10 do not meet the requirement, while the other 14 engines do.

Count effectiveness criteria, applied to entire population, sampling data

Now suppose the data in Table 5.5 represent a sample of 16 engines drawn from a larger installed population. Because these are not census data, it is no longer possible to say with certainty whether the population meets the requirement. We can estimate the probability that the population meets the requirement. The sample proportion of engines meeting the requirement is $7/8 = 0.875$. The standard error is $[7/(64 \cdot 16)]^{1/2} = 0.083$, so a two-sided 90% confidence interval for the probability that the requirement is met is

$$\left[0.875 - 1.96(0.083),\ 0.875 + 1.96(0.083)\right] = \left[0.713,\ 1.0\right].$$

The right-hand endpoint is actually 1.03 but is truncated at 1 because it is a probability. Another aspect of this problem bears examination. The way this requirement is written, it would take 10 years of data collection to determine compliance. This is impractical and undesirable. Is it possible to make any inference concerning compliance after, say, 1 year of data collection? Exercise 5 is an opportunity to try out a few ideas.

Count effectiveness criteria, applied to part of the population, census data
In this case, a requirement will look something like "at least 95% of the systems in service shall have no more than 5 failures in 10 years of operation." With census data, determining compliance is straightforward: if 95% of the observations are 5 or less, the requirement is met. In Table 5.5, the percentage of observations 5 or under is 68.8, less than 95, so the requirement is not met.

Count effectiveness criteria, applied to part of the population, sampling data
With data from only a sample of the installed population, only the probability that the requirement is being met can be estimated. The sample proportion from Table 5.5 that satisfies the requirement is 11/16 = 0.688, and the standard error of the estimate is 0.083. The probability that the requirement is met is approximately equal to the probability that a normal random variable having mean 0.688 and standard deviation 0.083 exceeds 0.95, and this probability is approximately 0.0008, so the requirement is almost certainly not met.

5.5.2.2 Count figure of merit requirements

Count figure of merit requirements usually apply to an entire population (e.g., the mean number of failures in the first year of operation shall not exceed 2). Count figure of merit requirements may also be expressed as a proportion, as in "the mean number of defective transactions per million opportunities shall not exceed 4." In this section, we will discuss mean and proportion count figures of merit when the data collected represent a census of the population or a sample of the population.

Count figure of merit, mean, census data
As always with census data, simply compute the figure of merit (in this case, the mean) stated in the requirement and compare the result with the value stated in the requirement. This supports a yes–no decision. Table 5.6 (Exercise 5) lists

TABLE 5.6 Marine Diesel Engine Failures

Engine No.	Failures in First Year
1	0
2	2
3	3
4	1
5	0
6	1
7	0
8	0
9	0
10	5
11	1
12	1
13	0
14	0
15	0
16	1

the number of failures for 16 marine diesel engines during their first year of operation. If the data in Table 5.6 are a census of the 16 engines in service, the mean number of failures in the first year in that census is 15/16 which is less than 2, so the requirement is met in this group of 16 engines.

Count figure of merit, mean, sampling data

The procedure here will be to estimate the population mean from the sample data and determine the probability that it is less than 2. The sample mean from the data in Table 5.6 is 0.94 and the standard error is 0.35. With 16 observations, the normal approximation should work well, and the probability that a normal random variable having mean 0.94 and standard deviation 0.35 is less than 2 is 0.999, so it is nearly certain that the requirement is being met.

Count figure of merit, proportion, census data

The service reliability requirements discussed in Chapter 8 are often written as limits on the proportion of transactions that fail in specified ways. In the telecommunications industry, these proportions are usually expressed as defects-per-million (DPM) opportunities. See Section 8.5.1.1 for a more complete review. Here we discuss statistical inference for requirements written as proportions.

> **Example:** A service reliability requirement that the mean number of defects per million transactions shall not exceed 3.4 is adopted. Two weeks' worth of busy-hour data are collected, with the following results: 0, 0, 4, 2, 0, 54, 1, 1, 3, 1, 0, 5, 0, 1 failures per million transactions. The mean number of defective transactions during these 14 hours is 5.2. If these 14 hours represent a census of the population, that is, these were the only 14 hours we were interested in, then the requirement is not met. In this data set, one might suspect that something unusual might have happened on day 6 (a "special cause" was in play). Eliminating day 6 from the data, we obtain a mean of 1.2, indicating the requirement is satisfied if there really is a special cause and whose elimination from the data is justified because it can be remedied. This analysis lends itself readily to graphical display using control charts [18] (see Exercise 5).

Count figure of merit, proportion, sampling data

Suppose now the data given in Section "Count Figure of Merit, Proportion, Census Data" is used as a sample from which the longer term reliability performance of this service is to be inferred. In other words, the data are a sample from a larger population of busy hours, only 14 of which have been measured. In this case, we estimate the probability that the requirement is met. The sample standard deviation from the data shown is 14.2, so the sample mean has approximately a normal distribution with mean 5.2 and standard deviation $14.2/\sqrt{14}=3.8$. The probability that the requirement is met is the probability that a normal random variable with mean 5.2 and standard deviation 3.8 is less than 3.4, which is approximately 0.32. If we were to eliminate day

6 from the data on the basis of there having been a special cause at play whose remediation would prevent many transaction failures from happening, then the sample mean of the remaining 13 days of observation is 1.38 and the sample standard deviation is 1.66. The probability that the requirement is met is 1 to 5 decimal places.

5.6 FAILURE REPORTING AND CORRECTIVE ACTION SYSTEM

The data that form the basis for the studies discussed in this chapter are most effectively obtained if there is systematic way to do this, that is, using a business process to gather these data. In reliability engineering, this process is called a FRACAS. The details of this process will vary depending on the type of product or service and the organization of the supplier's business, but all FRACAS processes are based on a simple idea. As a byproduct of the system maintenance plan, a flow of materiel (LRUs, etc.) around the locations of the supplier, manufacturer, customer, repair center(s), etc., is induced. A FRACAS places "transducers" at various points in this flow to acquire data on *numbers* and *times*. Data can consist of

- numbers of units flowing past the transducer per unit time,
- times that units spend in various states (operation, in repair, in a spares pool, etc.),
- numbers of units in certain locations at specified times,

and others. The data collected need to be tailored to the analyses that the organization needs to perform downstream; see Section 5.4 for the analyses that are most relevant and the types of data they need as raw material.

The remainder of this section is devoted to an example of a FRACAS for a system supplier who contracts out manufacturing but maintains internal repair facilities. This is a notional example, not drawn from any particular industry, but it illustrates important points about FRACAS applications.

The example here shows a FRACAS consistent with the remove-and-replace type of maintenance concept described more fully in Chapter 10. It concerns a system whose design is modular in the sense that the system is repaired when it fails by identifying the failed subassembly and replacing it by another like assembly drawn from a pool of spares. This plan induces a flow of the system's subassemblies through various locations at different times of their life cycle. Data enabling requirements conformance analyses can be gathered at different points in this flow. Figure 5.1 helps illustrate these ideas.

The solid lines in the diagram represent the flow of materiel and the dotted lines represent the flow of data. Assemblies are born in the factory and, in the example shown here, go to burn-in and test before being stored in a warehouse location. Assemblies that fail here are either queued for repair or scrapped if

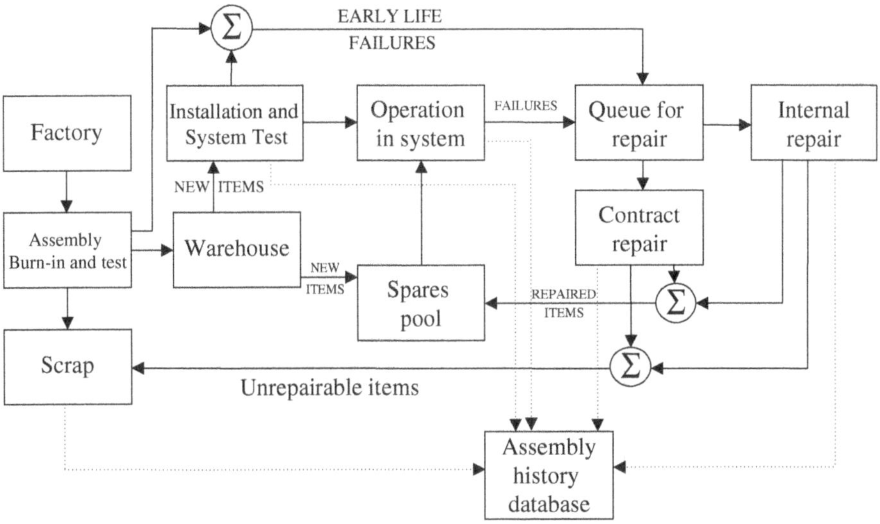

Figure 5.1 *Example of a FRACAS flow diagram.*

judged unrepairable. From burn-in, some assemblies are installed in new systems, while others stock a spares pool. Assemblies that fail while in service are queued for repair and repaired either by a contract manufacturer or internally by the supplier. Some assemblies may be unrepairable and will be scrapped. Those that are repaired are returned to the spares pool to be used as replacements for assemblies that fail in service. Some assemblies may pass all tests at the start of the repair process. These are sometimes called "no-trouble-found" assemblies and different industries treat these in different ways.

Data flows from various points in the diagram to an assembly history database. Data are most valuable when records are kept according to the assembly serial number which is facilitated by bar coding or other identification technology. When an assembly is installed into a system, either as a component of a new system or as a replacement for a failed assembly, the time at which it is installed is recorded in the database. When an assembly fails, the time of failure is sent to the database; from the time of its most recent installation, the current lifetime duration may be calculated. When the assembly is repaired, a listing of the components replaced during repair is also sent to the database; correlating the list with the assembly serial number allows computation of the lifetimes of those components.[6] When an assembly is scrapped, its serial number is recorded so that a complete history of that assembly can be compiled from its successive installation and failure times and component replacement records.

A FRACAS should also support corrective action based on fact. FRACAS can represent a significant expense, and simply collecting data may not present

[6] Depending on whether the components are replaced individually or in batches before testing, the data may contain block censoring which needs to be accounted for in estimating component lifetime distributions. See Ref. 2 for more information on block censoring.

a commensurate return on investment. Corrective action, based on what is learned during assembly test, system test, and repair, leads to design modifications and upgrades that should reduce the number of failures and duration of outages in the future. Corrective actions may include

- component changes,
- assembly redesign (e.g., circuit pack design for reliability as in Section 6.5.1)
- firmware redesign,
- environmental changes (e.g., adding a fan for cooling),

and others. While not underestimating the value of data collection and analysis, especially for determining the degree to which reliability requirements are achieved, a FRACAS without robust corrective action feedback into system design and configuration may be seen as not providing sufficient return on investment.

The FRACAS example shown here is not prescriptive, only illustrative. A successful FRACAS can only be instituted based on a solid understanding of the particular properties of the product or system, the supplier's business plan, the maintenance plan, and other factors affecting the flows of materiel and data in subassembly replacement. Correct FRACAS design will also include requirements to be instituted for activities performed by contractors (i.e., nonaffiliated repair shops) so that the FRACAS functions as a cohesive whole.

5.7 RELIABILITY TESTING

Reliability testing is the source of a great many interesting problems in test design, data collection, and analysis. The most prominent types of reliability testing are:

- component life testing,
- reliability growth testing, and
- software reliability testing.

Full treatment of any of these is beyond the scope of this book, but a brief introduction is useful for systems engineers. References are provided for those interested in further study.

5.7.1 Component Life Testing

The purpose of component life testing is to estimate the life distribution of a population of components under a variety of environmental conditions. For instance, if a circuit uses semiconductors built with 0.25-micron technology, reliability modeling for that circuit will require information about the life distribution of semiconductors built with 0.25-micron technology when operated under various temperature, humidity, and/or other environmental conditions. Even if it is possible to describe reliability of this technology from first principles (e.g., how long does it take for electromigration voids to appear in copper vias of a given thickness and width when a given current flows through it at a given operating temperature), integrated circuits (ICs) are complicated enough, containing numerous vias of different lengths and different current flows, that developing a reliability model for an entire integrated circuit from first principles may well be prohibitively expensive in time and resources. It may be more practical to carry out an accelerated life test for a population of ICs and their foundational technologies (using test structures such as capacitors to study dielectric breakdown, conductor patterns to study electromigration, and transistor structures to study hot carrier damage). The accelerated life test will subject some number of ICs and/or test structures to a variety of test conditions and record the time at which each fails (or if it is still alive at the end of the test period). Important questions include:

- How is an accelerated life test to be designed, given what may be known about the test subjects and the information desired from the test?
- What data should be gathered in the test?
- How should data from the test be analyzed?

Design and analysis of accelerated life tests has been extensively studied and is still a subject of active investigation. Some useful references include Refs. 9, 12–15.

5.7.2 Reliability Growth Testing

It is usually impractical to test an entire large system for reliability under its normal operating conditions. If carried out for a short period of time relative to its anticipated reliability, the test is likely to yield little or no data. For practical systems, whose reliability may be required to be high and for which times between outages will be long, a life test would be impractically long.[7] However, accelerated life testing and highly accelerated stress testing (HAST[8]) do work for some systems, and for these systems, a test, analyze, and fix (TAAF) procedure can be useful. This procedure is also known as reliability growth testing.

In reliability growth testing, a system in development is tested for some period of time. Root cause analysis is performed on any failures that may occur and corrective measures are instituted, creating a new version of the system. The new version is tested again, and this procedure repeats until satisfactory results are obtained. Popular methods for analyzing data from reliability

[7] A good rule of thumb is that a life test should run for at least 5 or more mean lifetimes in order to collect meaningful data.

[8] Testing a system beyond its specified environmental limits to deliberately stimulate failures.

growth testing include the Duane plot [16] and the AMSAA model [7]. The key point to remember in reliability growth testing is that each phase of testing deals with a different system: root cause analysis of the failures found in a phase of testing results in changes to the system that are intended to eliminate the responsible failure mechanisms, so the next version of the system is not the same as the previous one. Confusion sometimes results if this point is not clearly communicated.

5.7.3 Software Reliability Modeling

The most popular approach to software reliability modeling is reliability growth testing, specialized to the particular characteristics of software. The data on the times at which failures appear during testing are usually fit to a nonhomogeneous Poisson process. Decreasing intensity of the process is taken to indicate improvement in reliability of the software. This TAAF approach is discussed more fully in Section 9.4.1.

5.8 BEST PRACTICES IN RELIABILITY REQUIREMENTS COMPARISONS

5.8.1 Track Achievement of Reliability Requirements

This chapter has shown several comparisons that make sense in reliability engineering. The comparison between achieved reliability and reliability requirements should always be instituted because this is the best way to learn what customers are experiencing and where possible weak areas may exist in the design. The other comparisons may be added if resources and/or the business plan permit.

5.8.2 Institute a FRACAS

Most complex military, telecommunications, and other industrial systems will have a flow of materiel similar to that diagrammed in Figure 5.1, so a FRACAS is a natural outgrowth of this operation. The materiel is going to be moving around anyway, so it makes sense to exploit the opportunities it presents: control the flow by making its management a business process, instrument the flow by appropriate collection of data, and use the repair operations as an opportunity to collect the data needed to learn from failures and institute beneficial redesigns.

5.9 CHAPTER SUMMARY

Once reliability requirements are in place, it makes sense to ask whether they are being achieved in system operation. This chapter describes some tools that help make appropriate comparisons between

- reliability requirements and reliability performance,
- reliability performance and reliability modeling, and
- reliability modeling and reliability requirements.

Each of these is important for different aspects of the supplier's business. The chapter also describes an example of a failure reporting and corrective action system. The ideas presented in the example can be used to design other FRACAS for other system operations and maintenance plans.

5.10 EXERCISES

1. For each of the 15 helicopters in Section "Duration Effectiveness Criterion Applied to Part of the Population, Census Data", estimate the proportion of times between unscheduled outages that are less than 100 hours. Which helicopter(s) should be investigated for inadequate reliability?
2. For the group of 15 helicopters in Section "Duration Effectiveness Criterion Applied to Part of the Population, Census Data", estimate the proportion of times between unscheduled outages that are less than 100 hours. What does this tell you about the probability that future observations will be less than 100 hours, or the probability that this group will meet the requirement in future?
3. Complete the analysis described in Section "Duration Effectiveness Criterion Applied to Part of the Population, Sampling Data" by treating the 15 helicopters in Table 5.3 as a sample from a larger population of helicopters whose times between unscheduled outages have not been recorded. What is the probability that a helicopter drawn at random from the population will experience one or more times between unscheduled outages of less than 100 hours?
4. For the example in Section "Count Effectiveness Criteria, Applied to Entire Population, Sampling Data", find the largest confidence coefficient for which the right-hand endpoint of the corresponding confidence interval is 1 (see the second example in Section "Duration Figure of Merit, Applied to Entire Population, Sampling Data"). What is the corresponding confidence level? How would you express this to your manager?
5. Table 5.5 lists the number of marine diesel engine failures during their first year of operation (see Section "Count Effectiveness Criteria, Applied to Entire Population, Sampling Data"). What can you say about the probability that the requirement stated in Section "Count Effectiveness Criteria, Applied to Entire Population, Census Data" will be met after 10 years of operation? (Hint: some facilitating assumption is needed. Try treating each engine as experiencing failures according to a Poisson process whose intensity you can estimate from the data.)
6. Formulate the problem in the example in Section "Count Figure of Merit, Proportion, Census Data" as a control chart. What are some of the benefits and costs of this approach? How would you explain this to your manager?
7. Complete the example in Section "Count Figure of Merit, Proportion, Sampling Data."

REFERENCES

1. Barlow RE, Clarotti CE, Spizzichino F. *Reliability and Decision Making*. New York: Chapman and Hall; 1993.
2. Baxter LA. Estimation subject to block censoring. *IEEE Trans Reliab* 1995;44 (3):489–495.
3. Baxter LA, Tortorella M. Dealing with real field reliability data: circumventing incompleteness by modeling and iteration. Proceedings Annual Reliability and Maintainability Symposium; Piscataway, NJ: IEEE; 1994. p 255–262.
4. Berry DA, Lindgren BW. *Statistics: Theory and Methods*. 2nd ed. Belmont: Duxbury Press (Wadsworth); 1996.
5. Chung KL. *A Course in Probability Theory*. 3rd ed. New York: Springer; 2001.
6. Cox DR, Lewis PA. *The Statistical Analysis of Series of Events*. London: Chapman and Hall; 1966.
7. Crow LH. An extended reliability growth model for managing and assessing corrective actions. 2004 Annual Reliability and Maintainability Symposium. Piscataway, NJ: IEEE; 2004. p 73–80.
8. Cui L, Li H, Xu SH. Reliability and risk management. *Ann Oper Res* 2014;212 (1):1–2.
9. Elsayed EA. *Reliability Engineering*. 2nd ed. Hoboken: John Wiley & Sons, Inc; 2012.
10. Grigelionis B. On the convergence of sums of random step processes to a Poisson process. Theory Probab Appl 1963;8 (2):177–182.
11. Hoel PG, Port SC, Stone CJ. *Introduction to Statistical Theory*. Boston: Houghton Mifflin; 1971.
12. Lawless JF. *Statistical Models and Methods for Lifetime Data Analysis*. Hoboken: John Wiley & Sons, Inc.; 2011.
13. LuValle MJ, LeFevre BG, Kannan S. *Design and Analysis of Accelerated Tests for Mission-Critical Reliability*. Boca Raton: Chapman and Hall/CRC Press; 2004.
14. Meeker WQ, Escobar LA. *Statistical Methods for Reliability Data*. New York: John Wiley & Sons, Inc; 1998.
15. Nelson WB. *Accelerated Testing: Statistical Models, Test Plans, and Data Analysis*. New York: John Wiley & Sons, Inc; 1990.
16. Nelson WB. *Applied Life Data Analysis*. Hoboken: John Wiley & Sons, Inc; 2005.
17. Sauro J. 2005. Measuring usability: quantitative usability statistics and six sigma. Available at http://www.measuringusability.com/wald.htm#wilson. Accessed November 9, 2014.
18. Wadsworth HM, Stephens KS, Godfrey AB. *Modern Methods for Quality Control and Improvement*. New York: John Wiley & Sons, Inc; 2002.

6

Design for Reliability

6.1 WHAT TO EXPECT FROM THIS CHAPTER

Now that we have a good grasp of reliability requirements and some quantitative modeling supporting them, we turn to the question of how to arrange a design so that reliability requirements can be met. The development team needs to take deliberate actions to build reliability into the product or service. Without this attention, the product or service will have some reliability, but it will be just whatever you get by chance. You need to take control of system reliability and take positive steps to drive it in the direction you want, as summarized in the reliability requirements. This chapter discusses several techniques that you can use to build reliability into the product or service, an activity we call "design for reliability." These include:

- a thorough understanding of the reasoning process underpinning design for reliability,
- a CAD tool for design for reliability in printed wiring boards (PWBs),
- fault tree analysis (FTA),
- failure modes, effects, (and criticality) analysis,
- a brief introduction to design for reliability in software, covered in more detail in Chapter 9, and
- robust design as a reliability enhancement tool.

Reliability, Maintainability, and Supportability: Best Practices for Systems Engineers, First Edition. Michael Tortorella.

As with many of the ideas in this book, none of these receives an exhaustive treatment because they are each treated thoroughly elsewhere (references provided) as individual technologies in their own right. The primary intent here is to show how these apply in reliability engineering specifically and to give you the right material so that

- you can use these in your systems engineering practice and
- you are prepared to dig into any of these more deeply should you have need or interest.

6.2 INTRODUCTION

Our focus up to this point has been on writing appropriate and effective reliability requirements. When this task is completed, systems engineers still have a stake in the success of the project, and can add value by promoting efficient achievement of these (and other) requirements. In reliability engineering, the most effective tool available to do this is *design for reliability*. Design for reliability is the set of activities undertaken during product or service design and development to realize the product or service so that it meets its reliability requirements. In brief, design for reliability encompasses those actions taken during product or service design and development to **anticipate** and **manage** failures. "Manage failures" means to

- avoid those failures for which economically sensible countermeasures can be devised, and
- plan for how to react to failures when they do occur.

Design for reliability is a systematic, repeatable, and controllable process whose goal is to fulfill reliability requirements in an economically sensible way. This chapter covers not only the basics of design for reliability but also some deeper aspects of this process when applied in the specific domain of PWBs in electronic systems. Parts II and III of this book discuss maintainability and supportability procedures to minimize the duration of outages that take place when failures occur.

Design for reliability is most effective when it takes place early in product or service development. Because the product or service is not yet realized, a way is needed to assess the likelihood that its reliability requirements will be met so that product/service design actions to achieve reliability requirements can be guided. Accordingly, the first topic covered in this chapter is reliability assessment of a notional product or service, that is, one that is not yet real but whose development is starting or is partially completed. Reliability assessment is a way to tell where you are and provide feedback to the development team.

Design for reliability will not eliminate all possible failures. Design for reliability is a special kind of attempt to predict the future, and, as such, has the usual potential for errors of omission and commission. There are many examples of comprehensively executed failure mode and effects analyses (FMEAs) that described many potential failure modes in detail but completely missed a failure mode that caused many failures in service. One example is the Saturn seat recliner failure described in a 2000 National Highway Transportation Safety Administration case [22]. Because such omissions can easily occur, it is important to maintain a robust program of learning from past experiences. We return to this point in the design for reliability process description.

Design for reliability leads naturally into design for maintainability and design for supportability because it is important to provide the customer with means for dealing with failures that may occur. While design for reliability does provide a robust approach to failure prevention, it is misguided to imagine that all failures can be prevented. Failures will occur. The key question is how will the supplier also create the means for customers and users to minimize the impact on their operations of the failures that do occur. This is the content of design for maintainability and design for supportability that you will find in Chapters 11 and 13.

6.3 TECHNIQUES FOR RELIABILITY ASSESSMENT

There are important reasons for being able to assess product or service reliability throughout the life cycle. During design and development, we need a basis for design for reliability activities: it will be difficult to tell whether further improvement may be needed unless a reading of current status of reliability is available. During product manufacturing or service deployment, it is important to assure that related processes are not introducing into the product or service latent defects that may activate and cause failure later on. During operation and use by the customer, all parties are concerned with whether the product or service is meeting its reliability requirements. This section discusses quantitative reliability modeling and reliability testing as a means for assessing reliability during design, development, and manufacturing for products. Design for reliability for services is covered in Chapter 8. The all-important question of estimating reliability from data gathered during the customer's use of the product was covered in Chapter 5.

6.3.1 Quantitative Reliability Modeling

All of Chapters 3 and 4 were devoted to an exploration of quantitative reliability modeling for systems engineers. Quantitative reliability modeling is usually the domain of reliability engineering specialists, and these chapters go more deeply into this subject than is usually required of systems engineers. The material presented in Chapters 3 and 4 is to

- help guide systems engineers so that they can be good suppliers to, and customers of, reliability engineering specialists, development management, and other stakeholders,
- provide a solid foundation for further exploration of reliability modeling if that is needed or desired, and
- promote correct use of language and concepts that is so important for clarity of purpose and communications.

It is fruitful to incorporate a good customer–supplier model into the systems engineering process for dealing with sustainability specialists. While this is rarely done explicitly, all parties can benefit from understanding that there is a customer–supplier interaction talking place here and even an informal acknowledgment of that relationship can lead to more effective behavior and results. As detailed in Chapter 1, the system engineer supplies reliability requirements to the development team and is a customer for the reliability assessments and data analyses supplied by the reliability engineering staff. Common understanding and correct use of language is a necessary condition for this to be successful. Development management are also suppliers and customers for systems engineering. They provide development schedules and financial goals, insight into customer needs, and funding for project work. In turn, they need up-to-date and unvarnished information about current status of important project attributes, including reliability, maintainability, and supportability. Other stakeholders include customer representatives, with whom various negotiations go on concerning needs and likely outcomes in sustainability parameters, and executives on both sides of the table. Systems engineers are adept at managing all these interfaces, and the more knowledgeable they can be concerning these important factors, the more likely the system or service will be successful: desirable to the customer and profitable to the supplier.

We can use quantitative reliability modeling when a product is in the conceptual stage. No hardware or software or even prototypes need yet exist; modeling is the construction of a mathematical representation of the system and its constituent components that allows estimates or predictions about reliability of the components to be combined in a systematic way to yield estimates of the reliability of the system and its subsystems. Such modeling serves as one of the indicators that systems engineers can use to help decide whether resources need to be added (or may be taken away) to keep the anticipated reliability of the system at the level specified by the system reliability requirements. Information gained from reliability modeling should always be used by design teams to determine where designs may need to be strengthened when the modeling shows that there is significant chance that requirements will not be met.

Reliability modeling without use of the results by design teams is a wasted effort.

6.3.2 Reliability Testing

While testing as a means for reliability assessment can sometimes be helpful, there is usually little opportunity at the early stages of development for learning about the reliability of the system by testing. This is not to say that there are not valuable things that may be learned about a system by testing prototypes or early versions, or components and subassemblies that may be assembled into the system, but reliability testing specifically requires testing of a large number of units for a long time (typically, several multiples of the anticipated mean lifetime of the unit), and this is not practical during early development. This is one of the primary reasons why we urge systems engineers to employ reliability engineering and design for reliability techniques early in the system development process. The details of designing and interpreting reliability tests, including accelerated life tests, are beyond the scope of this book. Many excellent treatments are available, including Refs. 2, 14, 25, 29, 35, and others.

There is one notable exception to this principle. In contemporary practice, testing is the primary (though not the only) tool used for learning about reliability in software products or systems. Thorough discussion of this approach is contained in Chapter 9. What makes reliability testing viable for software products or systems is that the effects of software failure mechanisms are immediate. The standard model for software failures is that most root causes are faults in the code, and when a requirements-legal input condition[1] tendered to the software invokes that part of the code where a fault resides, an improper output results. "Improper" means "different from what it should be according to the requirements" so the improper output is a failure. This happens with essentially no delay whenever such an input condition is tendered, and it happens in the same way in every copy of the software (barring reproduction errors that are usually assumed to be rare or nonexistent). Reliability testing for software is practical because

- unlike in hardware, failure occurs instantly when a failure mechanism is activated,
- failure mechanisms are stimulated by specific input conditions, so testing may proceed by searching the space of possible input conditions, and
- faults may be corrected more quickly than they could be in hardware.

This kind of testing is facilitated by use of an *operational profile* which is a catalog of requirements-legal input conditions together with estimates of the probability that each will be encountered in service. More details are given in Chapter 9. See also Refs. 13, 36.

[1] One that is legitimate according to the software's requirements.

6.4 THE DESIGN FOR RELIABILITY PROCESS

The purpose of this section is to explain the abstract or generic thought process used in design for reliability. This process uses reasoning similar to FTA, and it can be convenient to use a tree-like diagram for summary of results and communication. See Ref. 8 for further comprehensive review of this topic.

The design for reliability (DFR) process that we describe here is a systematization of the idea that DFR proceeds by **anticipation** and **management** of failures. Each design choice has reliability consequences. In other words, each design choice introduces some potential failure mode(s) into the product or service. The DFR process catalogs those failure modes, determines the failure mechanisms and root causes associated with each failure mode, considers what preventive action(s) may be taken to prevent those failure mechanisms from becoming active, and lists the consequences of taking, or choosing not to take, the preventive action(s). This process may be visualized in the form of a tree (see Figure 6.1).

The diagram is necessarily incomplete because lack of space prevents explicit representation of all the failure modes, failure mechanisms, preventive actions, and consequences that stem from even the single design choice shown in the diagram. In practical cases, moreover, the number of developed failure modes from any given design choice is likely to be small, and a formal study of this nature, which may turn out to be costly if fully pursued, is likely to be reserved for truly critical design choices.

To use the DFR process, proceed from left to right in the figure. The first step is to determine the failure modes introduced into the system by each

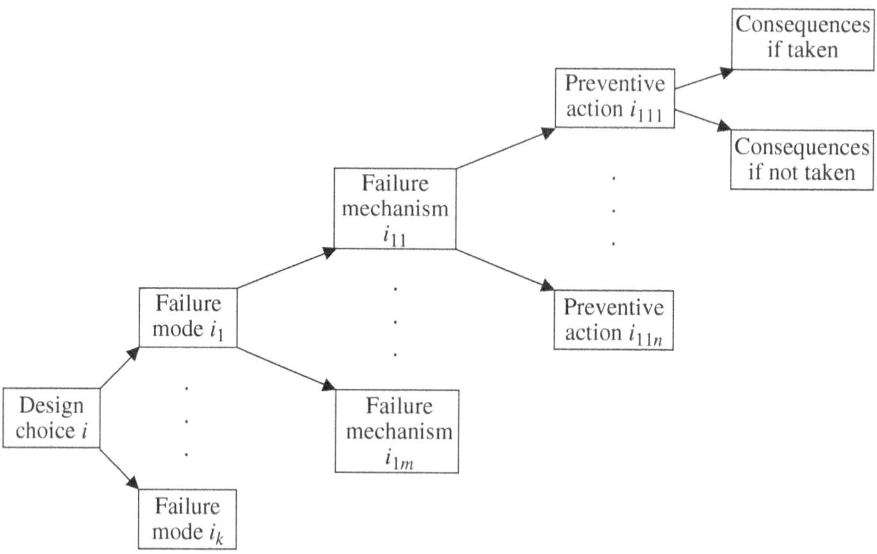

Figure 6.1 Design for reliability process tree.

design choice. This can be done systematically by reviewing each system requirement and listing the ways that the design choice can contribute to violation of the requirement. For example, the designer of a low-voltage power supply can choose aluminum electrolytic or tantalum electrolytic capacitors for filtering. For aluminum electrolytic capacitors, relevant failure modes include open-circuit failure, short-circuit failure, capacitance change, leakage current increase, open vent, and electrolyte leakage [23]. For these capacitors, $k=6$ in Figure 6.1. For tantalum electrolytic capacitors, relevant failure modes include short-circuit failure and thermal runaway failure, and in Figure 6.1, $k=2$ [26]. The design choice, which includes the voltage to be applied to the tantalum capacitor, may stimulate or suppress relevant failure modes in the capacitor, depending (in this case) on the voltage applied by the circuit.

The next step in the process is to associate with each failure mode listed in the first step the relevant failure mechanisms. There is often more than one failure mechanism for each failure mode. Failure mechanisms should be developed in enough detail so that root causes may be identified. That is, each statement about a failure mechanism should include the results of root cause analysis so that appropriate countermeasures may be discerned. For aluminum electrolytic capacitors, section 4 of Ref. 23 shows the failure mechanisms associated with each of the six failure modes listed earlier. For example, the short failure mode is caused by isolation failure in the dielectric film or a short between the electrodes. In turn, isolation failure in the dielectric film is due to localized defect(s) in the oxide film or dielectric paper weakness. This analysis continues until enough is understood about the failure mechanism so that countermeasure(s) can be identified. For the designer of the aluminum electrolytic capacitor, dielectric paper weakness may be prevented by instituting a suitable supplier management program with the supplier of the dielectric paper. A similar solution works for the defective oxide film failure mechanism.

However, our concern here is less with the designer of the capacitor than it is with the system developer or circuit designer, primarily to show how system developers may use the design for reliability process. That is, we are concerned not with the manufacturer of the capacitor but rather with the user of the capacitor in a circuit. For effective use here, we need to know how the factors that are within the scope of this user's control affect the reliability determined by the design choice, in this case, the electrolytic capacitor. The designer's scope of control will include

- circuit design, comprehending the electrical stresses that may be placed on the capacitor by the circuit design, and
- circuit physical layout, comprehending the mechanical (primarily thermal) stresses that may be placed on the capacitor by the physical design.

So root cause analysis needs to proceed as far as being able to identify how electrical and mechanical stresses affect the capacitor's reliability. Again referring to section 4 of Ref. 23, the list of mechanical and electrical stresses that can cause failures of aluminum electrolytic capacitors includes applied overvoltage, excessive ripple current, improper mechanical stresses, applied reverse voltage, halogen contamination, excessive charging or discharging, deterioration over time, and aging of seal materials. The user of this capacitor in a system needs to take action to ensure that these stresses are not present in the application, or at least, if present, are present at a low enough level that they do not cause activation of the related failure mechanisms (see Section 2.2.7). For this capacitor, appropriate design for reliability actions include choosing a capacitor with a voltage rating that is approximately twice the largest peak (for AC) or steady (for DC) voltage anticipated in the circuit, assuring that proper through-hole or surface-mount attachment procedures are followed, and choosing a capacitor whose seal degradation will not proceed far enough over the service life of the product that it will cause failures.

Section 6.5 describes a systematic approach to design for reliability using some of these ideas in the context of PWB design. Section 6.8 discusses how drift in related properties over time influences reliability of subassemblies using the components.

6.4.1 Information Sources

It should be clear from the foregoing discussion that design for reliability relies a great deal on stores of accumulated knowledge. Knowledge is needed about

- quantitative stress–strength relationships (Section 3.3.6), including
  - the stresses that affect the component in question,
  - the functional form relating stress value to the life distribution for the component (Section 3.3.5),
  - the parameter value(s) entering the functional form;
- the stress values that the application (circuit, mechanical design, etc.) places on the component, and
- effective means of arranging the circuit and/or mechanical design and component selection so that harmful effects of stress may be avoided.

These are substantial issues and, while much research has been devoted to their resolution, the important thing for systems engineers is to be able to provide design teams with practical, readily available information that helps them solve these problems quickly.

Reliability physics as a distinct discipline endeavors to determine from first principles the relationships between stress and reliability in a wide variety of devices, especially discrete and integrated semiconductors. The reliability physics literature is too large for most systems engineers to spend time distilling the information they need for day-to-day application, but the knowledge derived from these studies contributes to standard industry and academic databases and software that are available for use by reliability engineers.

In addition to reliability physics, other sources of information used in creating these databases include analysis of reliability data from operation of systems in customer environments. A properly designed FRACAS (Section 5.6) can yield significant amounts of usable data down to the component level. Analysis techniques for these kinds of data have reached an advanced stage of development [5, 6, 28, 31, 33], and many others. In Chapter 5, we have touched only on the basic techniques that would be most useful for systems engineers. The sources listed in the references offer greater flexibility and analytical power that may be useful in more complicated situations in which specialized reliability engineering and/or data analysis expertise may be called for.

Databases also rely on reliability testing for additional information. As often noted, reliability testing is time-consuming and expensive. It is usually reserved for new technologies, for situations in which the response of a component to a particular type of stress is needed, or for high-consequence systems (Chapter 7). Extended discussion of reliability testing technology (accelerated life testing, experimental designs, data analysis, etc.) is beyond the scope of this book. Many resources cover this area, including Refs. 14, 17, 29, 31, 38.

An extremely important function of a reliability database is the acquisition and archiving of reliability lessons learned from experience with products or services previously designed by the organization.[2] No one can foresee every possible failure mode and failure mechanism that may pertain to a new system. Without an easily accessible source of information, even if only anecdotal, about past failures and related root cause analyses, it is easy to forget valuable lessons. Furthermore, formal attention to this need boosts institutional memory which is important because individuals move on to other opportunities. Without a systematic means for capturing their knowledge and experience, it will be lost to the organization until a similar failure scenario brings back a painful situation that need not have been reexperienced.

Finally, an important characteristic of information sources is that they be designed for usability. The general principle is that the information should be available to the designer, in a form that is convenient and consistent with the design process, at the time and the location where it is needed. Usually, this means that printed documentation is not adequate, because the extra step needed to locate the document and physically leaf through it to find the exact information needed adds time and the possibility of error. In extreme cases, the inconvenience may even deter designers from using the resource at all. A more fruitful approach is to integrate the database with the design workflow so that when the designer is working with, say, capacitors, during schematic capture or circuit layout, reliability information for capacitors automatically appears, perhaps, in another window on the same CAD workstation.

[2] Ideally, of course, it would be useful to have access to this kind of experience information regardless of source, but many organizations are loath to share negative experiences publicly. While understandable for competitive reasons, this does make more designer's task more difficult.

6.5 HARDWARE DESIGN FOR RELIABILITY

In this section, we will design PWBs for reliability by describing a tool that exposes the stresses components on the board are subjected to and using the stress–strength relationship (Section 2.2.7) to select components and arrange the physical layout of the board so that the number of component failures due to those stresses is minimized. Recall that the postulate of the stress–strength relationship is that a device fails when it is subjected to a stress that is greater than its strength. Therefore, design for reliability for PWBs consists primarily of efforts to prevent overstresses[3] from reaching devices on the board. Similar ideas apply in other contexts and material systems, and this section is not intended to be a complete catalog of techniques. The ideas presented here can form the basis of a practical DFR system for PWBs and should stimulate you to research similar practices when you are faced with different material systems. We also discuss design for reliability in more complex systems.

6.5.1 Printed Wiring Boards

Line-replaceable units (LRUs) are often configured as individual PWBs, also called printed circuit boards (PCBs) or circuit packs, or assemblies of several PWBs. PWBs typically consist of one or more layers of copper circuit lines ("vias" or "traces"), separated by some insulating material such as FR-4 glass epoxy, phenolic, or similar, with individual components soldered (either through-hole or surface-mount) to the copper traces. A typical double-sided (having wiring traces on both sides of the substrate) PWB is pictured in Figures 6.2 (a and b).

For PWB design for reliability, the key considerations are the stress–strength relationships germane to each of

- the board material,
- the electronic and mechanical components themselves, and
- the solder attachments.

There may also be mechanical attachments, like fasteners, especially for large components (transformers, backplane connectors, etc.). If these are torqued to proper values[4] during assembly, they will not be a cause of failure later on. The design for reliability implication is that these torque values need to be known to the assembly designer, which again points to the need for adequate information supporting DFR.

6.5.1.1 *PWB design for reliability: board material*

Torsional and bending stresses are placed on the PWB substrate in normal operation. The relevant strength is the PWB substrate material and thickness. These should be chosen so that cracks, breakage, trace delamination, or other related

[3] A common abbreviation for stresses exceeding the strength of the device in question.

[4] "Proper values" are determined by the need to resist shock and vibration stresses that may appear in operation.

(a) (b)

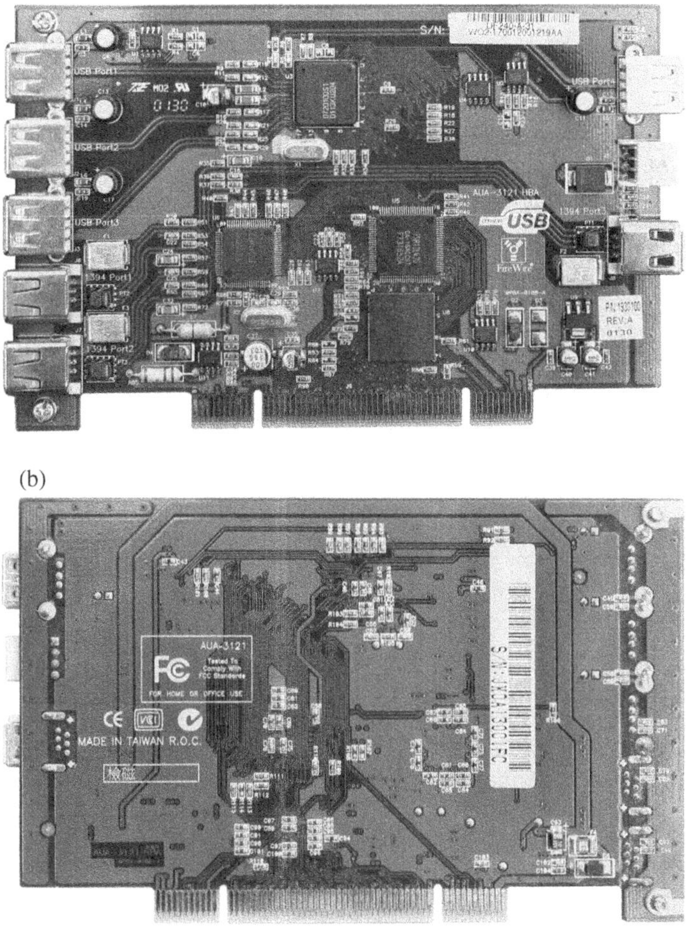

Figure 6.2 Example of a printed wiring board (a) top view (b) bottom view. Photo courtesy of A. G. Blum.

failures do not occur over the intended service life of the equipment [24]. PWB substrate material should be covered by the same component sourcing management program that is used for other components. See Section "Early life failures."

6.5.1.2 PWB design for reliability: assemblies of electronic and mechanical components

Following the model for component reliability discussed in Section 3.3.4.4, we may approach design for reliability in assemblies using electronic and mechanical components by addressing each of the three lifetime phases individually. To recall, these three phases are

1. early life failures caused primarily by failure of components having defects introduced during their manufacture,

2. mid-life failures caused primarily by overstresses applied to otherwise "normal" components (i.e., ones manufactured according to their design intent and containing no manufacturing-introduced defects), and
3. end-of-life failures caused primarily by wearout failure modes.

Early life failures

In the "standard model" of component failure (Section 3.3.4.4), the so-called early life or "infant mortality" failures are caused by defects introduced into components during their manufacture. Components having these defects cause their strength to be less than the majority of the population of components not having manufacturing defects. If we were to draw the strength density for this population at time 0, it would be bimodal: most of the components in the population have strength clustering around the larger mode, and some of the population (those with manufacturing defects) clustering around the smaller mode. For example, a metal oxide semiconductor (MOS) requires an oxide of a specified thickness. If the oxide in a particular device is thinner than specified, that device may fail when subjected to "normal" voltage stresses that an oxide of the specified thickness may easily be able to withstand. If some devices with this defect escape the manufacturer's process controls and make their way into the population of devices sold to users, the users will suffer these "premature" failures.

This model postulates that most of the components in the population that have manufacturing defects eventually will have failed and are no longer in use. When the mechanism is as described in the model, these failures should happen relatively rapidly and the time by which all (or most) of the defective devices have failed should be relatively short. The key to managing the number of early life failures, given this understanding, is to ensure that components are sourced from reputable suppliers who practice systematic quality engineering with effective process controls. Economics usually dictates that reliability testing and "burn-in" of sourced components before assembly is impractical except in some high-consequence systems like nuclear weapons, nuclear power plants, satellites, undersea cable telecom systems, etc.[5] The details of supplier management programs for component reliability are beyond the scope of this book. A good place to start exploring these programs is the American Society for Quality publications [7, 30].

Overstress failures

For components in the "normal" part of the strength distribution, most failures are caused by occasional imposition of stresses beyond their strength. Consequently, the key design for reliability principle applicable to prevent component failures in this phase is to ensure that stresses in excess of the components' strengths are not applied in routine operation.[6] For instance, the

[5] See Chapter 7 for some techniques applicable to incoming reliability inspection used in high-consequence systems.

[6] That is, operation within the environmental limits specified in the reliability requirements.

TABLE 6.1 Components and Stresses

Component	Primary Failure-Causing Stresses
Capacitor	Voltage
Resistor	Dissipated power
Inductor	Current
Electromechanical relay	Contact current
Semiconductor	Reverse voltage, forward current

voltage rating of a capacitor selected for a particular circuit application should always exceed (usually by a factor of at least 2) the voltage that exists in the circuit at that point. To apply similar reasoning to all the components on the PWB would require that the relevant stresses impinging on each of the components be discernible. Table 6.1 lists some important electrical and mechanical components included on many PWBs and the primary stresses that cause failure for each.

It is still possible that extraordinary overstresses may reach circuits because of exogenous shocks like lightning strikes, power surges, cosmic rays, etc.; and depending on the economics, systems engineers may choose to adopt requirements that force designs to be more resistant to these known stressors. It is rare that a stress that cannot be shielded against; the determination of how much preventive action should be taken to protect against known stresses is a matter for reliability economics and risk management [3, 11, 34]. There is also a significant issue with unanticipated stresses. Design for reliability is necessarily an attempt to forecast the future and, therefore, has better or worse outcomes depending on the tools and information available (see Section 6.4.1) and the abilities of the people who use them, so beyond the standard catalog of known stresses there may be stresses that simply were not anticipated, perhaps because of a not-yet-fully understood new technology application or an unanticipated customer environment. The conventional advice for dealing with unknowns of this kind is to allow more margin in the stress–strength analysis where it is possible to do so reasonably. Implementing this advice requires knowledge of stresses and strengths: what is the current state of the design with respect to known stresses, and how much risk are suppliers and customers willing to assume regarding unknown stresses.

The catalog of stresses affecting electronic components on PWBs includes, but is not necessarily limited to,

- heat, including thermal cycling,
- power dissipation,
- voltage,
- electrostatic discharge,
- current,
- humidity,
- shock, and
- vibration.

Quantitative models for the effect on reliability of at least some of these are found in Section 3.3.5. The design for reliability idea here is to enable designers to use these models in a convenient way during schematic capture (PWB electrical design) and PWB layout (PWB physical design) to ensure that the circuit and physical designs do not present overstresses to components. In particular, a design for reliability procedure for PWBs must identify the stresses on the components in the current iteration of the PWB design, identify those that are overstresses, and offer the designer options by which those overstresses can be reduced or removed. In this section, we describe the basic outline of a procedure of this kind that covers thermal and electrical stresses.

Requirements tip: Refer to the component reliability model discussed in Section 3.3.4.4. The idea underlying the PWB design for reliability procedure is that the components on the PWB have reached the constant hazard rate period of their lifetimes, for this is where the influence of randomly occurring overstresses on hazard rate is postulated to occur (see Exercise 1 of Chapter 3). Therefore, effective use of this tool requires that a component acquisition management program be implemented for the purpose of eliminating from the incoming population of components any that may have manufacturing defects and may be subject to early life failure. The postulate of the model that components have not yet reached their end-of-life phase is also to be respected when using this procedure. Reliability requirements should be constructed with this in mind when designing PWBs for reliability using this procedure.

Two main ideas underlie this technology:

1. Include electrical and thermal stresses in the analysis.
2. Integrate the procedure with the computer-aided design process.

These requirements determine what analyses and information will be needed and how they should be interconnected. Electrical stresses (voltage, current, and power dissipation) are discernible from circuit simulation. Once the power dissipated by each component is known, a thermal analysis can be run to determine the distribution of heat across the PWB and find the temperature of each component on the PWB. Once the temperature and electrical stress for a component are known, its reliability may be estimated from an accelerated life model or other appropriate stress-life model (Section 3.3.5). Figure 6.3 shows a schematic of the analyses, databases, and their interconnections (information flows) that can be used to construct a design for reliability procedure for PWBs.

The thermal impedances for each power-dissipating component needed by the thermal analysis tool can be stored in the reliability database or some other database accessible to the thermal analysis program. The schematic capture and physical layout analyses are typically part of a computer-aided design (CAD) system so the design for reliability procedure easily integrates with CAD, allowing for early insight into potential reliability problems. Based on the PWB reliability estimate output, the designer may select other components to provide greater margin

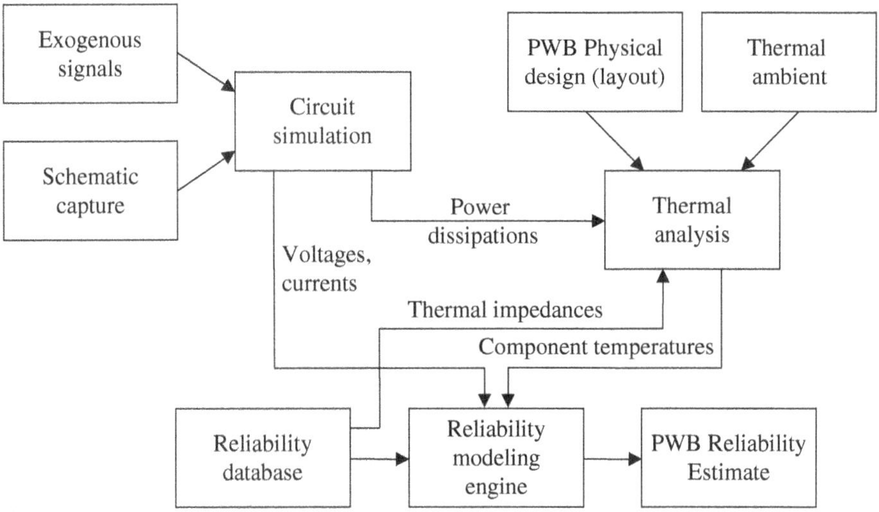

***Figure 6.3** Design for reliability procedure for PWBs.*

between stress and strength or rearrange the board layout to minimize hot spots and decrease the temperature on sensitive components. These changes should result in improved PWB reliability. The use of this type of design for reliability procedure is consistent with the idea of incorporating reliability engineering considerations into the system design from the earliest possible moment.

A brief summary of operation of the procedure follows. Once schematic capture is complete, circuit simulation may be run with whatever inputs come from elsewhere in the system. While the primary purpose of circuit simulation is to gain greater understanding of the circuit's performance against its functional requirements, for purposes of this procedure it is used to also develop voltages, currents, and power dissipations pertinent to each component. Voltages and currents combine with stress-life models (Section 3.3.5) to yield reliability estimates for those components. Power dissipations are combined with ambient temperature and thermal impedances to determine individual component temperatures that serve as inputs to thermal analysis of the entire PWB. Thermal analysis is typically a finite element approximate solution of the heat equation on the PWB using the component temperatures as sources; it produces a thermal profile of the entire PWB so that potential overheating of neighboring components may be detected and remedied by rearrangement of the circuit layout on the PWB. While component reliability estimates resulting from this analysis are important for a reliability model for the entire PWB, an equally important purpose of the procedure is to highlight components that may be overstressed because of circuit design, component selection, and/or physical layout. The design for reliability aspect of this work is that it enables these overstresses to be detected early in the design process when changes to remove the overstresses can be made with minimal disruption.

End-of-life failures

It is particularly important to avoid end-of-life failures because the increasing hazard rate characteristic of those failures means that a many failures may accumulate with increasing frequency over a population of installed systems. While this is a statement about a population of nonrepairable items, a system whose life distribution is increasing and which is repaired according to the revival protocol will experience an increasing failure intensity (Section 4.4.3) (Ascher and Feingold's "sad system" [1]; see Exercises 2 and 3).

For electronic components, a strategy that is often used is to select components so that, with the stresses applied by the circuit, the end-of-life (increasing hazard rate) period does not begin until after the service life of the system is over. This strategy is one of the foundations of the life distribution model described in Ref. 21; see also Section 3.3.4.4. This strategy is effective when it is possible to use it. Sometimes, new or unproven technology must be used in a cutting-edge system, and it is still necessary to protect against the risk of premature activation of a wearout failure mode. The AT&T SL-280 undersea fiber-optic cable communication system used optical transmitters containing laser diodes that had a known wearout failure mode [32]. To meet a very demanding system reliability requirement of no more than three failures in 25 years of operation, it was necessary to provide cold standby redundancy: for each laser transmitter in the system, three cold standby spares were provided along with an innovative optical relay switching system that enabled the spares to be inserted into the optical path as needed. This example illustrates the value of an appropriate redundancy strategy as a means for dealing with known wearout failure modes. See Chapter 7.

For mechanical components, in addition to (or instead of) these two approaches, there is the possibility of preventive maintenance. A good example of this is the internal combustion engine. Auto manufacturers recommend a schedule of engine oil changes to guard against two wearout failure mechanisms: first, unlubricated or poorly lubricated sliding friction of piston rings against cylinder walls and bearings against journals produces mechanical wear; second, engine oil deteriorates in use and its lubricating properties diminish. So far, no one has invented an engine oil whose wearout mechanism (physical/chemical deterioration) does not activate until after the vehicle's service life (potentially, many hundreds of thousands of miles) is over, so renewing the oil periodically is the only sensible preventive maintenance option. Doing so also forestalls the sliding friction wearout failure mode in the cylinders and bearings.

6.5.1.3 PWB design for reliability: solder attachments

The through-hole and surface-mount components are attached to the copper circuit traces on the PCB with solder, usually using a wave-soldering machine. Therefore, in addition to managing the reliability of components on the PWB, it is also necessary to manage the reliability of the solder connections. The major factors contributing to solder connection failure are

- defective solder joints (e.g., "cold" solder joints),
- thermal cycling, and
- shock and vibration.

The incidence of defective solder joints introduced during manufacturing is minimized through use of appropriate process controls. The degree of susceptibility of solder joints to thermal cycling, shock, and vibration depends on the solder material (usually an alloy of tin, lead, bismuth, antimony, etc.) and the shape and size of the solder joint. Solder attachment reliability has been widely studied [15, 16], but new materials are constantly being developed, especially in response to recent reduction of hazardous substances (RoHS) regulations. A thorough design for reliability process for PWBs should incorporate the latest understanding of the solder attachment system used.

6.5.2 Design for Reliability in Complex Systems

At higher levels of assembly, other factors besides component failures contribute to system failures and outages. These may include unanticipated interactions (i.e., timing mismatches in digital circuits) between or among subassemblies, connector and cable failures, operator errors, software faults, etc. Good design for reliability endeavors to anticipate as many of these factors as possible and to design the system so that consequent failures are minimized, and properly managed when they do occur. Quantitative system reliability modeling is useful in this task, and should be undertaken with the motivation of using it to discover weak spots in the design which then can be strengthened (within the economic constraints that prevail) so that reliability requirements will be met.

It is possible to use mathematical optimization techniques to help design for reliability in complex systems, either by minimizing cost subject to a reliability requirement (used as a constraint), or by maximizing reliability within a given cost constraint. The literature on these techniques is found under the topic of "reliability optimization" and is extensive. Important studies that may be used as a starting-off point in this literature include but are not necessarily limited to Refs. 10, 27. In practical reliability engineering applications, these methods can often provide qualitative insight, especially with regard to architecture selection (redundancy), but rarely can they provide specific design solutions. The methods require information about the cost and reliability of all design alternatives, such as LRU architecture and design. There is not usually a continuum of reliability *versus* cost in components or entire LRUs. Most often, there may be only a few components that may be suitable for a particular application, which implies that the reliability/cost function for those components is discrete. For example, in decades past, integrated circuits (ICs) were manufactured in plastic cases for commercial applications and in ceramic, hermetic cases for military applications. The ceramic-packaged ICs were considerably more expensive than the corresponding commercial versions, and while their reliability was commonly assumed to be better than the plastic-packaged versions, later understanding determined this not to be so, and the manufacture of ceramic-packaged ICs was discontinued. Nonetheless, the point of the example is that, for these ICs, there were only two choices available to the designer: two reliabilities and two corresponding costs. A continuum of reliabilities and costs did not exist for these parts, and reliability optimization studies became straightforward (choice between two alternatives) or impossible (for the lack of a continuum of costs and reliabilities).

All is not lost, however. Design for reliability for complex systems is facilitated by two qualitative techniques that have found wide applicability and good success over many kinds of systems, including high-consequence systems like nuclear power plants (Chapter 7). The next section discusses FTA and failure modes, effects, and criticality analysis (FMECA) techniques that can be applied at any stage in design but, as always, are recommended for use as soon as possible after the design concept has been established.

6.6 QUALITATIVE DESIGN FOR RELIABILITY TECHNIQUES

In this section, we introduce two techniques for anticipating and managing failures. FTA and FMECA help the systems engineer catalog the possible failure modes in a system or service, understand how the user may use or misuse the system and how such use or misuse can contribute to system failures, and assess whether proposed countermeasures will be cost-effective as well as effective in preventing failures.

6.6.1 Fault Tree Analysis

6.6.1.1 Introduction

FTA is a disciplined approach to discovering the failure mechanisms and failure causes associated with a failure mode. It is a form of the root cause analysis practiced in quality engineering. It is often referred to as a "top-down" approach because the starting point of an FTA is a system failure, a negative result we wish to avoid to the extent possible within the economics of the system or service. This requirements violation is pictured as the root event of a tree diagram and is referred to as the "top event" in the tree. The analysis proceeds by diagramming as branches emanating from an event the causes of each event in the tree, starting at the top event and working down the tree. The Boolean operators "and," "or," and "not" are used when there are multiple causes for an event. Page connectors are used when it is desirable to divide the diagram into smaller pieces for clarity or to fit to available space. The root cause analysis continues until it reaches a stage where it is possible to identify a reasonable countermeasure for the event at that stage. The reasoning is then that preventing the events at the "bottom" of the tree prevents the occurrence of all the undesirable events above it in the tree, including the final "top event" that represents system failure. Usually, some code or abbreviation is employed in the diagram to identify the events without having to take up space by writing out their entire description in the limited space available in the diagram. FTA, therefore, is a deductive approach to root cause analysis, supported by a simple graphical representation of a hierarchy of causes. Its aim is to find, for each system failure mode studied, a primitive cause or causes that can be prevented through identifiable actions.

A fault tree as we have described it is oriented toward negative outcomes, or failures, and the causal relationships between them. The top event is a violation of some system requirement. A thorough design for reliability analysis can be conducted systematically by cycling through all the system requirements and starting a fault tree for each identifiable violation. This is likely to result in an enormous amount of work and only rarely is so comprehensive a fault tree study conducted. The exceptions are cases where the consequences of failure are catastrophic, such as in a nuclear power plant, or where repair is not possible while a long useful life is desired, such as for a satellite. One can also conduct a similar analysis using successes instead of failures. In this analysis, the tree is oriented toward positive outcomes and the causal relationships between them. For each requirement, a top event can be formed as a successful operation according to the requirement, and the contributing events are events that cause proper operation rather than failure. Such an analysis might more properly be called a "success tree analysis." The success tree is less adapted for design for reliability because the very events that you need to uncover—namely, those involving failures and their causes—are left implicit in the success tree. So the choice between an FTA and a success tree analysis is not only a question of volume (if there are many ways in which a requirement can be violated but only a few in which it is fulfilled, a success tree analysis might be less time-consuming, and conversely). Most reliability engineering specialists use FTA, but you should be aware that alternatives exist.

FTA also may be used quantitatively to find the probability of occurrence of the top event if reasonable probabilities can be assigned to each of the primitive, or root, causes. The explicit use of Boolean operators in the tree allows direct application of the calculus of probabilities to "roll up" the probabilities from the lowest level of the tree to the top event. Usually, all the required independence and disjointness properties of the events in the tree are assumed, but this should always be examined. Sometimes, a fault tree will contain multiple connections between branches or the same event may appear in more than one place in the tree. In these cases, simple calculus of probabilities is not adequate to obtain the probability of the top event. A cut-set analysis (Section 6.6.1.3) is used instead.

Before we undertake a more detailed description of FTA, let's consider a small example.

6.6.1.2 Example: passenger elevator fault tree

For a passenger elevator, the occurrence of a free fall of the elevator car is a serious event with the potential for serious injuries or loss of life. Presumably, there is a safety requirement stating that the elevator car is not to fall freely at any time under any conditions. To illustrate how FTA works, we build a fault tree for the top event "Elevator car falls freely." To construct a fault tree for this event, we need to know something about how an elevator operates, so the next few paragraphs provide a brief description of passenger elevator operation. This description is sketchy and incomplete but provides enough information so that the principles involved in construction of a fault tree can be discerned. A realistic fault tree for a real passenger elevator installation will necessarily be more detailed and comprehensive. Do not mistake this simple illustration for a fault tree that is thorough enough to be used for a real elevator installation.

The three primary assemblies that affect the elevator operation are the control unit, the drive and suspension unit, and the brake unit. The control unit contains a microprocessor that responds to users' signals to move to a desired floor. The control unit starts the drive unit that moves the car to the desired floor and opens the entry doors when the car has come to a stop. The control unit also receives signals from switches in the elevator shaft so that it knows where the car is at all times. The drive and suspension unit holds the car suspended in the shaft and moves the car in response to signals from the control unit. The drive and suspension unit is supposed to be inactive (i.e., the car should not move) unless a signal is received from the control unit. The brake unit operates on the motor in the drive and suspension unit to hold the car motionless when power is removed from the motor and to allow the motor to turn when power is applied to it.

We recognize that there are three possible causes for the car to fall freely: the system does not hold the car, the suspension cable breaks, or the suspension cable slips off its pulley. Label the top event "1" and the causing events as "2," "3," and "4," respectively. Then Figure 6.4 is the start of the fault tree.

In Figure 6.4, the top event "1" and event "2" are drawn as rectangles, indicating that they will be further analyzed to discover their more basic causes. The diamond-shaped boxes indicate events that will not be further analyzed. These are considered to be root causes for which effective countermeasures can be applied. An "or" connector is used to indicate that if any of events "2," "3," or "4" occurs then the top event occurs. It may be possible to further decompose events 2 and 3 into root causes, but we shall not do so in this simple example. Again, the decision to seek deeper root causes rests on whether effective countermeasures can be devised for the event(s) at the bottom of the diagram. In this example, for instance, determining how to prevent the suspension cable from breaking may require additional information about the causes of a cable break.

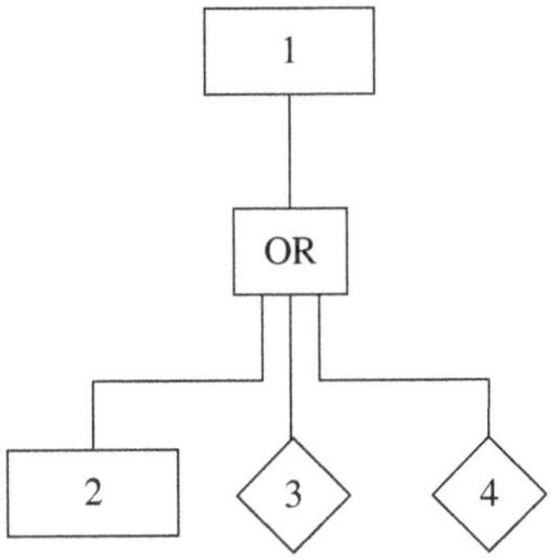

Figure 6.4 Beginning a fault tree for passenger elevator example.

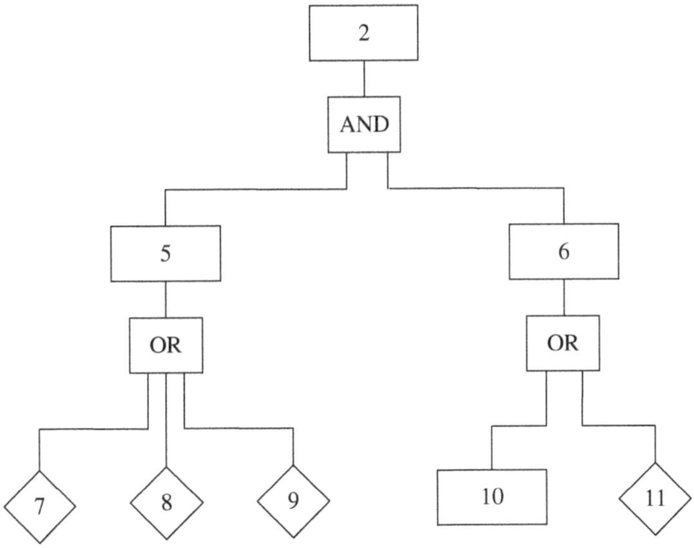

Figure 6.5 Sub fault tree for "system does not hold car" event.

We may now look for causes of event "2," the system does not hold the car. Two things have to take place for the system to not hold the car when the suspension cable is not broken or off the pulley: the brake fails ("5") and the motor turns freely ("6"). We identify three causes for "brake fails": loss of friction material ("7"), the brake solenoid sticks in the "brake off" position ("8"), or the control unit erroneously disengages the brake ("9"). There are two causes for the motor to turn freely: either there is no power to the motor ("10"), or the motor has failed ("11"). These events and their causes are diagrammed in Figure 6.5.

Note that each of the three causes of lack of braking is considered sufficiently analyzed (at least for purposes of this simple example) that countermeasures may reasonably be applied to them. In a real FTA for this system, it

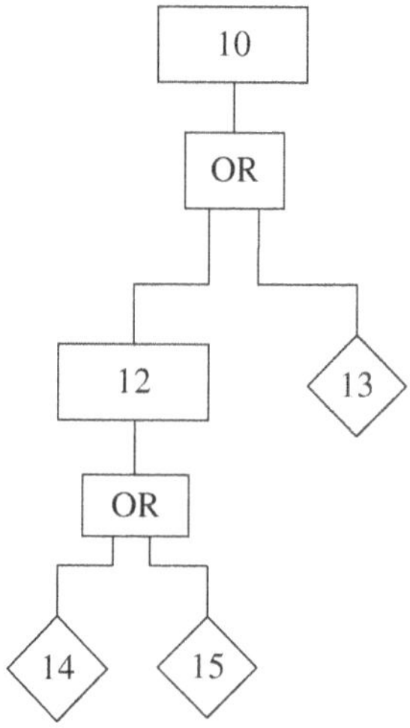

Figure 6.6 Sub fault tree for "no power to motor" event.

would be necessary to identify further root causes for event 7, such as improper preventive maintenance allowing the brake linings to wear beyond a safe point.

Loss of power to the motor may be caused by either the controller erroneously turning off power to the motor ("12") or by a total loss of system power ("13"). The erroneous behavior of the controller may be caused by a hardware failure ("14") or a software failure ("15"). These events and their causes are shown in Figure 6.6. Again, in a realistic FTA, the events 14 and 15 would most likely be analyzed further because "hardware failure" and "software failure" are not specific enough to be able to apply reasonable countermeasures.

We may include estimates of the probabilities for each of the elementary events (those in the diamond-shaped boxes) and use the calculus of probabilities to obtain an estimate of the probability of the top event. The elementary events are 3, 4, 7, 8, 9, 11, 13, 14, and 15. Let p_i denote the probabilities of the events in the tree, $i = 1, \ldots, 15$. Then the probabilities of the events in the tree are as follows:

- $p_{12} = p_{14}p_{15}$
- $p_{10} = p_{12} + p_{13}$
- $p_6 = p_{10} + p_{11}$
- $p_5 = p_7 + p_8 + p_9$
- $p_2 = p_5 p_6$
- $p_1 = p_2 + p_3 + p_4$.

The probability of the top event, written in terms only of the probabilities of the elementary events, is, finally, $p_1 = (p_7 + p_8 + p_9)(p_{14}p_{15} + p_{13} + p_{11}) + p_3 + p_4$. Of course, to write the probability of the top event (and the intermediate events) in this way, we need to assume that all events entering an "and" gate are stochastically independent, and all events entering an "or" gate are disjoint. In general, this will not be true and evaluation of the probability of the top event cannot be done with the ordinary calculus of probabilities. A useful alternative is provided by the method of cuts.

6.6.1.3 Cuts and minimal cuts in fault trees

Calculating the probability of the top event in a fault tree is often more complicated than shown in the earlier example because the same event may appear in different branches of the tree. This is a reflection of the fact that the same event may be the cause of several different effects. For example, consider the following modification of the fault tree in Figure 6.6.

Here, the top event can be caused by either event A or event B. C and D act together to cause event A, and C and E act together to cause event B. Let the probabilities of these events be $p_A, \ldots, p_E$. Then the probability of the top event is

$$\begin{aligned}
P(\mathrm{A} \cup \mathrm{B}) &= P(\mathrm{A}) + P(\mathrm{B}) - P(\mathrm{A} \cap \mathrm{B}) \\
&= P(\mathrm{C} \cap \mathrm{D}) + P(\mathrm{C} \cap \mathrm{E}) - P(\mathrm{C} \cap \mathrm{D} \cap \mathrm{C} \cap \mathrm{E}) \\
&= P(\mathrm{C} \cap \mathrm{D}) + P(\mathrm{C} \cap \mathrm{E}) - P(\mathrm{C} \cap \mathrm{D} \cap \mathrm{E}) \\
&= p_C p_D + p_C p_E - p_C p_D p_E = p_C (p_D + p_E - p_D p_E) \\
&= P(\mathrm{C})[P(\mathrm{D} \cup \mathrm{E})].
\end{aligned}$$

Notice how the repeated appearance of event C requires reduction of the intersection in the second line. This phenomenon appears frequently in realistic fault trees. Most realistic fault trees are more complicated than this example, so a simpler means of computation is desirable.

The use of paths, minimal paths, cuts, and minimal cuts to model system reliability was introduced in Section 3.4.7. The key idea was that for reliability block diagrams, especially those not having a series-parallel structure, another approach to developing an expression for the system reliability is offered by the method of paths and cuts. The cut technique is well adapted to the computation of the top event probability in a fault tree. To illustrate the use of cut-set methods in fault trees, we will list the cuts and minimal cuts in the passenger elevator fault tree example in Section 6.6.1.2.

Recall that a cut in a graph is a set of nodes and links whose removal from the graph causes the graph to become disconnected. In the fault tree case, the links serve only as connectors and may be disregarded. Two useful rules for cuts in fault trees are

- if some events are connected to a higher event by an AND gate, only one of those events need be part of a cut containing the higher level event and
- if some events are connected to a higher event by an OR gate, all those events need to be part of a cut containing the higher level event.

For example, in Figure 6.7, the cuts are {A, C, B, C}, {A, C, B, E}, {A, D, B, C}, {A, D, B, E}, {A, C, E}, {A, D, E}, {B, C, D}, {B, D, E}, {C, D}, {C, E}, {D, E}, and {C}.

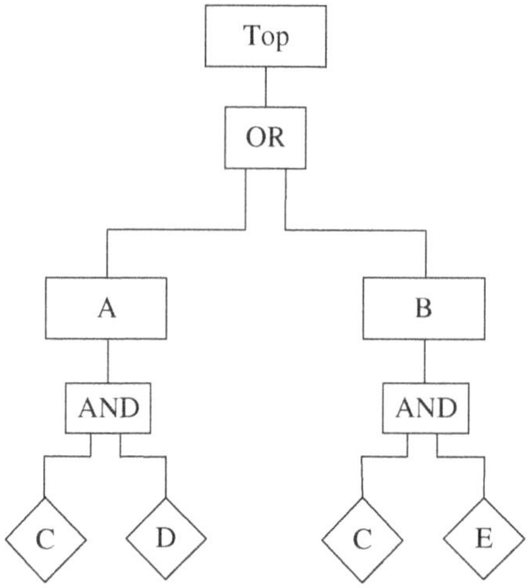

Figure 6.7 *Fault tree illustration with a repeated event.*

The minimal cuts are {C} and {D, E}, so the minimal cut set is {{C}, {D, E}}. Following the reasoning of Section 3.4.7, the probability that the fault tree fails to function (i.e., the top event does NOT occur) is given by the probability of the tree's minimal cut set. Repeating: in the case of a fault tree, "the system (the fault tree) fails" means "the top event does not occur" because the "system" in this case is "successful occurrence of the top event." As in Chapter 3, we (ab) use the same letter for the event and for the indicator that the event occurs, so the probability that the top event does not occur is $P(\{C=0\}$ or $\{D=0$ and $E=0\})$. Using $p_C = 1 - P\{C=0\} = 1 - P(\sim C)$, etc., we obtain

$$
\begin{aligned}
P(\sim \text{Top}) &= P(\sim \text{C} \cup (\sim \text{D} \cap \sim \text{E})) = P(\sim \text{C} \cup \sim (\text{D} \cup \text{E})) = P \sim (\text{C} \cap (\text{D} \cup \text{E})) \\
&= 1 - P(\text{C} \cap (\text{D} \cup \text{E})) = 1 - p_{\text{C}}(p_{\text{D}} + p_{\text{E}} - p_{\text{D}}p_{\text{E}})
\end{aligned}
$$

which is the same as the expression following Figure 6.7. Note the confusion that can easily occur here. The top event is one of the failure modes of the original system, that is, a system failure. Using the cut-set reasoning, our target is the top event NOT occurring, that is, the "system" represented by the fault tree fails. In other words, "failure of the fault tree 'system'" means "the top event does not occur," which is equivalent to saying "the (original) system[7] does not fail." Fortunately, these difficulties occur mainly when constructing a fault tree by hand and are obviated by the availability of many software packages for fault tree construction and analysis (one of the earliest of these is [40]; since then, many professional fault tree packages have appeared in the commercial marketplace).

[7] The system for which the fault tree was constructed.

We close with the observation that **the minimal cut set is the catalog of actions that can be taken to forestall or prevent the top event (system failure when so defined) from occurring.** For example, in Figure 6.7, if you can prevent C from happening, or if you can prevent both D and E from happening, then the top event does not occur. In this way, a useful by-product of fault tree construction is a simple procedure for cataloguing countermeasures.

6.6.1.4 Application: system reliability budget

We began a discussion of reliability budgeting in Section 4.7.3 in which a solution based on mathematical optimization was proposed. Sometimes, it may not be possible or desirable to develop a formal reliability budget using the optimization ideas shown there. But it is still a best practice to create a reliability budget so that all development teams know what reliability targets they need to reach. For a more informal approach to reliability budgeting, consider that an FTA can be a viable approach. To do this, the elements of the fault tree need to correspond to the items that will appear in the reliability budget. A fault tree approach provides additional flexibility in that the elements of the fault tree need not be subassemblies or subsystems only, but may also be events of significance to system reliability. To use this approach, assign probabilities (or life distributions, availabilities, etc., as appropriate) to each element of the tree and use the standard fault tree computation to determine the appropriate reliability effectiveness criterion or figure of merit for the system. If the tree is small enough, trial and error may work well. For larger trees, one would need to link the output of the fault tree construction software to the input of a budgeting optimization routine.

6.6.2 Failure Modes, Effects, and Criticality Analysis

FTA provides a systematic way of discovering the causes of system failures. Its premise is that, starting with a list of system failure modes, which comes from reviewing the system's attribute requirements and cataloguing the different ways they can be violated, reasoning can be applied to discern the intermediate and root causes of these system failures. The results are displayed in a tree diagram with the system failure as the top (root) node and the intermediate and root causes branching from that, with the links indicating causality relationships. FMEA, by contrast, provides a systematic way of discovering the consequences of failures of parts of the system. It begins by postulating failures of parts of the system and determines the consequences of those failures in a sequence whose terminus is a system failure. So in a sense, FMEA is an opposite, or complementary, reasoning process to that of FTA. The most common graphical display of FMEA results is as a table or spreadsheet as illustrated in the next section.

> **Language tip:** Note that the use of "failure" in FMEA is different from that in FTA. In FTA, "failure" refers to system failure, violation of a system

requirement. FMEA, by contrast, concerns failures of components of a system and how the effects of these failures propagate throughout the system to cause (one or more) system failure(s).

Language tip: In this book, we will use the acronym FMECA to mean FMEA with the criticality component, FMEA to mean failure modes and effects analysis specifically excluding the criticality component, and FME(C)A to indicate either FMEA or FMECA generically.

Requirements tip: To know what "failure" means for system components, we need to know the requirements for those components. For simple components like resistors, bearings, and the like, these requirements are usually limited to operation of the component within its specified parameters. For example, a resistor may be required to have a resistance value of between 950 and 1050 Ω and to be able to dissipate 0.25 W of power. A resistor with these parameter values is selected for use in a system if these values allow it to perform its circuit function(s) in the system. Usually, no additional requirements for isolated, single components like these are imposed. FME(C)A asks what are the consequences for circuit operation if the resistor no longer meets these specifications. More complex subassemblies may have additional or more functional requirements; in the FME(C)A context, failure of a subassembly means violation of one or more of these.

6.6.2.1 Failure modes and effects analysis

We begin by examining FMEA, a purely qualitative technique. We discuss two types of FMEA: concept FMEA which is suitable for use early in the design of a system, before any specific hardware or software have been specified, and design FMEA which captures the additional detail that is beneficial when the design is more complete. Section 6.6.2.2 adds a quantitative dimension to FMEA by incorporating the notion of criticality as a way of ranking the importance of the failures studied.

Concept FMEA

Concept FMEA is useful first at the stage of a design where it is possible to identify a system functional decomposition and the major design elements that will fulfill the identified functions. The system functional decomposition is a good place to begin a concept FMEA because constituent elements of the system are identified in the functional decomposition and the FMEA reasoning process can begin with these elements. The FMEA reasoning is a process of inquiring for the consequences of failures, and the consequences of those consequences, etc., continuing until a system failure is identified as a final consequence of a chain of failures beginning with the system's components or subassemblies. Concept FMEA begins by postulating failure of one of the system's functional elements and listing the consequences of that failure. Each of those consequences, in turn is subjected to the same reasoning. The effect of a failure of an element of the system could be the cause of failure of another element of the system, etc. The process stops when a chain of consequences, or failures, ends at a system failure. The design for reliability aspect is that preventing the first element failure in the chain prevents the others, and the consequent system failure.

Example: Concept FMEA for a home alarm system. A home alarm system typically includes these features:

- *Intrusion detection,*
- *Carbon monoxide,*
- *Smoke detection,*
- *Alerting of a central monitoring facility when an alarm event (unauthorized access, excess CO, smoke/fire, etc.) takes place,*
- *Battery backup during periods of AC power failure, and*
- *Facility for user-initiated testing.*

It contains a sensor for each location where access is possible (usually doors and first-floor windows), carbon monoxide and smoke detectors, connection to a telephone line, a local CPU, and a control panel. A functional decomposition for a (generic) home alarm system could be developed (see Figure 6.8).

AD-1 through AD-*n* are the access detection switches. The system can be in any of three states at any time: armed, ready-to-arm, and failed. Concept FMEA can begin with the blocks in the functional decomposition. Table 6.2 aids in completing a concept FMEA for this example.

Obviously, this example is far from complete. Other failure modes are not included; some failure causes are not specific, etc. The purpose of this example is not to provide a complete or generic concept FMEA for a home alarm system. Rather, it is to illustrate the reasoning process that is used to complete the

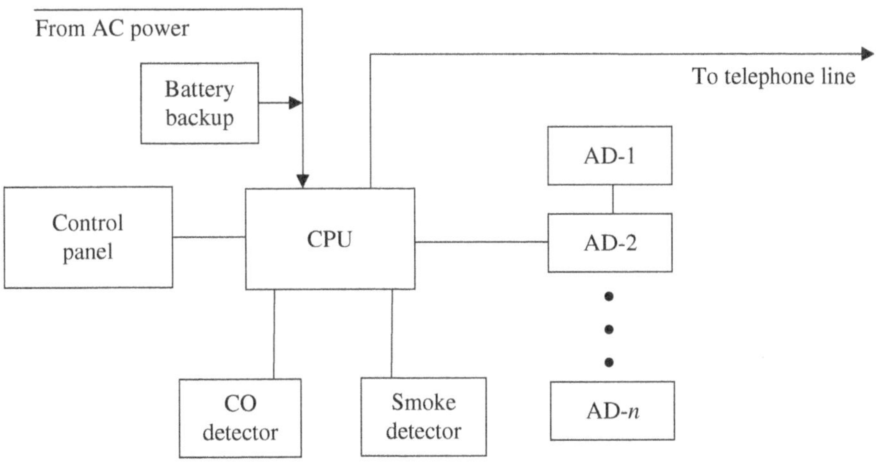

Figure 6.8 Home alarm system functional decomposition.

TABLE 6.2 Concept FMEA Table

Item	Function	Failure Mode	Failure Cause(s)	Effect of Failure
1	CPU	PWB failure	Component failure	System totally inoperative
			Solder attachment failure	
			Power surge	
2	Telephone dialer	Does not detect dial tone	Loose or broken telephone line connection	Cannot dial out to alarm HQ
		Does not generate DTMF tone(s)	Solder attachment failure	
			Component failure	
3	Access detector	False positive	Defective switch (short)	False intrusion alarm
		False negative	Defective or dirty switch (open)	Alarm event missed
4	CO detector	False positive	Defective component, CO sensor element, or solder attachment, or other circuit failure	False CO alarm
		False negative		Missed CO event, possible loss of life
5	Smoke detector	False positive	Defective component, smoke sensor element, or solder attachment, or other circuit failure	False fire alarm
		False negative		Missed fire event, possible property damage or loss of life
6	Control panel	Does not respond to inputs	Defective keypad	Alarm system inoperative
		Illumination failure	LED failure	Difficult to use in the dark
7	Power	No power to system	AC power and backup battery failures	System totally inoperative

table. The completeness of the system functional decomposition determines to a great extent the quality of a concept FMEA. It is good practice to include each block in the system functional decomposition in a concept FMEA.

Design FMEA

The distinction between concept FMEA and design FMEA is that design FMEA can be more detailed, and therefore more specifically helpful for design for reliability, because more of the design is committed. Specific hardware and software have been identified that will perform the functions in the system. In other words, the functional blocks in the system functional decomposition are fleshed out with specific hardware and software to perform the functions. Design FMEA can begin with more specific system component information and so promotes more effective design for reliability. Specific design for reliability decisions can

TABLE 6.3 Design FMEA Example

Item	Function	Failure Mode	Failure Cause(s)	Effect of Failure
1	Connector	Open	Corrosion, improper assembly, not correctly seated	Excess CO may not be alarmed, possible loss of life
		Short	Improper assembly	CPU sees CO detector as defective
2	CO sensor element	Open	Broken lead due to improper assembly	Cannot detect excess CO, possible loss of life
		Short	Manufacturing defect	False CO alarm
		Oversensitive		
		Undersensitive	Accumulation of dirt, corrosion, manufacturing defect	Cannot detect excess CO, possible loss of life
3	CO detector circuit	False positive	Component or solder attachment failure	False CO alarm
		False negative		Excess CO may not be alarmed, possible loss of life
4	Wiring to CPU	Open	Improper initial assembly, damage from vermin	Missed CO event, possible loss of life
		Short	Defective insulation	CPU sees CO detector as defective

be taken on the basis of design FMEA. For instance, the functional decomposition for a domestic forced-air heating system will contain a block for "air impeller." A concept FMEA explores the consequences of air impeller failure. Design FMEA can be undertaken when a specific blower motor and fan model is chosen; the failure characteristics of that motor and fan model add understanding to the concept FMEA and allow the system designer to decide whether this specific blower motor model is appropriate for use in the system.

Table 6.3 illustrates a design FMEA for the CO detector in the home alarm system discussed in Section "Concept FMEA."

Again, this design FMEA is incomplete. It would normally be executed for a specific CO detector circuit and failure causes could therefore be more specifically identified. Circuit simulation would provide an indication of how the distribution of sourced component values and drift of component values over time may affect operation of the detector when new and as it ages. Specific countermeasures could then be devised and their economics evaluated. This is the value of design FMEA in design for reliability.

Design FMEA is related to robust design (Section 6.8). Some of the same tools are used. For example, circuit simulation is used not only to determine requirements for individual circuit elements but also to help understand how operation of the circuit outside its specification limits may contribute to failures further up the design hierarchy.

6.6.2.2 *Failure modes, effects, and criticality analysis*

FMECA adds a quantitative component, *criticality*, to FMEA. Criticality is a number attached to each outcome of the chain of causality. Criticality, also called *risk priority number* (RPN), is arrived at by multiplying three numbers:

- Probability of occurrence of the failure,
- Severity of the failure, and
- Probability that the failure will affect the customer/user.

Unless a detailed reliability model for the system exists, the probability of occurrence of the failure is subjectively assessed. A 1–10 scale is commonly used in accordance with the Table 6.4.

Table 6.4 provides an illustration of a possible FMECA probability scale. The values and criteria shown are not universal; other probability scales are sometimes used. The scale is somewhat arbitrary and other choices also make sense because FMECA aims to provide a relative ranking of failures rather than an absolute measurement of failure impact. At the stage of development at which FMECA is normally used, it is usually not possible to be very precise about these probabilities. A good deal of subjective judgment, based on experience, is used. The absolute values of the criticalities or risk priority numbers are not, and are not intended to be, reliable. The relative ranking of these is used to help assess which failures should receive attention first. The output of an FMECA is displayed as a Pareto chart (ordered bar chart) with the risk priority numbers in descending order from left to right. It is possible to tell at a glance the priority in which potential failures should be addressed.

TABLE 6.4 Example of a FME(C)A Probability Scale

Scale Quantity (Rank)	Qualitative Criteria	Failure Occurrence per 100,000 Units	Failure Occurrence per Year
1	Remote possibility of occurrence Unreasonable to expect that failure would occur	Negligible	
2	Relatively low likelihood of occurrence. Failure would be surprising	5	1/3
3		10	1 (annually)
4	Moderate likelihood of failure; occasional failure but not in major numbers	50	2 (biannually)
5		100	4 (quarterly)
6		500	12 (monthly)
7	High likelihood of failure; comparable to products/processes that have caused problems before	1,000	52 (weekly)
8		5,000	365 (daily)
9	Very high likelihood of failure, almost certain that many failures will occur	10,000	10,000 (hourly)
10		50,000	40,000

TABLE 6.5 Example of a Severity Scale

<table>
<tr><th>Rank</th><th>Qualitative Description</th></tr>
<tr><td>1</td><td rowspan="3">Minor—A failure not serious enough to cause injury, property damage, or system damage, but which will result in unscheduled maintenance or repair</td></tr>
<tr><td>2</td></tr>
<tr><td>3</td></tr>
<tr><td>4</td><td rowspan="3">Marginal—A failure may cause minor injury, minor property damage, or minor system damage which will result in delay or loss of availability or mission degradation</td></tr>
<tr><td>5</td></tr>
<tr><td>6</td></tr>
<tr><td>7</td><td rowspan="2">Critical—A failure can cause severe injury, major property damage, or major system damage which will result in mission loss</td></tr>
<tr><td>8</td></tr>
<tr><td>9</td><td rowspan="2">Catastrophic—A failure can cause death or complete system loss</td></tr>
<tr><td>10</td></tr>
</table>

TABLE 6.6 Sample Scale for Probability of Affecting Users

<table>
<tr><th>Rank</th><th>Qualitative Description</th><th>Probability of Affecting Users</th></tr>
<tr><td>1</td><td>Remote likelihood that the defect will not be detected before occurrence—it will not affect the user.</td><td>0–0.05</td></tr>
<tr><td>2</td><td rowspan="2">Low likelihood that the defect will not be detected before occurrence—it will probably not affect the user.</td><td>0.06–0.15</td></tr>
<tr><td>3</td><td>0.16–0.25</td></tr>
<tr><td>4</td><td rowspan="3">Moderate likelihood that the defect will not be detected before occurrence—it may affect the user.</td><td>0.26–0.35</td></tr>
<tr><td>5</td><td>0.36–0.45</td></tr>
<tr><td>6</td><td>0.46–0.55</td></tr>
<tr><td>7</td><td rowspan="2">High likelihood that the defect will not be detected before occurrence–it probably will affect the user.</td><td>0.56–0.65</td></tr>
<tr><td>8</td><td>0.66–0.75</td></tr>
<tr><td>9</td><td rowspan="2">Very high likelihood that the defect will not be detected before occurrence—it will affect the user.</td><td>0.76–0.85</td></tr>
<tr><td>10</td><td>0.86–1</td></tr>
</table>

The second component of the risk priority number is the severity of the failure. A scale is imposed on a set of qualitative criteria so that a risk priority number can be developed. Table 6.5 gives an example of a possible severity scale.

Again, the scale shown is arbitrary, and other similarly useful scales are in common use.

Finally, the third component of the risk priority number is the probability that the failure will be detected before it affects the system user or customer. The reasoning here is that a defect can be detected and prevented from affecting system users should rate lower on a risk priority scale than one that is not prevented and does affect system users. Accordingly, Table 6.6 illustrates a sample user impact scale.

Sometimes, it is helpful to think of this scale in reverse terms, that is, as a scale measuring how readily the effect may be prevented. Effects for which there is a readily available preventive measure receive lower ranks. Effects for which preventive measures are not available or difficult or expensive to implement receive higher ranks.

The risk priority number resulting from use of these scales is a number from 1 to 1000, the product of the three rank numbers from each of the three factors. The RPN has no absolute meaning, but RPNs provide a relative ranking of the importance of each of the defects considered in the FMECA. The Pareto chart is an effective way to display RPN information, allowing users to discern at a glance the most critical defects.

The fundamental structure of FMECA lends itself readily to incorporation in spreadsheet software. FMEA tables are augmented with four additional columns in which the criticality information is displayed. The first three columns contain the rank information from Tables 6.4 to 6.6, and the fourth column contains the RPN, the product of these three numbers.

Other FME(C)A applications

FME(C)A is also effective as a design for reliability tool for less-tangible objects like processes, services, and software. For processes, replace the system functional decomposition with a process flow diagram so that the downstream effects of improper operation of any process step can be determined. For services, the service functional decomposition (Section 3.4.2.3 gives an example; see also Section 8.3) serves the same purpose. An FME(C)A for software is facilitated by data flow and control flow diagrams [13].

6.6.2.3 Summary

FME(C)A and FTA are complementary design for reliability tools. At any block in the system functional decomposition, it is possible to start an FME(C)A or an FTA as long as requirements for that functional block exist. FTA will look from a particular functional block down the design hierarchy, into the supporting blocks, inquiring how failures in those supporting blocks may cause failure in the study block. FME(C)A will look from a particular functional block up the design hierarchy, into the blocks it supports, inquiring how failures in the study block may cause failures in functional blocks it supports. Both points of view are useful. Limited resources may force a choice because of lack of staff or time to carry out both procedures. One important factor influencing the choice is that FTA requires more detailed information about the design than does FME(C)A. A concept FME(C)A can begin earlier in the design process because only the system functional decomposition is needed. To be useful, FTA needs more-detailed information about components and subassemblies of the system and how they operate so that cause and effect may be determined in specific, rather than general, terms.

Many resources are available to help reliability engineers and associated staff carry out FME(C)A studies. Standards include MIL-STD-1629A for the defense industry, SAE J-1739 for the automotive industry, IEC 60812 for the electronics industry, and others. Even though FME(C)A is easily enough automated with simple spreadsheet software, many dedicated software resources for FME(C)A are available in the commercial marketplace.

6.7 DESIGN FOR RELIABILITY FOR SOFTWARE PRODUCTS

Almost every contemporary technological system contains software—lots of software. Many notable failures are caused by software problems. In the past, the primary means for assuring reliability of software-intensive products was through a test, analyze, and fix (TAAF) cycle based on a sequence of development releases that underwent testing and removal of any errors found in that testing. This procedure is depicted schematically in section 6 of Ref. 18 and is the basis for many of the software failure intensity models reviewed in Chapter 9. More recently, it has been recognized that design for reliability can be a more effective means of assuring reliability even in software-intensive products [4, 18]. So the usual procedures we use to design for reliability need to be adapted to the characteristics of software products. In particular, this requires understanding of the failure modes and failure mechanisms peculiar to software products. Section 9.5 discusses failure modes and failure mechanisms for software.

Most system reliability requirements are written in terms of behaviors that are discernible to the user (see Section 9.5.1). As such, it would be unusual for there to be reliability requirements at this level specifically for software. What usually happens is that software contributes one or more causative events at the second level or below of a fault tree. When these cause-and-effect relationships are discerned, system reliability requirements can be allocated to its software component(s). See Section 9.3.1.

Finally, it is important to note that safety and security often come to the fore in discussions of software products. These do not look overtly like reliability issues, but the point of view we adopt in this book is that if a requirement exists for a system, a violation of that requirement is a failure. Even if safety and security requirements are not explicit (and it is a bad idea to leave them as tacit requirements only), most users and customers would regard a system operation that causes injury or loss of life to be a failure. The key idea is that issues like safety and security, even if they do not have explicit requirements, can be treated using the reliability engineering and design for reliability technologies: identify the safety or security failure modes and subject them to the design for reliability process (Section 6.4), use FTA and FME(C)A (Section 6.6) to discover the relevant failure mechanisms, and devise countermeasures tailored to these failure mechanisms. We repeat a point we made in Chapter 1, which is that while the consequences of safety (or security) failures may be more serious than those of other system failures, they may be treated by the same DFR and reliability engineering techniques that we would apply to any other system requirement. In software-intensive products, the constituent materials may be different, but the basic reliability ideas are the same.

6.8 ROBUST DESIGN

The design process yields an "ideal" design in the sense that all design elements (architecture, components, etc.) are abstract until realized in physical space. Any such realization differs from the "ideal" design because of random factors such as component values differing from those specified, drift of component values over time, operating environments differing from those anticipated, and the like. When these random factors are at play, properties of the realized design (and there may be various copies, at least in the case of hardware) differ to some extent from corresponding properties of the "ideal" design. One would like to know in advance whether these potential differences could be significant. The discipline of *robust design* has grown up to, among other benefits, provide a means for the development team to anticipate the variation in realized versions of the ideal design and adopt corrective measures if necessary. The robust design technology was introduced by Taguchi [37] as a means of improving quality while reducing cost. In this section we will discuss the robust design technology as a reliability-improving activity that prevents failures while avoiding overdesign and decreasing cost.

One of the hallmarks of robust design is the categorization of the factors affecting the operation of the system as *control factors* and *noise factors*. Control factors are those that the system development team and system users are able to manipulate to achieve desired effects in a deterministic (or, at least, statistically predictable) fashion. These include design parameter nominal values, material composition, etc. Noise factors are those over which the development team and users have little or no control. They are usually represented as random influences that come from the environment, from the manufacturing process, from variation in design parameters in realization, or from other sources. For instance, while the designer may specify a range of values for a certain component, a physically realized component may still take a random value within (or outside of) that range. Robust design aims to adjust the control factors so that the response of the design to noise factors is minimal, or damped. For instance, a circuit may be designed to have a certain performance. When the circuit is physically realized as a chip or as a PWB with components, the random variation of the chip manufacturing process or of the values of the components on the PWB may cause the performance of the realized circuit to vary from its nominal target.

Robust design is specifically a quality improvement technique. It is not the purpose of this book to conduct a thorough review of the robust design technology. However, it is reasonable to ask whether a design that is robust (little or no response to noise factors) is also reliable, or, to put it another way, can robust design be used as a reliability improvement technique also? The purpose of this section is to show how the notion of drift of design attributes over time unifies robust design and reliability engineering. Paying appropriate attention to these drifts is an important aspect of design for reliability. In brief, the aim of robust design as a design for reliability technique is to maintain a suitably small number of component failures stemming from drift outside specification limits by anticipating the total amount of that drift at the end of the system's service life and ensuring that this amount of drift can be tolerated by the system [9, 19, 20, 39].

We postulate that certain system attributes drift from their nominal values as time passes. To fix ideas, consider a simple electronic component like a 1000-Ω resistor. Two important ideas are at play here:

1. Not every resistor in the population of (nominally) 1000-Ω resistors is exactly 1000 Ω. Rather, the value of a resistor drawn at random from that population has a distribution, usually taken to be approximately normal, with mean 1000 Ω and some standard deviation σ Ω [12].
2. As time passes, the values of the resistors in the population change gradually, or drift, so that after the passage of some time the population distribution may retain its original form (i.e., may still be approximately normal) but with a different mean and standard deviation—indeed, in this formalism these are functions of time.

Electronic components, like these resistors, also usually specify a tolerance (1%, 5%, etc.). These tolerance values usually indicate that at least six standard deviations of the component's value fit within the tolerance percentage. For instance, for a 1000-Ω, 5%-tolerance resistor, this would indicate that $6\sigma \leq 100$ (=1050 − 950), so we may infer that $\sigma \leq 16.6$ Ω for this population. For higher tolerance components, such as 0.1%, components are sometimes specially selected to fit within the tolerance interval (for 0.1%, this is 999–1001 Ω) and the resulting population is no longer normally distributed. See Exercise 1.

In light of these ideas, the design for reliability actions using robust design are twofold:

1. The response (performance) of the circuit at time of manufacture should be evaluated using the distributions of the values of all the circuit elements. This will help determine whether the response of the circuit to this noise factor (e. g., the varying resistance value across the population of resistors) can be satisfactorily damped when the circuit is new. For electronic circuits, this evaluation can be accomplished with a circuit simulation program like SPICE.
2. The response of the circuit at future times should be evaluated using knowledge about the drift of the resistance value with time. This will give an idea of how many circuits will have failed by the end of the system's service life.

For example, consider that a circuit using this resistor will perform satisfactorily if the value of the resistor is between 950 and 1050 Ω. If the population of new resistors is normally distributed with a mean of 1000 Ω and a standard deviation of 30 Ω, then approximately 9.5% of the circuits built will not perform satisfactorily (clearly this is not a realistic example). This situation is illustrated in Figure 6.9.

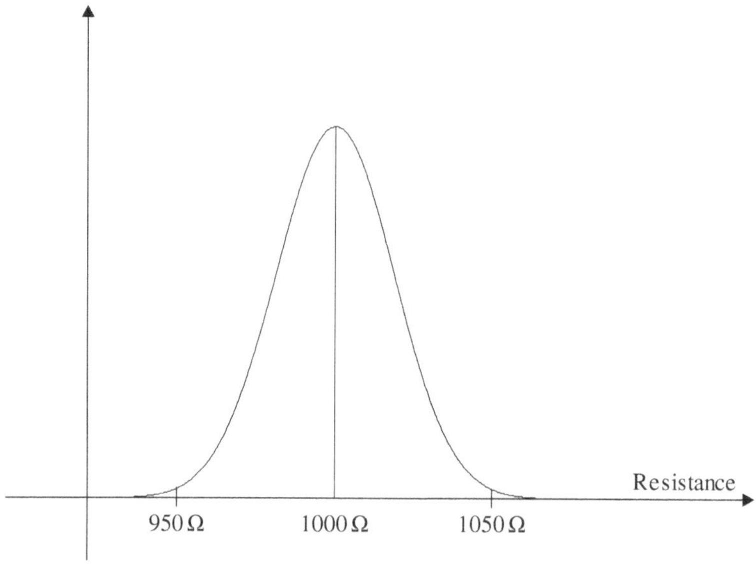

Figure 6.9 *Normal density of resistor values in new population.*

To illustrate the notion of drift, suppose that the mean resistor value increases (linearly) at a rate of 10 Ω per year and the standard deviation increases (linearly) at 2 Ω per year. Then after t years, the resistor population has a mean of $1000+10t$ and standard deviation $30+2t$. After 10 years, the population mean is 1100 Ω and standard deviation is 50 Ω. At that time, 16% of the circuits do not operate satisfactorily. The drift of the resistor values has caused additional circuits to fail.

We turn to a general framework for solving design for reliability problems using the robust design method. The key point is that this technology is appropriate for dealing with failures due to design characteristics drifting over time outside of an acceptable range. To promote consistent use of terminology, we will call the objects in the population "items." These may be electronic or mechanical components, or subassemblies or entire systems. An item may have one or more failure modes, but at least one of these is of the nature of a drift of a characteristic that, when the characteristic changes enough, causes a failure of the system of which the item is a part. An item is described by one or more characteristics. For example, a resistor is described by its resistance, power dissipation rating, material content, physical size, mounting (leaded or surface-mount), etc. Let $X(t, \omega)$ denote for item ω at time t the vector of characteristics whose drift we want to focus on. The notation explicitly allows for this vector to depend on time. Denote by $F(x, t)$ the cumulative distribution function (cdf) of $X(t)$, that is, $F(x, t)=P\{X(t) \leq x\}$, and by $f(x, t)$ the corresponding density function. Suppose that the dimension of X is d and that there is a region $A \subset \mathbf{R}^d$ that represents the region of acceptable operation for these items.

The region A is usually determined by external criteria. For example, a pump may be considered failed when it can no longer pump as much as 60 gph, where the 60 gph requirement comes from the application of the pump in some larger context. Then the proportion of the population that is in the working state (not failed) at time t is

$$\int_A f(x,t)\,dx.$$

In particular, if we let $L(\omega)$ denote the lifetime of item ω, that is, the time it takes for item ω's characteristics to drift outside of the region of acceptable operation, then

$$P\{L > s\} = \eta\left(\int_A f(x,s)\,dx\right),$$

where η represents the lower envelope function

$$\eta(g(s)) = \begin{cases} g(s) \text{ if } g \text{ is decreasing at } s \\ g(\bar{s}) \text{ if } g \text{ is increasing at } s \end{cases}$$

where $\bar{s}$ is the nearest point to the left of s at which g is decreasing,[8] or

$$\bar{s} = \max\{u : u \le s \text{ and } g \text{ is decreasing at } u\}.$$

Alternatively, $\eta(g) = \max\{h : h \text{ is increasing and } h \le g\}$. The use of the lower envelope function means that we do not allow an item whose characteristics have drifted into A to "recover," or once an item has failed in this fashion, it remains failed. This is consistent with the way a system failure is treated: if drift of an item's characteristics causes a system failure, remediation of the failure usually begins immediately without waiting to see whether the system heals itself by virtue of the item's characteristics drifting back into the acceptable region.

Example: Suppose a population of items has a (one-dimensional) characteristic $R(t)$ that has, for each t, a normal distribution with mean $0.2 + 0.002t$ and standard deviation $0.03 + 0.001t$, where t is measured in years. Then we can write $R(t) = 0.2 + 0.002t + (0.03 + 0.001t)Z$ where Z is a random variable having a standard normal distribution. For definiteness, it may help to think of $R(t)$ as the rise times, measured in nanoseconds, of a population of digital integrated circuits. Suppose the failure criterion is that an item is failed if its rise time exceeds 4 ns. Then the lifetime $L(\omega)$ of item ω is

$$L(\omega) = \min\{t : R(t,\omega) > 0.4\}.$$

[8] Throughout, by "decreasing" we mean "nonincreasing" and by "increasing" we mean "nondecreasing."

The survivor function of L is

$$
\begin{aligned}
P\{L>t\} &= P\left\{\sup_{0\le s\le t} R(s)\le 0.4\right\} = P\{R(s)\le 0.4, 0\le s\le t\} \\
&= P\{30Z+(2+Z)s\le 200, 0\le s\le t\} \\
&= P\{30Z+(2+Z)s\le 200, 0\le s\le t, 2+Z\le 0\} \\
&\quad + P\{30Z+(2+Z)s\le 200, 0\le s\le t, 2+Z>0\} \\
&= P\{Z\le -2\} + P\{30Z+(2+Z)t\le 200 \mid Z>-2\}\,P\{Z>-2\} \\
&= 0.023+0.977\Phi\left(\frac{200-2t}{30+t}\right).
\end{aligned}
$$

This expression allows us to obtain the value of the life distribution for this population of items at any time t (measured in years). For instance, the probability that the lifetime of an item from this population exceeds 20 years is $0.023+0.977\Phi(3.2)=0.023+0.997(0.9993)=0.99932$.

Although this example is contrived, it does help point out some useful ideas in the application of the robust design technology to design for reliability. First, distributions of population characteristics can spread (or shrink), as well as shift, as time passes. Second, life distributions derived from this formalism can be defective. For instance, in the example, as $t\to\infty$, 4.5% of the population is still alive. Also, $P\{L=0\}=0.977[1-\Phi(20/3)]$ which is slightly positive.

The robust design technology may be applied more widely in design for reliability than in only the parameter drift scenario. Operational requirements for engineering systems are often written in terms of certain performance parameters such as, for example, the power output of an RF amplifier or the propagation delay across a digital circuit. The value of the performance parameter in any physical realization of the design is influenced by many noise factors. Those that are important for purposes of this engineering technique are

- unit-to-unit variation in system construction,
- unit-to-unit variation in performance of a system's components and subassemblies,
- drift of physical, electrical, or chemical characteristics of items with time, and
- variation in the environmental conditions in which the systems operate.

In this technology, we postulate that satisfactory system operation requires that the system's performance parameters remain within some acceptable set of values that we will continue to call A. We consider those failure modes in which performance parameters falling outside this set means that the system has failed. The time of first occurrence of this event is the *failure time* and the period of satisfactory operation between commencement of system operation and the failure time is the *lifetime*. We now introduce a formalism that may be used to solve design for reliability and robust design problems of this nature that may incorporate any or all of the four factors listed earlier.

Let $X(t, u(t))$ denote the vector of system performance parameter values at time t when the environment (the collection of all extrinsic noise factors that influence system performance) is described by a function $u(t)$ of time. u may be modeled as a deterministic function or as a stochastic process. For simplicity, we will assume that there is only one parameter of interest ($d=1$) so $A \subset \mathbf{R}$, and is usually an interval $[a, b]$. The case $d>1$ develops similarly. The failure time is given by $L=\inf\{t : X(t, u(t))<a \text{ or } X(t, u(t))>b\}$. The environment could be a vector function. For example, the power output of an RF amplifier is affected by its ambient temperature and supply voltage. In this example, $u(t)=(u_1(t), u_2(t))$ where $u_1(t)$ is a temperature and $u_2(t)$ is a voltage. This formalism assumes there is no "environmental memory" in the sense that the value of the performance parameter at any particular time depends only on the value of the environment at that time and not on any previous values of the environment or on what path the environment variable may have taken to reach its value at that time. Further development of this model rests on assumptions about the distribution(s) of the noise factors included in the model. For instance, the earlier development in this section covered the third item in the earlier list, the drift of physical, electrical, or chemical characteristics of items with time. In many realistic studies, the distribution of the failure time L may be difficult to obtain analytically. In this case, simulation can be an effective approach.

6.9 DESIGN FOR RELIABILITY BEST PRACTICES FOR SYSTEMS ENGINEERS

Rarely will the systems engineer also be the engineer charged with carrying out design for reliability activities herself. However, the systems engineer is vitally concerned with the likelihood that the product's reliability requirements will be met when its development is complete. Consequently, he should put in place a program that will

- allow estimation of this likelihood throughout the design and development process so that appropriate feedback can be provided to design and development staff to guide improvements where necessary,
- conduct reliability testing where necessary and justifiable, and
- engage appropriate design for reliability practices (FTA, FMECA, etc.) at the right time in the development cycle.

6.9.1 Reliability Requirements

Good design for reliability cannot proceed without correct reliability requirements. Systems engineers should review all reliability requirements to ensure that they meet the criteria established in Chapter 2. All relevant reliability requirements should be explicitly stated. It is especially important to include reliability requirements for safety and security so that failures in these key areas are addressed and not overlooked.

6.9.2 Reliability Assessment

Make a quantitative reliability model as soon as a system functional description can be defined. Continually update the model as the design develops. Use the results of the modeling to improve the design.

6.9.3 Reliability Testing

Reserve reliability testing for those systems that use a lot of new or unproven technology and for which the cost of reliability testing can be justified by either the criticality of the system or the possibility of highly profitable sales (through volume or margin or both). Reliability testing for software-intensive products will be discussed in Chapter 9.

6.9.4 DFR Practices

Use FTA to systematically walk through all system requirements, extract the relevant failure modes, and look for root causes for which countermeasures can be deployed. Use FME(C)A to develop the consequences of user actions, environmental interactions, and potential system misuses so that a good understanding of the full range of system operation is obtained. Use the PWB DFR procedure whenever circuits are built up from discrete or integrated components. Note that this process also can be used for other types of circuits, such as those that are point-to-point wired, as long as the stress-reliability relationships are available.

Systems engineers always balance the cost of undertaking a design action with the benefits presumed to flow from that action. In many cases, cost precludes full implementation of FTA or FME(C)A techniques for all possible failure modes. However, even for consumer products that have a short service life before obsolescence, it is important to be able to establish that the product cannot be used in an unsafe manner or a manner that may cause property damage, injury, or loss of life. Auto manufacturers know well the costs of a vehicle recall to repair a design defect that may cause a fire, an accident, or other unsafe condition. FTA and FME(C)A focused on safety should always be part of a product development.

6.10 SOFTWARE RESOURCES

Time was when fault tree analyses and FME(C)A would have to be documented by hand on large sheets of paper. This is no longer necessary because many commercial and open-source software packages for carrying out these tasks are available. Most office suites contain a facility for drawing cause-and-effect (Ishikawa) diagrams.

6.11 CHAPTER SUMMARY

Design for reliability is the process of anticipating and managing failures by application of preventive measures within the economic constraints of the system development. The design for reliability process described in Section 6.4 is based on the stress–strength model. It takes the user through a systematic consideration of the stresses created by the electrical and mechanical designs and recommends actions to take that prevent those stresses from activating the susceptible failure mechanisms. In cases in which certain preventive actions are deliberately not taken, systems engineers are urged to quantitatively understand the consequences of this omission in terms of an increased number of system failures over what would have been had the preventive action been taken. The raw material supporting the design for reliability process is information, so it is vitally important to maintain a robust information repository comprising not only knowledge acquired from external sources such as supplier-provided material and the research literature but also from internally produced information such as lessons learned from prior experience with similar systems.

Two specific designs for reliability techniques are introduced in Section 6.6: FTA and FME(C)A. FTA is a deductive, top-down approach to determining the root cause(s) of system failures. It is supported by a simple graphical aid called the fault tree which depicts the causal relationships that allow us to connect simple events, for which countermeasures may be readily devised, to the complex event of system failure. FTA can be used quantitatively by attaching estimated probabilities to each of the elementary events in the tree and using the calculus of probabilities or cut set analysis to derive an estimate of the probability of system failure. FME(C)A is an inductive, bottom-up approach to determining the consequences of certain design choices. When the criticality analysis is omitted, the technique is called FMEA. FMEA is a purely qualitative tool that still has value in promoting a disciplined approach to uncovering failure modes and their effects, and helping organize countermeasures.

6.12 EXERCISES

1. Suppose that 1000-Ω resistors from a population having 10% tolerance are selected for 0.1% tolerance. What is the distribution of the values of the survivors of this selection? How much of the original population is discarded?

2. Show that a system whose life distribution has an increasing hazard rate and which is repaired according to the revival model has an increasing failure rate.
3. What can be said about the system in Exercise 2 if it is repaired according to the renewal model?
4. Consider the fault tree example for the passenger elevator in Section 6.6.1.2.
   - List the cuts in the fault tree.
   - What is the minimal cut set?
   - Write the probability of the top event in terms of the minimal cut set.

REFERENCES

1. Ascher H, Feingold H. *Repairable Systems Reliability: Modeling, Inference, Misconceptions, and Their Causes*. New York: Marcel Dekker; 1984.
2. L. J. Bain and M. Engelhardt (1991), *Statistical Analysis of Reliability and Life-Testing Models: Theory and Methods* (Vol. 115). Boca Raton, FL: CRC Press.
3. Barlow RE, Clarotti CE, Spizzichino F. *Reliability and Decision Making*. New York: Chapman and Hall; 1993.
4. Bauer E. *Design for Reliability: Information- and Computer-Based Systems*. New York: John Wiley & Sons, Inc./IEEE Press; 2011.
5. Baxter LA. Estimation subject to block censoring. IEEE Trans Reliab 1995;44 (3):489–495.
6. Baxter LA, Tortorella M. Dealing with real field reliability data: circumventing incompleteness by modeling and iteration. Proceedings of the Annual Reliability and Maintainability Symposium; January 24–27; Anaheim, CA; 1994. p 255–262.
7. Bossert JL, editor. *The Supplier Management Handbook*. 6th ed. Milwaukee: American Society for Quality; 2004.
8. Chan CK, Tortorella M. Design for reliability: processes, techniques, information systems. 2000 Annual Reliability and Maintainability Symposium, Tutorial Volume, 1–13; January 24–27; Los Angeles, CA; 2000.
9. Chan CK, Mezhoudi M, Tortorella M. A theory unifying robust design and reliability engineering. Presented at Joint Research Conference on Statistics in Industry and Technology; New Brunswick, NJ; 1997.
10. Coit DW, Smith AE. Reliability optimization of series-parallel systems using a genetic algorithm. IEEE Trans Reliab 1996;45 (2):254–260.
11. Cui L, Li H, Xu SH. Reliability and risk management. Ann Oper Res 2014;212 (1):1–2.
12. Davis PJ. Fidelity in mathematical discourse: is one and one really two? Am Math Mon 1972;79 (3):252–263.
13. DeMarco T. *Concise Notes on Software Engineering*. New York: Yourdon Press; 1979.
14. Elsayed EA. *Reliability Engineering*. 2nd ed. Hoboken: John Wiley & Sons, Inc; 2012.
15. Engelmaier W. Surface mount solder joint long-term reliability: design, testing, prediction. Soldering Surf Mt Technol 1989;1 (1):14–22.
16. Engelmaier W. solder attachment reliability, accelerated testing, and result evaluation. In *Solder Joint Reliability—Theory and Applications* (J. H. Lau, ed.). New York: Van Nostrand Reinhold; 1991. p 545–587.

17. Escobar LA, Meeker WQ. A review of accelerated test models. Statist Sci 2006;21 (4):552–577.
18. Everett WW, Tortorella M. stretching the paradigm for software reliability assurance. Softw Qual J 1994;3 (1):1–26.
19. Field D, Meeker WQ. An analysis of failure-time distributions for product design optimization. Qual Reliab Eng Int 1996;12:429–438.
20. Hamada M. Reliability improvement via taguchi's robust design. Qual Reliab Eng Int 1993;9:7–13.
21. Holcomb D, North JR. An infant mortality and long-term failure rate model for electronic equipment. AT&T Tech J 1985;64 (1):15–38.
22. http://www.autosafety.org/node/32435
23. http://www.elna-america.com/tech_al_reliability.php
24. Jawitz MW, Jawitz MJ. *Materials for Rigid and Flexible Printed Wiring Boards*. Boca Raton: CRC Press; 2006.
25. D. Kececioglu (2002), *Reliability and Life Testing Handbook* (Vol. 2). Lancaster: DEStech Publications, Inc.
26. Kemet Corporation Application Notes for Tantalum Capacitors. http://www.kemet.com/kemet/web/homepage/kechome.nsf/weben/08114D8D1402B2D6CA2570A500160901/$file/F3100_TaLdPerChar.pdf
27. Kuo W, Prasad VR. An annotated overview of system-reliability optimization. IEEE Trans Reliab 2000;49 (2):176–187.
28. Lawless JF. *Statistical Models and Methods for Lifetime Data Analysis*. Hoboken: John Wiley & Sons, Inc; 2011.
29. LuValle MJ, LeFevre BJ, Kannan S. *Design and Analysis of Accelerated Tests for Mission Critical Reliability*. Boca Raton: CRC Press; 2004.
30. Lynch GS. *Single Point of Failure: The Ten Essential Laws of Supply Chain Risk Management*. Milwaukee: American Society for Quality; 2009.
31. Meeker WQ, Escobar LA. *Statistical Methods for Reliability Data*. New York: John Wiley & Sons, Inc; 1998.
32. Nash FR, Joyce WB, Hartman RL, Gordon EI, Dixon RW. Selection of a laser reliability assurance strategy for a long-life application. AT&T Tech J 1985;64 (3): 671–716.
33. Nelson WB. *Applied Life Data Analysis*. Hoboken: John Wiley & Sons, Inc; 2005.
34. Samaniego F, Cohen M, editors. *Reliability Issues for DoD Systems: Report of a Workshop*. Washington: National Academies Press; 2002.
35. Sinha SK. *Reliability and Life Testing*. New York: John Wiley & Sons, Inc; 1986.
36. Smidts C, McGill J, Rodriguez M, Lakey P. Operational profile testing. In: Marciniak JJ, editor. *Encyclopedia of Software Engineering*. Abingdon: Taylor and Francis; 2010.
37. Taguchi G. *Introduction to Quality Engineering*. Tokyo: Asian Productivity Organization; 1986.
38. Viertl R. *Statistical Methods in Accelerated Life Testing*. Göttingen: Vandenhoeck and Rupprecht; 1988.
39. Wasserman GS. A modeling framework for relating robustness measures with reliability. Qual Eng 1996;8 (4):681–692.
40. R. R. Willie (1978), Computer-aided fault tree analysis. Defense Technical Information Center AD-A066567.

7

Reliability Engineering for High-Consequence Systems

7.1 WHAT TO EXPECT FROM THIS CHAPTER

A high-consequence system is one in which the consequences of failure are so severe that the tradeoff between prevention cost and external failure cost almost always leads to a decision to strongly emphasize prevention cost. This is not to say that in high-consequence systems, money is no object in the prevention cost budget, but rather the bias for prevention in such systems is so strong that extraordinary measures are usually easily justified. This chapter discusses reliability engineering practices that may not be fully implemented in ordinary cases but are appropriate for high-consequence systems.

7.2 DEFINITION AND EXAMPLES OF HIGH-CONSEQUENCE SYSTEMS

7.2.1 What is a High-Consequence System?

Much of modern life is made possible by systems whose proper functioning is usually taken for granted by lay persons but whose failure would have severe consequences that may range from relatively benign problems, such as extreme

Reliability, Maintainability, and Supportability: Best Practices for Systems Engineers, First Edition. Michael Tortorella.

expense to repair, to very malign events, possibly even including social collapse. We refer to these systems in this book as *high-consequence systems*.[1] High-consequence systems have one or more of these attributes:

- Extreme consequences of failure to users of the system:
  - Many injuries,
  - Loss of life,
  - Social unrest, disruption, or collapse;
- Extreme consequences of failure to the owner of the system:
  - Loss of profitability to an extent threatening the survival of the organization;
- Extreme difficulty in repairing the system when it fails:
  - Remote or inaccessible location,
  - Need for specialized and expensive equipment for repairs.

Because of these characteristics, high-consequence systems are usually of very high cost and often only a small number are deployed.

These properties bring the need for additional reliability engineering efforts. In a high-consequence system, one may say that the reliability requirements trump all others, and reliability almost always wins in trade studies with other needs. This chapter discusses some additional strategies for meeting reliability requirements in high-consequence systems. These strategies are normally not all used in ordinary systems because they may be too expensive or time-consuming for routine use, but this additional expense and time may be justified by the need for extremely high reliability in high-consequence systems.

7.2.2 Examples of High-Consequence Systems

7.2.2.1 Critical infrastructures

Advanced societies rely on certain critical infrastructures to accomplish tasks that are large in scope and vital to social functioning. These include electric power systems, water, oil, and gas distribution systems, mass transit systems, road networks (including bridges), fire prevention and response systems, etc. In most of these systems, short or localized outages may be tolerable, especially if a backup scheme is in place. However, a long, widespread outage can be deadly or provoke a degree of social unrest or disruption that it is fortunate that it is experienced only rarely.

Let's consider electric power systems in more detail. At the risk of oversimplifying, electric power systems comprise generating plants and distribution (transmission) networks. Localized outages are not uncommon, and there have been a few notable widespread outages in recent decades. Many institutions and individuals plan for these outages and have backup systems in place. Almost all hospitals have local generators to provide backup electric power

[1] Some authors refer to these as mission-critical systems.

when their provider has an outage because loss of electric power in a hospital can readily lead to loss of life. Telephone central offices use 48-V batteries to provide power for about 8 hours after an outage begins, and larger offices also use on-site generators to provide power after the batteries have discharged. Other users of backup power schemes include wireless service providers, Internet service providers, some food retailers, etc. Some individual homeowners also use generators in the event of an outage.

Localized outages and outages of relatively short duration are manageable by these schemes. However, a widespread outage of long duration, such as the power outages caused by hurricanes Katrina and Sandy in recent years, have more serious consequences. Sometimes, gasoline for generators cannot be obtained because electric power to run pumps at filling stations is not available. Such outages entail complicated, expensive, and sometimes slow emergency response by federal and state governments and private aid agencies. Imagine if the Northeast blackouts of 2003 [12] or 1985 [13] had lasted weeks instead of hours. Major social disruptions could easily have resulted. Even in the (relatively) short outages due to hurricanes Katrina and Sandy, day-to-day life was severely disrupted, especially for the less-fortunate, and even after power was restored life did not return to normal for many people until a long time afterward.

7.2.2.2 Commercial aircraft

Commercial aircraft used in passenger service are clearly high-consequence systems [25]. An onboard failure in flight can easily lead to many injuries or lives lost. Even a fire on the ground can be a serious incident. Should the failure be due to a design flaw (as opposed to, say, a faulty component), the airframe manufacturer may find their business at risk as well. Recently, the Boeing 787, a new aircraft, has experienced several fires due to lithium-ion battery thermal runaway [14] and miswiring of a fire suppression system [16]. While no injuries or deaths resulted from these incidents, you may be certain that the manufacturer is working vigorously to correct these problems.

7.2.2.3 Satellites and undersea cable telecom systems

Satellites and undersea cable telecommunications systems may be considered high-consequence systems because of the impossibility, or great difficulty and expense, of repairing them when failures occur. Until recently, repair of a satellite[2] was indeed unthinkable. Astronauts based at the International Space Station have made repairs and upgrades to the Hubble Space Telescope, so repair of satellites is no longer impossible, but it can hardly be considered a routine exercise. It is still breathtakingly complicated and expensive. Undersea cable telecommunications systems are based on earth; but at the bottom of the ocean, still another relatively unexplored frontier. The technology for grappling a cable to the surface and splicing in a new repeater or cable section has

[2] Apart from remote software maintenance and reboots.

existed since the earliest deployment of electronic undersea cable systems in the 1950s, but time has little simplified or made less expensive this endeavor. In the balance between reliability and maintainability for such systems, it is easy to justify expending more design and development resources to prevent failures in preference to reacting to failures when they occur.

7.2.2.4 Other high-consequence systems

You can cite many other examples of high-consequence systems, including

- spacecraft,
- hazardous materials shipping and handling systems,
- medical systems, and
- nuclear reactors,

and more. In all cases, the social and economic costs of the catastrophic consequences of failure of these systems far outweigh the amounts needed to be spent on failure prevention. We might envision this as something like a 1–10–1,000,000 rule instead of a 1–10–100 rule.

We may also contrast with systems that are clearly not high-consequence. Some of these include

- consumer entertainment products,
- systems whose operation affects only a small number of people and for which most failures are benign (e.g., garage door openers, ball-point pens),
- lighting fixtures,

and the like. Even these seemingly simple examples can have serious failures (a lighting fixture can catch fire and cause a house to burn down, for example). Tragic as these may be for the persons involved, these systems do not rise to the level of high consequence as defined here because failures do not cause mass casualties or major social disruption or infrastructure damage.

7.3 RELIABILITY REQUIREMENTS FOR HIGH-CONSEQUENCE SYSTEMS

The special properties of high-consequence systems make it possible, if not obligatory, to adopt a different approach to reliability requirements. Because the consequences of failure in high-consequence systems are much more serious, and because there are sometimes only a few systems deployed, it may make sense to adopt reliability requirements limiting the number and duration of failures directly, rather than requirements limiting the mean, median, or other abbreviation of the quantities involved. Because the consequences of failure are severe, this number will be small; and when there is only a small number of systems involved (perhaps only one), it is preferable to adopt requirements for the reliability effectiveness criterion itself. Reliability models for these systems are constructed to assess the probability of meeting the requirement. Adopting this point of view leads naturally to a holistic model of reliability risk management in the context of decision theory. The details of this approach are beyond the scope of this book, but interested readers may consult Refs. 1 and 6 for further information. Analysis of data to determine compliance with the requirement is a simple matter when a census of the population of installed systems is available, and requires only a little more work when only a sample of the installed population is available. See Chapter 5 for more details.

For example, the reliability requirement for the TAT-8 undersea cable telecommunications system was that there should be no more than three failures requiring a ship repair[3] over the anticipated 25-year life span of the system [8]. The requirement was intended to cover only failures caused by malfunctions within the system itself, and did not apply to events such as damage due to earthquakes,[4] shark attacks, cable breaks caused by fishing trawlers, etc. Reliability models were constructed from which the probability of four or more failures requiring a ship repair in 25 years was estimated to be less than 0.05. There is only one TAT-8 transatlantic system, although the SL-280 and SL-560 fiber-optic systems were used in other undersea applications in subsequent years. These other applications, such as TAT-9, TPC-4, and others, had their own reliability requirements: the point is that the reliability requirement is a property of the application, and it is the responsibility of the system supplier (in this case, the SL-280 and SL-560 manufacturer, which was a joint effort of AT&T, Standard Telephone and Cables of the United Kingdom, and Submarcom of France) to assure the purchaser that the reliability requirement for the application will be met by the proposed system solution.

Another consequence of there having been only one TAT-8 system is that there is no need for complicated data analysis to determine whether the requirement has been met or not. There is only one system, so it is easy to count the number of failures requiring a ship repair and see whether this number is greater than 3 or not. The TAT-8 system was installed in 1988, and was removed from service in 2002 because systems with far larger capacity had been by then installed, and it was no longer economical to maintain a system with such small capacity (40,000 voice channels). Unfortunately, data on the number of internal system failures is always hard to come by as suppliers usually hold this as confidential. This example also points out a disadvantage of requirements written on effectiveness criteria: it is difficult to determine whether the requirement

[3] "Ship repair" is the industry term of art for a failure requiring the dispatch of a cable repair ship to grapple the cable to the surface, perform whatever replacements are necessary, and lay the cable back down to the ocean floor.

[4] Earthquakes, wildfires, floods, etc. may cause failures in all kinds of infrastructures, including high-consequence systems. Our primary intent in this chapter is to study measures to prevent failures. Disaster recovery is outside the scope of this book.

is met until the entire period over which it is to apply has elapsed. Prospective models can be created using data from a limited period of time, but the conclusions expressed from these will properly be probabilistic only.

We have previously emphasized that there are three vital components of a good reliability requirement: the numerical requirement itself, the (operating) conditions under which the requirement applies, and the length of time over which the requirement applies. Reliability requirements for high-consequence systems are different only in that the numerical requirement applies to the reliability effectiveness criterion (e.g., number of failures over the system lifetime) rather than some related reliability figure or merit (e.g., mean number of failures per year).

7.4 STRATEGIES FOR MEETING RELIABILITY REQUIREMENTS IN HIGH-CONSEQUENCE SYSTEMS

7.4.1 Redundancy

Adding redundancy is perhaps the most basic reliability improvement strategy. Redundancy can be quite effective provided proper attention is paid to the reliability of switching circuits and devices. The disadvantages of redundancy are complication and expense.

Redundancy adds complication to operations and maintenance. Operators need additional training to help them avoid errors when working with redundant equipment. For example, in fully duplicated electronic switching systems, it is possible to erroneously remove a line-replaceable unit, or circuit pack, from the active side of the switch instead of from the side of the switch that is failed. Instances of this are rare but not unheard of. If this removal takes place while the switch is in a brink-of-failure state because one-half of the switch is out of service due to a corrective or preventive maintenance activity, then the entire switch will fail. More comprehensive operator training helps avoid such errors. Technological solutions have also been proposed [20].

The clear disadvantage of redundancy is added cost, both initial cost for acquiring redundant units and recurring cost for ongoing operation. In a high-consequence system, it may be that added cost for reliability improvement is readily justifiable, but it would still be unusual if explicit cost and reliability tradeoffs were not performed. Even in a high-consequence system, systems engineers should be prepared to contend for higher reliability solutions that may raise prevention costs because the advantages of early investment in design for reliability strategies may not always be fully understood by all stakeholders. Thorough preparation is important.

The TAT-8 application of the SL-280 undersea cable fiber-optic telecommunications system provides an instructive example of redundancy and its costs and benefits. The undersea portion of the system comprises a six-fiber

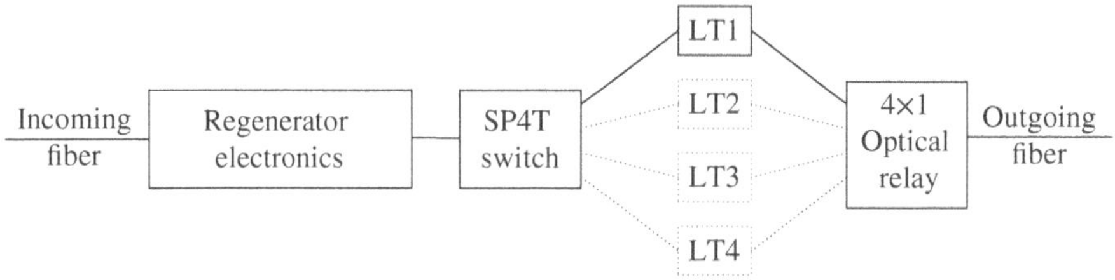

Figure 7.1 *Regenerator section system functional decomposition.*

cable and discrete repeaters spaced approximately every 35 km. In the TAT-8 application, three fibers are used for eastbound traffic and three for westbound traffic. Of the three, two are active and one is a spare. The repeaters contain six regenerators, one per fiber. The regenerator is an electronic amplifier that compensates for the attenuation of signals in a fiber section (a fiber section is a length of fiber between two repeaters). It requires an optical-to-electrical conversion at its input and an electrical-to-optical conversion at its output. The spare fiber, with its associated regenerator(s), can be switched into service in individual sections. Each regenerator includes a laser transmitter for which three cold standby laser transmitters are provided. All switching takes place within the repeater but is controlled remotely from shore terminals. Figure 7.1 shows a functional decomposition of one regenerator section within a repeater for the TAT-8 application of the SL-280 system.

The regenerator section provides the amplification for one fiber. The regenerator electronics contains an O–E (optical-to-electronic) conversion circuit and an amplifier. The output of the regenerator electronics is directed to one laser transmitter (denoted LT1 through LT4 in Figure 7.1) in a cold standby ensemble of four laser transmitters. E–O conversion takes place in the laser transmitter. Laser transmitters LT2 through LT4 are shown in the diagram with dotted lines and borders to indicate that they are cold standby units.[5] The (active) laser transmitter output is switched onto the outgoing fiber by an optical relay having four inputs and one output. One direction of transmission within a repeater (a "repeater half") comprises a two-out-of-three cold standby system of regenerator sections. The entire repeater comprises two halves, that is, two directions of transmission. The supervisory circuits, required to monitor performance, detect failures, and switch redundant units, are not shown in Figure 7.1 although they are an integral part of the redundancy scheme. Also not shown is the powering arrangement.

The system operates at full capacity with two fibers for each direction of transmission and with a single laser transmitter operating in each active regenerator. Purely to increase reliability to the point where the requirement could be met, the cost of the system was multiplied by a factor of 1.5 to provide the extra fiber and regenerators for the third, redundant, path in each direction.

[5] This is not a universally accepted diagramming scheme, but it can be helpful in simple situations.

The regenerator cost was increased by approximately a factor of 4 to provide for the spare laser transmitters and the optical relay. The example shows that this expense was willingly incurred as a reasonable tradeoff against the potential revenue loss and repair expense that would flow from an excessive number of failures requiring ship repair. While the SL-280 and SL-560 technologies are now obsolete, undersea cable telecommunications systems still rely on redundancy as a basic reliability improvement strategy.

7.4.2 Network Resiliency

Many high-consequence systems operate as networks. Electric power generation and distribution, water distribution, oil and gas transport, telecommunications [7], and public transportation are but a few examples. Certainly some of these are high-consequence systems. The electric power system is a multilayer high-consequence system as not only is the system as a whole high-consequence but also important components of it—e.g., nuclear power plants—are themselves high-consequence systems.

For reliability improvement in network-like infrastructures, the concept of network resiliency [21, 23] is helpful. Network resiliency is used to describe two (related) properties of a network:

1. The network continues to provide the service(s) for which it was deployed in a satisfactory manner despite the occurrence of network element failures;
2. When network element failures disrupt service delivery, the network can be returned to normal functioning in a short period of time.

Fundamental to the concept of network resiliency is the idea that the network has a purpose, namely, to deliver some service(s).[6] The effectiveness of the network is measured in terms of a *delivery function* that measures the aggregate flow in the network from all origination nodes to all destination nodes.[7] In the first case, network resiliency is measured as the change in the delivery function resulting from a change in the network's capacity matrix. This is a sensitivity measure, or importance measure, akin to the Birnbaum reliability importance measure discussed in Section 3.4.8. In the second case, rapid return of a network to normally functioning condition is promoted by added attention to supportability and maintainability. Establishment of principles and design practices promoting network resiliency is still a field of active study. Some of this work may be found in Refs. 2, 10, 21, and the references therein.

[6] In Chapter 8, we explicitly conceptualize networks as "service delivery infrastructures" to emphasize their role as enablers of desired services.

[7] For more specialized studies, subsets of the origin and destination nodes may be considered within the same framework.

7.4.3 Component Qualification and Certification

When replacement of failed components is complicated and/or expensive, or the consequences of component failure are serious, as is the case in most high-consequence systems, it is worth employing measures to prevent component failures. The basic strategies for preventing system failures that stem from component failures have been discussed in Chapter 6. To control end-of-life failures, use components whose wearout failure mechanisms do not activate until after the system's planned service life is over. To control failures stemming from application at random times of stresses exceeding a component's strength, use the thermal and electrical stress analysis procedures discussed in Section 6.5.1.2 to minimize the occurrence of "random" failures during component mid-life. In this section, we discuss a strategy to mitigate

- failures due to intrinsic component failure mechanisms, and
- the problem of sourced components having manufacturing defects that may appear after a system is put into service.

This strategy is called *qualification and certification* and proceeds by special reliability testing and screening of sourced components. You may feel that we are here advocating an activity that is commonly considered inefficient, costly and sometimes ineffective, but in the context of high-consequence systems where external failure costs are very high, additional investment in prevention is justified.

In brief, qualification is a test scheme, focused on a component's intrinsic failure mechanisms, that is intended to ascertain whether a population of components is capable of surviving the majority of stresses that may be applied to it during operation of a system containing them. Qualification usually consists of a number of accelerated life tests focused particularly on the component's known failure modes as well as accelerated life tests intended to uncover possibly unknown failure modes. It answers a question about an entire population of (like) components[8]: given the components, the system service life, and the stresses that component is likely to see during operation of the system, will enough of the components in the population survive beyond the end of the system service life for use of this population in the (high-consequence) system to make economic sense? Qualification is a procedure that leads to a decision about an entire population of components, and, as such, is subject to the usual type I (reject a population that should be accepted) and type II (accept a population that should be rejected) errors.

Certification, by contrast, is a procedure used to identify individual components for reliability sufficient to guarantee that, with high probability, that component will not fail in operation throughout the useful life of the (high-consequence) system. Certification is also a decision procedure: for each individual component, a decision to accept or reject it for use in the system is made. As such, the decision is also subject to type I and type II errors.

[8] Sometimes specially selected by the component manufacturer.

This section is devoted to determination of the type I and type II error probabilities for qualification and certification and the life distributions of the survivors of qualification and certification testing. Much of this material was previously published by the author in case 6 of Ref. 9 and is reprinted here with permission.

7.4.3.1 Qualification as a decision process

Qualification is the process of ascertaining whether a population of components can be sourced that is economically viable after certification is carried out. That is, qualification is intended to determine whether the proportion of components satisfying the certification criteria is large enough that enough components pass the certification testing for the overall cost of acquiring a certified population of components to be reasonable. In the context of an undersea telecommunications cable system in which the reliability requirement is for a 25-year system life [19, 22], we want to determine whether there are in the population enough components whose lifetimes are greater than 25 years (the anticipated service life of the TAT-8 system) so that a sensible certification (screening of individual components; see Section 7.4.3.2) can be implemented without compromising the system profit (clearly there is nothing essential about 25 years; henceforth, we replace it by T where T is the system useful life requirement in years). If F is the life distribution of the population to be used, then qualification attempts to determine the value of $F(T)$ with the hope that it is close to 0. In particular, let us suppose that there is some number θ, $0<\theta<1$, for which a sufficient condition for economical system deployment is $F(T)\leq\theta$. That is, the definition of a population being qualified is that its life distribution satisfies $F(T)\leq\theta$. θ is arrived at by considering cost and technology tradeoffs. The choice of qualification criterion in this form reflects the notion that for a population to be qualified means that a large enough proportion of it $(1-\theta)$ has sufficiently long lifetimes that the cost of qualification and certification does not unduly affect negatively the overall economics of the system. As qualification makes a judgment about $F(T)$ in relation to θ, it is possible that this judgment could be incorrect. This section shows how the quality of this decision influences the life distribution of the survivors of qualification.

Qualification may be construed as a decision process: it is the collection and analysis of data to support a judgment about whether $F(T)\leq\theta$ or $F(T)>\theta$. As such, the decision is subject to type I and type II errors. The magnitude of these errors has an effect on the life distribution of the survivors of qualification and certification. This Section explores the role of type I and type II errors in the qualification decision. Section 7.4.3.2 concerns the role of type I and type II errors in the certification decision.

Qualification proceeds by a sequence of alternating reliability tests and product redesigns.[9] After each reliability test, a judgment is made as to whether the population is qualified. Represent by Q and N the states of a population's being qualified and not qualified, respectively, using these letters both for the

[9] Note the similarity to "reliability growth testing" [5].

state of nature (the "actual," unknowable state of the population), and for the judgments made following reliability testing. Define also S_k to be the state of nature after the $(k-1)^{st}$ product redesign and J_k to be the judgment rendered after the kth reliability test, $k=1, 2, \ldots$. S_k and J_k then take on the values N or Q. Finally, define F_k to be the life distribution of the population after the $(k-1)^{st}$ product redesign, with $F_1 = F$ being the original population life distribution (before any testing or redesign is performed).

Necessarily, the sequence of judgments comprises some number of N values followed by a Q because once the population is deemed qualified, the sequence of product redesigns and reliability tests halts. Any judgment may individually be mistaken, including the last one, and so we need to examine not only the individual type I and type II errors at each step, but the aggregate or overall type I and type II errors in the final qualification decision.

For the qualification decision (after, say, k steps), the <u>overall</u> type I error is rejecting the conclusion that $F_k(T) \leq \theta$ when it is in fact true (this is the producer's risk) and the <u>overall</u> type II error is accepting the conclusion that $F_k(T) \leq \theta$ when it is in fact false (this is the consumer's risk). Define $\alpha_Q(k)$ to be the probability of overall type I error and $\beta_Q(k)$ be the probability of overall type II error when the decision is taken at the kth step. Then, $\alpha_Q(k) = P\{\text{Decide } F_k(T) > \theta \mid F_k(T) \leq \theta\}$ and $\beta_Q(k) = P\{\text{Decide } F_k(T) \leq \theta \mid F_k(T) > \theta\}$. For purposes of this study, which focuses on the magnitudes of the possible decision errors that can be made, the details of the qualification testing, data collection, and the use of the information developed thereby to make the decision are not relevant. Obviously, in practice we would like $\alpha_Q(k)$ and $\beta_Q(k)$ to be as small as possible, subject to whatever time and resource constraints may apply to the qualification undertaking. The present study concerns only how the values of $\alpha_Q(k)$ and $\beta_Q(k)$ influence the life distribution of the final survivors of certification that is performed on (what was decided to be) a qualified population. Table 7.1 may help illuminate the definitions of the overall type I and type II errors in this context.

Qualification proceeds through a sequential process of testing, decision, and modification of the product if the population is judged to be not qualified, until a decision is reached that the (suitably modified) population is qualified. Let us assume that there are $\nu - 1$ "not qualified" decisions followed by a "qualified" decision, at which point the modification process stops. While the distribution of ν is unlikely to be geometric, because the decisions are likely to be stochastically dependent, we are going to assume that the individual-step type I and type II errors are independent from one trial to the next and that their probabilities (α and β, respectively) remain the same throughout. This may be reasonable provided the type of testing done to qualify the population is

TABLE 7.1 Qualification Decision Errors

State of Nature	$F(T) \leq \theta$	$F(T) > \theta$
Qualification accepts population	Correct decision	Type II error
Qualification rejects population	Type I error	Correct decision

substantially the same after each modification, the same personnel are involved in each decision, etc. Define $\sigma_k(N)$ (resp., $\sigma_k(Q)$) to be the number of N (resp., Q) states in $\{S_1,\ldots, S_k\}$; then $\sigma_k(N)+\sigma_k(Q)=k$. Recall that if $J_k=Q$, then $J_1=\cdots=J_{k-1}=N$. Then we have

$$P\{J_k=Q \mid S_k=Q\}=\sum_{n=1}^{k}(1-\alpha-\beta)^n \alpha^{k-n} P\{\sigma_{k-1}(N)=n-1, \sigma_{k-1}(Q)=k-n\}$$

and

$$P\{J_k=Q \mid S_k=N\}=\sum_{n=1}^{k}(1-\alpha-\beta)^{n-1} \alpha^{k+1} \beta P\{\sigma_{k-1}(N)=n-1, \sigma_{k-1}(Q)=k-n\}$$

with $P\{\sigma_1(N)=1, \sigma_1(Q)=0\}=1-P\{\sigma_1(N)=0, \sigma_1(Q)=1\}$, the probability that the initial population is not qualified, being given. Then the overall type I and type II error probabilities at the kth step are

$$\alpha_Q(k)=P\{J_k=N \mid S_k=Q\}=1-P\{J_k=Q \mid S_k=Q\}$$

and

$$\beta_Q(k)=P\{J_k=Q \mid S_k=N\}.$$

These are computed recursively from the equations above.

Effect of type I and type II errors on the life distribution

If $J_k=Q$, then we refer to the life distribution of the population judged qualified at step k as the "final" life distribution F_k. To say that $J_k=Q$ means that, as far as we know, $F_k(T)\leq\theta$ (which is equivalent to $S_k=Q$). In this section, we study the quality of our knowledge about $F_k(T)$ based on the overall type I and type II errors in qualification. Accordingly, we wish to examine $P\{F_k(T)\leq\theta \mid J_k=Q\}$. We have

$$\begin{aligned}
P\{F_k(T)\leq\theta \mid J_k=Q\}&=P\{S_k=Q \mid J_k=Q\}\\
&=\frac{P\{J_k=Q \mid S_k=Q\}P\{S_k=Q\}}{P\{J_k=Q\}}\\
&=\frac{P\{J_k=Q \mid S_k=Q\}P\{S_k=Q\}}{P\{J_k=Q \mid S_k=Q\}P\{S_k=Q\}+P\{J_k=Q \mid S_k=N\}P\{S_k=N\}}\\
&=\frac{\left[1-\alpha_Q(k)\right]P\{S_k=Q\}}{\left[1-\alpha_Q(k)\right]P\{S_k=Q\}+\beta_Q(k)P\{S_k=N\}}.
\end{aligned}$$

This equation represents a "degree of belief" in whether $F_k(T)$ is greater than or less than θ when a judgment is made at the kth step that it is. If both $\alpha_Q(k)$ and $\beta_Q(k)$ are equal to zero, then $P\{F_k(T) \le \theta \mid J_k = Q\} = 1$ and our judgment accurately reflects the state of nature. If either $\alpha_Q(k)$ or $\beta_Q(k)$ are positive, then our judgment is flawed, and the larger they are, the less accurate is our judgment.

To complete the computation, a model for the distribution of $\{S_1, \ldots, S_k\}$ is needed. A very accurate model would require deep knowledge of the particular processes of testing and redesign in question, so the following remarks should be taken as illustrative only. A simple model for $\{S_1, \ldots, S_k\}$ is a two-state Markov chain having $p_{QQ} = P\{S_{j+1} = Q \mid S_j = Q\} = 1 - p_{NQ} =$ "large" (close to 1) and $p_{QN} = 1 - p_{NN} =$ "medium." A model like this would allow computation of the terms involving probabilities of events in the σ-field determined by $\{S_1, \ldots, S_k\}$ and so allow completion of computations following Table 7.1. In practice, k is usually rather small, on the order of at most 2 or 3, so computations in this model would not be too onerous. Other than this simple suggestion, further exploration of this issue is beyond the scope of this discussion because of the strong dependence on particular program details.

7.4.3.2 *Certification as a decision process*

Because certification makes a separate decision for each component regarding whether or not its lifetime exceeds T years, it is important to consider the possibility that the decision may be made incorrectly in particular cases. Let L denote the (random) lifetime of a given component from the population judged to be qualified that has survived the certification tests. That is, $L(\omega)$ is the lifetime of component ω, an element of the sample space that describes the population of components entering certification that survive the certification tests. Further, we divide this population into two parts: $\Omega_A = \{L > T\}$ and $\Omega_U = \{L \le T\}$. These names are meant to call to mind that lifetimes exceeding T years are Acceptable and those not exceeding T years are Unacceptable. For each ω, let $C(\omega) = A$ (resp., U) if certification places component ω in Ω_A (resp., Ω_U). That is, $C(\omega)$ is the result of the certification decision on component ω.

If $\omega \in \Omega_A$ and $C(\omega) = A$, or if $\omega \in \Omega_U$ and $C(\omega) = U$, then the certification decision is correct for ω. If, on the other hand, $\omega \in \Omega_A$ and $C(\omega) = U$, or if $\omega \in \Omega_U$ and $C(\omega) = A$, then the certification decision is incorrect for ω. In the first case, we have an example of *producer's risk*, or type I error, in which an acceptable component is incorrectly discarded. In the second case, we have an example of *consumer's risk*, or type II error, in which an unacceptable component is incorrectly retained. For purposes of most high-consequence systems, type II error is much more significant because a component on which a type II error is committed is one that is used in the system assembly and that will likely fail before T years. The probability of a type I error is $P\{C = U \mid \Omega_A\} = \alpha_C$ and the probability of a type II error is $P\{C = A \mid \Omega_U\} = \beta_C$. Table 7.2 may help clarify the situation.

TABLE 7.2 Certification Decision Errors

State of Nature	Component is in A, $L>T$	Component is in U, $L \le T$
Certification marks component in A	Correct decision	Type II error
Certification marks component in U	Type I error	Correct decision

7.4.3.3 Life distributions

Qualification is usually accomplished through some accelerated life test(s). This means that components accumulate a certain amount of age during qualification, and we reflect this in the life distribution model by postulating a time τ_Q that represents the age consumed during qualification. Usually, the survivors of qualification testing are not used in downstream production; only the untested portion of the population (that is judged to be) qualified is used; in this case, we would have $\tau_Q=0$. However, in cases where the survivors are used in downstream production, then $\tau_Q>0$ and if $J_k=Q$, then the life distribution of the survivors of qualification is given by $(F_k(t+\tau_Q)-F_k(\tau_Q))/(1-F_k(\tau_Q))$ for $t \ge 0$ (here and below we use a new time origin for the population of survivors to be consistent with the actions taken in practice, where the survivors are considered a new, distinct population having a new life distribution that is zero at the time origin). In addition, we know that F_k satisfies the equation in Section "Effect of type I and type II errors on the life distribution".

$P\{L \le t \mid C=A\}$ is the life distribution of the components that have been selected for use by the certification procedure. It is easier to work with the survivor function, so we have

$$\begin{aligned} P\{L>t \mid C=A\} &= \frac{1}{P\{C=A\}} P\{L>t, C=A\} \\ &= \frac{1}{P\{C=A\}} \left[P\{L>t, C=A, \Omega_A\} + P\{L>t, C=A, \Omega_U\} \right] \end{aligned}$$

where $P\{C=\text{A}\}$ is given by

$$\begin{aligned} P\{C=A\} &= P\{C=A \mid \Omega_A\} P(\Omega_A) + P\{C=A \mid \Omega_U\} P(\Omega_U) \\ &= (1-\alpha_C) P\{L>T\} + \beta_C P\{L \le T\} \\ &= (1-\alpha_C)\left[1-F_Q(T)\right] + \beta_C F_Q(T). \end{aligned}$$

We now assume that, given A (or given U), the lifetime value and the certification decision are conditionally independent. This reflects the idea that the decision maker does not know the lifetime of the device exactly. This is, of course, only an approximation because the certification decision is made based on some testing which may lead to an estimate of the device's lifetime, but this would result in a more complicated model that is beyond the scope of this discussion and is a suitable subject for future research. Then

$$\begin{aligned}
P\{L>t,\ C=A,\ \Omega_A\} &= P\{L>t, C=A \mid \Omega_A\}P(\Omega_A) \\
&= P\{L>t \mid A\}P\{C=A \mid \Omega_A\}P(\Omega_A) \\
&= (1-\alpha_C)P\{L>t, L>T\} \\
&= (1-\alpha_C)\left[1-F_Q(t\wedge T)\right] \\
&= \begin{cases} (1-\alpha_C)\left[1-F_Q(T)\right], & 0\le t\le T \\ (1-\alpha_C)\left[1-F_Q(t)\right], & t>T \end{cases},
\end{aligned}$$

where the conditional independence is used at the second step. Similarly,

$$\begin{aligned}
P\{L>t, C=A, \Omega_U\} &= P\{L>t, C=A \mid \Omega_U\}P(\Omega_U) \\
&= P\{L>t \mid U\}P\{C=A \mid \Omega_U\}P(\Omega_U) \\
&= \beta_C P\{L\le t, L>T\} \\
&= \begin{cases} 0, & 0\le t\le T \\ \beta_C\left[F_Q(t)-F_Q(T)\right], & t>T \end{cases}.
\end{aligned}$$

Altogether, we obtain

$$P\{L>t \mid C=A\} = \begin{cases} \dfrac{(1-\alpha_C)\left[1-F_Q(T)\right]}{(1-\alpha_C)\left[1-F_Q(T)\right]+\beta_C F_Q(T)}, & 0\le t\le T \\[2ex] \dfrac{(1-\alpha_C)\left[1-F_Q(t)\right]+\beta_C\left[F_Q(t)-F_Q(T)\right]}{(1-\alpha_C)\left[1-F_Q(T)\right]+\beta_C F_Q(T)}, & t>T \end{cases}.$$

This is the desired survivor function of the population of components that pass the certification screen.

Note that when $\alpha_C=\beta_C=0$, that is, the certification decision is always correct, then $\{C=A\}=\{L>T\}$, and we obtain

$$P\{L>t \mid C=A\} = \begin{cases} 1, & 0\le t\le T \\[1ex] \dfrac{1-F_Q(t)}{1-F_Q(T)}, & t>T \end{cases}$$

which is the same as $P\{L>t \mid L>T\}$ as it should be.

As α and β increase, $P\{L > t \mid C = A\}$ decreases for each fixed t, indicating that the consequences of incorrect certification decisions become more costly as the probability of incorrect decision increases. In effect, what incorrect certification decisions do is increase the number of sub-T-year lifetime components in the population of components that survive qualification and certification, with the consequence that more failures will occur in service due to these components.

7.4.3.4 Section summary

A review of recent component qualification and certification studies and textbooks [3, 4, 10, 11, 18, 22, 24] shows that the importance of decision errors on reliability predictions for qualified and/or certified components has not yet been fully appreciated, particularly during planning exercises in which the reliability of qualified and/or certified components must be predicted. Our intention here is not to provide a complete and comprehensive guide to decision processes in qualification and certification, but rather to provide you with a basis for implementing a qualification and certification program that accounts for the decision-making nature of these procedures.

7.4.4 Failure Isolation

Sometimes, despite best efforts, failures do occur. In distributed high-consequence systems covering large geographical areas or large populations, the strategy of failure isolation can be effective in limiting damage. Failure isolation refers to the idea of preventing a failure that occurs in a system from causing failures in other systems that the failed system may be connected to or communicating with. Sometimes this condition is called "cascading failures."

> **Language tip:** The phrase "failure isolation," or more often "fault isolation," is also used to refer to the process of locating within a system the component or subassembly that has failed, causing a system failure. To avoid this confusion, we consistently refer to this activity as "fault location." This is discussed at length in Chapter 12.

7.4.4.1 Example

In January 1990, the AT&T Signaling System 7 (SS7) transaction management data network suffered an extensive outage because a software failure in one of the SS7 switching nodes propagated over a large area when nodes connected to the failed node went out of service because they received improper status information from the failed node. The failures cascaded in less than 10 minutes to over 100 SS7 nodes, resulting in a nine-hour outage that affected 60,000 customers and cost AT&T approximately $60 million in lost revenue [15]. Had flawed status information not been passed on to neighboring nodes, the scope of the outage would have been greatly reduced.

7.4.4.2 Failure Isolation Strategies

While distributed systems offer many advantages in autonomous provision of services over wide geographical areas and/or to diverse populations of customers, some coordination is always required to ensure that, for example, all records belonging to the same customer are coordinated regardless where they may be stored. Communication across elements of a distributed system is essential, but can also serve as a medium for undesired propagation of errors throughout a system. Therefore, whenever systems must share information across a communications channel, good software engineering practices should be followed:

- Before a message is transmitted, ensure that
  - the message format is correct (i.e., is in the format that the receiver is expecting),
  - the message content is correct,
  - all return codes are checked and verified.
- When a message is received, and before it is acted on,
  - verify that the error-detection code result is correct.

Many relevant practices are a part of good software design for reliability.

Failure of an element of a distributed system or network may increase the load (stress) places on neighboring system or network elements. For example, in a fluid distribution network, failure of a pump may cause backflow in the pipes entering the pump, and may place extra load on pumps feeding those pipes. A design review of a distributed system or network should include a section focused especially on discovery of unanticipated stress overloads caused by system or network element failures, with particular emphasis on whether these overloads can cause failures of other neighboring or communicating system or network elements. Structurally, this examination is facilitated by use of the FMECA tool (Section 6.6.2.1).

7.5 CURRENT BEST PRACTICES IN RELIABILITY ENGINEERING FOR HIGH-CONSEQUENCE SYSTEMS

The designation of a system as "high-consequence" is one of those situations where a difference of degree becomes so great that it is really a difference of kind. All systems exist on a continuum from failures being nothing more than nuisances (e.g., entertainment equipment) to failures having life-threatening consequences (e.g., medical devices—for a specific example, see Ref. 17). So the first step in reliability engineering for high-consequence systems is the determination of whether the system in question is high-consequence. The import of a "yes" answer is that it is justifiable to spend heavily on prevention because external failure costs are so high. Consequently, a vigorous design for reliability program (Chapter 6) should be undertaken. In systems that are not high-consequence, some reliability engineering tasks may be omitted because their cost-to-benefit ratio may not justify their use. In this kind of system, it is appropriate to justify the inclusion of activities into its reliability engineering program. For example, it may not be desirable to undertake both a fault tree analysis and an FMECA for a simple, low-impact system like a DVD player. In a high-consequence system, a reliability engineering program ought to require justification for exclusion of any tasks.

The following are some additional ideas to consider in reliability engineering for a high-consequence system:

- Reliability requirements development: It is of vital importance that a high-consequence system has appropriate and explicit reliability requirements. All system requirements should be formally examined for the ways in which they can be violated, that is, for their associated failure modes. Reliability requirements should be subjected to a design review with representatives of all areas of the development team.
- Reliability modeling: A system reliability model can also function as a means for keeping track of design for reliability activities. As activities are completed, their results are folded into the system reliability model, providing the systems engineers with a current view of projected reliability at all stages of development.
- Design for reliability: High-consequence systems should receive formal design for reliability attention. Reasons for omission of any design for reliability tasks discussed in Chapter 6 should be carefully studied and completely understood.
- Network resiliency: In high-consequence systems that are realized as infrastructure networks, such as electric power distribution, the emerging science of network resiliency offers design principles and practices that enable the network to continue to function satisfactorily (i.e., deliver the service(s) that it is designed to provide) despite failure of one or more network elements.

7.6 CHAPTER SUMMARY

High-consequence systems are those in which failures have extremely serious consequences. In these systems, spending heavily on prevention costs is often justifiable because external failure costs are so high. Reliability engineering strategies for high-consequence systems include

- Write reliability requirements directly in terms of reliability effectiveness criteria rather than reliability figures of merit.
- Employ redundancy where possible.
- Consider component qualification and certification as an essential part of supplier management.

- Employ failure isolation practices in distributed systems.
- Use network resiliency concepts to discover particularly vulnerable locations in a network and redesign the network to mitigate these vulnerabilities.
- Institute formal design reviews focused on reliability to gain the benefit of varied experiences available in cross-functional teams.

While design for reliability can be very effective in high-consequence systems, it is also necessary to be prepared in case failure does occur. See Chapters 11 and 13 for design for maintainability and design for supportability in high-consequence systems.

7.7 EXERCISES

1. List five high-consequence systems and five systems that are not high-consequence (other than those given in Section 7.2.2). Discuss.
2. Refer to the example in Section 7.4.1.
   a. Using suitable notation, write an expression for the lifetime of a regenerator section in terms of the lifetimes of the incoming and outgoing fibers, the regenerator electronics, the SP4T switch, the laser transmitters, and the optical relay.
   b. If the life distribution of the laser transmitters is exponential with a mean of 40,000 hours, what is the life distribution of the cold standby ensemble of four laser transmitters? What is the mean life of the ensemble? How do your answers differ if the life distribution of the laser transmitters is lognormal with mean 40,000 hours and a spread factor 2.3?
   c. Would hot standby have been an effective redundancy strategy in this example? Consider how hot standby would be implemented and compare this with how cold standby is implemented.
3. Consider a multiprocessor computing system. Draw a cause-and-effect (fishbone, Ishikawa) diagram identifying root causes of cascading failures. Suggest countermeasures for the three most important root causes.

REFERENCES

1. Barlow RE, Clarotti CE, Spizzichino F. *Reliability and Decision Making*. New York: Chapman and Hall; 1993.
2. Baroud H, Ramirez-Marquez JE, Barker K, Rocco CM. Stochastic measures of network resilience: applications to waterway commodity flows. Risk Anal 2014;34 (4):1317–1335.
3. Booker JM, Ross TJ, Rutherford AC, Reardon BC, Hemez FM, Anderson MC, Doebling SC, Joslyn CA. An engineering perspective on UQ for validation, reliability, and certification (U). 2004. Los Alamos National Laboratory report no. LA-UR-04-6670.

4. Boydston A, Lewis WD. Qualification and reliability of complex electronic rotorcraft systems. Presented at AHS Specialists Meeting on Systems Engineering; October 15–16, 2009; Hartford, CT; 2009.
5. Crow LH. Tracking reliability growth. 1974. Defense Technical Information Center document no. AD-0785614.
6. Cui L, Li H, Xu SH. Reliability and risk management. Ann Oper Res 2014;212:1–2.
7. Dolev D, Jarmin S, Shavitt Y. Internet resiliency to attacks and failures under BGP policy routing. Comput Netw 2006;50 (16):3183–3196.
8. Easton RL, Hartman RL, Nash FR. Assuring high reliability of lasers and photodetectors for submarine lightwave cable systems: introduction. AT&T Tech J 1985;64 (3):661–670.
9. Elsayed EA. *Reliability Engineering*. 2nd ed. Hoboken: John Wiley & Sons, Inc; 2012.
10. Hallberg Ö, Eriksson B, Francis R, Hjortendal R, Lindberg L-I, Saevstroem B. Hardware reliability assurance and field experience in a telecom environment. Qual Reliab Eng Int 2007;10 (3):195–200.
11. Harry CC, Mathiowetz CH. ASIC reliability and qualification: a user's perspective. Proc IEEE 1993;81 (5):759–767.
12. http://en.wikipedia.org/wiki/Northeast_blackout_of_2003. Accessed November 11, 2014.
13. http://en.wikipedia.org/wiki/Northeast_blackout_of_1965. Accessed November 11, 2014.
14 http://en.wikipedia.org/wiki/Boeing_787_Dreamliner_battery_problems. Accessed November 11, 2014.
15. http://www.informit.com/library/content.aspx?b=Signaling_System_No_7&seqNum=19. Accessed November 11, 2014.
16. http://www.bloomberg.com/news/2013-08-14/ana-finds-wiring-defects-in-dreamliners-as-jal-plane-scraps-trip.html. Accessed November 11, 2014.
17. Leveson NG, Turner CS. An investigation of the THERAC-25 accidents. Computer 1993;26 (7):18–41.
18. Nash FR. *Estimating Device Reliability: Assessment of Credibility*. New York: Springer-Verlag; 1993.
19. Runge PK. Undersea lightwave systems. AT&T Tech J 1992;71 (1):5–13.
20. Tortorella M. Electromagnetically locking latch to prevent circuit pack removal. US Patent 6,312,275. 2001.
21. M. Tortorella (2011), Design for network resiliency. In JJ Cochran, LA Cox, Jr., P Keskinocak, JP Kharoufeh, JC Smith *Encyclopedia of Operations Research and Management Science*, vol. 2, 1364–1381. Hoboken: John Wiley & Sons, Inc.
22. P. Trischitta, M. Colas, M. Green, G. Wuzniak, J. Arena (1996), The TAT-12/13 cable network. IEEE Commun Mag February, 24–28.
23. Whitson JC, Ramirez-Marquez JE. Resiliency as a component importance measure in network reliability. Reliab Eng Syst Saf 2009;94 (10):1685–1693.
24. Wright MW, Franzen D, Hemmati H, Becker H, Sandor M. Qualification and reliability testing of a commercial high-power fiber-coupled semiconductor laser for space applications. Opt Eng 2005;44 (5):054204.
25. Yeh YC. Unique dependability issues for commercial airplane fly-by-wire systems. Presented at 18th IFIP World Computer Congress; Toulouse, France. 2004. Available at http://webhost.laas.fr/TSF/IFIPWG/Top3/08-Yeh.pdf. Accessed November 11, 2014.

8

Reliability Engineering for Services

8.1 WHAT TO EXPECT FROM THIS CHAPTER

Services are becoming an ever-larger part of the world economy. There will be an increasing number of service consumers across the globe. These consumers will expect that the services they purchase will be reliable. In this chapter, we will discuss what it means for a service to be reliable and some of the engineering techniques used to make them so. We cover always-on services and on-demand services but focus more on on-demand services because reliability for always-on services is equivalent to the reliability of the infrastructure used to deliver them.

8.2 INTRODUCTION

So far, we have worked with reliability engineering for *systems*, by which we meant tangible, physical objects designed and assembled to fulfill certain purposes. Now we turn to examination of those purposes and how to ensure that users of the systems receive the services they desire from those systems, or how well those systems fulfill their purpose(s) from the point of view of the users. In an important sense, this is an even more fundamental viewpoint because it finally encompasses the entire value chain from supplier to user, enabling the

Reliability, Maintainability, and Supportability: Best Practices for Systems Engineers, First Edition. Michael Tortorella.

systems engineer to achieve a holistic view of system development from concept to user. While many publications concerning service reliability may be found in the literature (e.g., Refs. 3, 5, 9, and many others), the systematic study of service reliability as a distinct discipline was begun in Refs. 10, 11.

Service reliability engineering differs somewhat in details, but not in fundamental principles, from product reliability engineering. What is required, as is always the case with reliability engineering study, is a good understanding of the failure modes and failure mechanisms of the service. To achieve this understanding, it helps to begin by defining the concepts and primitives connected with services. We consider two types of services, those that are delivered by means of discrete transactions, and those that are intended to be "always on."

8.2.1 On-Demand Services

An *on-demand service* comprises actions in which a *server* accomplishes some deed or takes some action in response to a request from a *user* or *customer*. Some examples of this kind of service are voice telephony, auto repair, retail sales, package delivery, business processes, etc. In an on-demand service, a user makes a request of a server (a caller dials digits, a car owner requests that a certain repair be accomplished, a customer buys a refrigerator, a sender contracts with the postal service to deliver a package, an insurance claim is filed, etc.), the server performs some actions to carry out the request (the telephone network sets up a (possibly virtual) connection to the called party, the car repair shop assigns a service technician who works on completing the repair, the retail store vends and delivers the refrigerator, the postal service forwards the package on to its destination, an insurance adjuster inspects the damaged property and makes a determination about payment, etc.), and the action has an identifiable completion upon which the server is dismissed and is able to accept new transaction requests. On-demand services are characterized by an interaction between a user and a server that is called a *transaction*. We think of the transaction as the basic unit of on-demand services. Informally, service reliability engineering for on-demand services is concerned with the delivery of successful transactions to all users throughout the useful life of the service. We consider the useful life of a service to be over (in general) when the service provider ceases offering the service or (for a particular service purchaser) when the service contract or agreement expires. We also consider the useful life of a service to be over when the service provider changes the requirements for the service. In that case, the service is changed, and it is different from the previous version; from a reliability engineering point of view, one should treat this as a new service.

Delivery of services to users is accomplished with the aid of certain equipment and processes. These "background" resources are the *service delivery infrastructure* (SDI). In a package delivery service, the SDI includes the service provider's transportation network comprising vehicles, airplanes, routing software, etc., the customer interfaces such as service counters, local carriers, etc., and billing and payment mechanisms (over-the-counter, via the Internet, etc.). Understanding the SDI is important because it is the source of service failure mechanisms. In other words, we trace the failure mechanisms associated with each service failure mode back to events taking place in the SDI. This extra step is required in fault tree analysis (Chapter 6) for services. This approach underlies all the analyses in this chapter.

In addition, reliability engineering for a PC- or smartphone-based application is effectively accomplished using the concepts and methods of service reliability engineering. The user requests an action, such as bringing up an e-mail reader, starting a game, paying a bill by near-field communication, etc., to be performed by the PC or smartphone. The SDI in this case includes the hardware of the PC or smartphone and the software running on it; such applications may also require accessing resources over a remote network such as the Internet, and in these cases, the remote network becomes part of the SDI as well.

Service reliability can readily be understood, then, as the ability to continually deliver (the service provider's concern) or carry out (the user's concern) successful transactions in the service. Note how this is a consistent adaptation of the standard definition of reliability (the continued satisfactory operation of a system, according to its requirements, under specified conditions, for a specified period of time) to the particular properties of a service. Here is the formal definition:

> **Definition:** Service reliability is the ability to deliver satisfactory transactions in the service, according to its requirements, under specified conditions, throughout the useful life of the service.

A transaction is considered satisfactory if it meets all the requirements of the service. As with systems or products, every requirement contains within it one or more failure modes, ways in which the requirement can be violated. This is completely analogous to the definition of reliability for systems that we have used so far. It incorporates the notion that the service has a set of requirements that are to be satisfied in order that (a transaction in) the service be deemed successful. This is analogous to the notion of successful operation of a system: all requirements for the system are met in successful operation. An instance of not meeting a requirement is a failure; reliability engineering for products or systems is the process by which we endeavor to make the product or system as free from failures as is economically reasonable. The same is true of services. Reliability engineering for services is the process by which we endeavor to make the service as free from failures as is economically reasonable. Accordingly, we gather in detail the knowledge available about service failures. In particular, this means we need to study service failure modes and failure mechanisms.

8.2.2 Always-On Services

In addition to services that are delivered by discrete transactions, many important services are of the nature that they are (supposed to be) always ready for use. Prominent examples are utilities: electric, natural gas, water, etc.

Users expect that these will always be ready to use at any time. Always-on services may be accommodated in a transaction-based framework in two ways:

1. The service may be thought of as a single transaction that began at a time in the past and is to continue into the indefinite future. In this interpretation, electricity service may be considered a single transaction that began when the current user first requested service to begin at his/her premises and ends when that user requests that service be discontinued at that premises. The primary concern for this user will be service continuity (Section 8.4.2.2), which concerns interruptions and instances of electricity being provided that is outside of the utilities' requirements for voltage, frequency, etc.
2. The service may be conceptualized as being delivered through transactions in which each transaction is a request by the premises owner to access the service; this would make each attempt to turn on a lamp, run a machine or appliance, etc., into a transaction, and the transaction-based reliability engineering models discussed in this chapter can be used without modification.

Either approach can yield useful results. The former approach is better adapted to the utility (service provider) view of the service, while the latter reflects more faithfully the user's point of view. It is most likely that the service provider will be doing these analyses, and the user's service fault tree analysis will rapidly reduce to the utility's service continuity because when the utility service is accessible, the causes of a user's individual transaction failures will be local to the user's premises. Therefore, it usually suffices for the service provider to carry out the former analysis.

8.3 SERVICE FUNCTIONAL DECOMPOSITION

You can create a functional decomposition for a service in the same way that a functional decomposition can be created for products and systems. In most cases, the service will be provided through the use of some SDI—an arrangement of hardware and software owned by the service provider and used by the service consumer to access and use the service. A service functional decomposition comprises a sequence of operations that delivers the service. Those operations are carried out on a hardware and software platform, or SDI, which has to configure itself in certain defined ways to successfully deliver the service. The configuration of the underlying SDI is a key part of the service functional decomposition.

Example: The service we consider in this example is off-site backup of enterprise computer data "in the cloud." This is an example of a cloud computing service. The service user contracts with a provider of memory space at a remote location to store certain files from various users at the remote location.

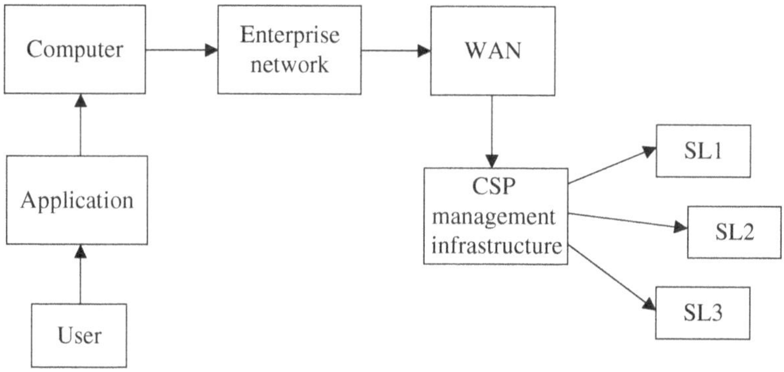

Figure 8.1 *Cloud backup service functional decomposition.*

This may be thought of as an on-demand service (Section 8.2.1): in addition to scheduled times at which files may be transferred from users to the cloud storage location, users may asynchronously request files from the backup and/or send files to the backup. Users do this by running an application locally. We may construct a functional decomposition for this service as Figure 8.1.

The user invokes an application on her computer that forms a request to access some data stored remotely by the cloud service provider. The request travels through the user's enterprise network and through a wide-area network to the service provider's management infrastructure (computers, software, billing, etc.). The desired data are transmitted back to the user through the same means. More detail could be added to this service functional decomposition if necessary. For instance, it does not show the sequence of operations during any of the interactions depicted by the arrows (user with application, etc.). These interactions involve various requests and replies, usually mediated by some software residing in different parts of the platform. The decomposition presented in Figure 8.1 is adequate for a high-level reliability model in which there may exist some broad-based reliability estimates for each of the parts of the decomposition. If more detail is necessary or desired, the operations should be part of the decomposition also. Section 3.4.2.3 presents an example of a detailed service functional decomposition including the specific messages among the three entities in the diagram.

8.4 SERVICE FAILURE MODES AND FAILURE MECHANISMS

8.4.1 Introduction

Consistent with the considerations of Section 8.2.2, we will focus on reliability engineering for on-demand services for the remainder of this chapter.

Service failure modes derive from the service's attribute requirements in exactly the same way system failure modes derive from the system's attribute (functional, performance, physical, and safety) requirements. Any instance in which a requirement is not met is a failure, and a failure mode is an overt indication that this has taken place. For example, in voice telephony service, when one or the other party can no longer hear the other, a "cutoff call" has occurred. That these are rare events (except possibly in wireless telephony) speaks to the advanced state of development of telephony infrastructure in the developed world. This is an instance of a service failure, and the failure mode is the cutoff call. Many additional examples of service failure modes for telecom services are found in Ref. 10. Sufficiently complex services operate nearly 100% of the time in some degraded state for some transactions or users. For instance, folklore has it that about 5% of the routers in the Internet are failed at any given time. The robustness of IP helps ensure that these failures are hardly noticeable by users, although if demand increases enough, added congestion in the access network will be noticeable.

To facilitate the examination of service failure modes and failure mechanisms, it helps to analyze the concept of transaction in greater detail. As a transaction has an initiation phase, a proceeding phase, and an ending phase, we may classify service failure correspondingly as

- service accessibility failures,
- service continuity failures, and
- service release failures.

Service accessibility failures are any failures that are connected with the inability to set up or initiate a transaction. Service continuity failures are any failures connected with the ability to carry on a transaction to its completion, given that the transaction started correctly. Service release failures are any failures that are connected with the inability to dismiss a transaction that has completed its proceeding phase correctly. Some detailed examples in the context of telephony services are found in Ref. 10.

As with products or systems, reliability engineering requires that the failure mechanisms associated with each failure mode be understood. Here the study of service reliability requires an additional step. The key observation in service reliability engineering is that the <u>service failure mechanisms are events in the SDI</u> that cause transaction failures. This additional step is needed in service reliability engineering because services are intangible: an action can only fail if something required to carry out that action does not (by action or omission) complete the steps needed to further the action. In other words, service failures are caused by failures in the SDI, and the type of service failure seen is determined by the type of infrastructure failure. A good example of this is again drawn from telecom: infrastructure failures that occur while a call is being set up lead to service accessibility failures and infrastructure failures that occur during the stable phase of a call lead to service continuity failures.

Definition: <u>Service accessibility</u> is the ability to set up a transaction at a time desired by the user. <u>Service continuity</u> is the ability to carry on a transaction, without interruption and with satisfactory quality, until the desired completion of the transaction, given that it has been successfully initiated. Service release is the ability to successfully dismiss a transaction when it has been completed. This three-part breakdown is called the standard transaction decomposition.

Language tip: As usual, the same words are used for both the definition as given earlier and for the associated probabilities. So, for instance, we may speak of "service accessibility" both as the corpus of events associated with setting up or initiating transactions and as the probability of being able to successfully set up or initiate a transaction. Care should be taken to distinguish which meaning is intended if it is not clear from the context.

The service provider sets requirements for services using the same process of understanding customers' needs and performing economic trade-offs that is used in setting requirements for tangible products and systems. When any requirement is not met in a particular transaction, that transaction is failed, and a service failure has occurred. Service failure modes are the overt signs that a service requirement is not met. It is helpful to classify service failure modes according to the standard transaction reliability analysis given earlier.

Requirements tip: Sometimes, what looks like a single service to a customer is a composite of two or more services from different service providers. For example, purchasing goods over the Internet from a specific seller involved two SDIs controlled by two different providers: the seller's servers and associated software and the Internet service provider's wide-area and access networks. The seller cannot set a requirement for any service failure modes without an understanding of the ISP's performance and possibly some agreement with the ISP regarding carriage of the seller's traffic. Under net neutrality rules, the ISP can make no special provision for the seller's traffic, and any service reliability requirements from the seller need to be harmonized with the ISP's service reliability performance. Should net neutrality rules be modified, service providers like Netflix may be free to write agreements concerning the contribution of the ISP to their service reliability. This book takes no position on net neutrality, noting only that systems engineers need to account for all segments and contributors to an SDI when crafting service reliability requirements.

8.4.2 Service Failure Modes

8.4.2.1 Service accessibility

Service accessibility failures are those occurring in the process of starting, or "setting up," a transaction. Some examples of service accessibility failure modes are failure of a requested www page to load, excess delay in completing a telephone call (from receipt of dialed digits to the start of ringing at the far end) to the distant party (if the service provider has a requirement for delay), or failure of a home heating oil delivery to meet an appointed time. The typical service accessibility failure mode is excess delay: the customer makes a request to initiate a transaction, and service provider's SDI does not respond in a timely manner (provided the service provider's service accessibility requirement(s) contain a limit on the amount of time a transaction setup should take). A complete failure to set up a transaction would be an instance in which no matter how long the user were to wait, the transaction would not begin. An excess delay may look to an impatient user like a complete failure to set up a transaction, but the user's balking probability must be taken into account. See Exercise 3.

In some cases, particularly for telecom and datacom services, service accessibility failures may be the result of congestion in the service provider's network, even if all elements of the network are working properly. This is caused by an excess of demand above the network's capacity.[1] To the user, this looks like a failure, while to the service provider, this may be normal behavior. Here is an example of a service reliability requirement in action. The service provider may have established a service accessibility requirement that says "No more than 1.5% of transaction initiation attempts will experience a delay of greater than 3 seconds."[2] A user experiencing more than 3 seconds of delay may be one of those caught in the 1.5% when the service provider's SDI is operating nominally. It is also possible that this requirement is not being met—more than 1.5% of users are experiencing delay over 3 seconds at that time. To discern which is the case, measurements need to be taken, data collected, and analyzed. We have discussed interpretation of requirements for products and systems in Chapter 5, but some additional ideas are needed when dealing with services. See Section 8.5.2.

8.4.2.2 Service continuity

Service continuity failures are those that take place during the time a transaction is being conducted, after it has been properly set up or initiated. Some examples of service continuity failures are failure to deliver a package that was properly accepted by the postal service and paid for by the user, excess distortion in the voice or video of a video teleconference call (if the service provider has a requirement limiting distortion), and delivery of a refrigerator model different from the one that was sold. It is sometimes useful to distinguish two types of service continuity failures. One type comprises those failures in which a transaction is interrupted and does not resume at all. The other comprises those transactions that carry on to completion but fail to adhere to one or more of the service provider's quality requirements. The cutoff call is a good example of the first type. The parties are talking and suddenly "the connection disappears" and neither party hears the other. This condition persists indefinitely, that is, no matter how long the parties were to wait, the conversation would not

[1] When network elements fail, additional congestion may result from the same level of demand that may lead to normal congestion when all network elements were working properly.

[2] Old telephone engineers may recognize this as the standard Bell System dial tone delay requirement.

be restored. An example of the second type is furnished by the video teleconference in which audio and/or video are distorted beyond the limits allowed for by the service provider's requirements. The conference has not ceased entirely, but its quality fails to meet the relevant requirement(s). Hoeflin [6] refers to the latter type of service continuity failures as "service fulfillment failures."

8.4.2.3 Service release

Service release failures are those connected with the release of the service provider's resources after a transaction is complete. They also include any post-transaction actions that are connected with the specific transaction, such as billing, that affect the user's perception of transaction success. Service release failures tend to occur less often than service accessibility or service continuity failures, but they do have economic consequences for the service provider who should, therefore, include them in a service reliability plan. Some examples of service release failures include an incorrect bill for a completed shipment, refusal of a videoconference to terminate after the terminate signal was given, etc.

8.4.3 Service Failure Mechanisms

As noted in Section 8.1, identifying the failure mechanisms connected with a service failure mode requires understanding what takes place in the service provider's SDI to create the particular type of transaction under study. Failure of these actions to take place, or completion of the actions in an inadequate way, leads to transaction failures. Two important features of service failure mechanism analysis are

1. It is essentially a fault tree analysis (Chapter 6) with the additional step of incorporating how elements and processes in the service provider's SDI work together to deliver a transaction in the service.
2. As we saw in the product/system case, the fault tree can be developed down to very detailed events. The analyst needs to determine the amount of detail that is included and how small the probability of an event should be for it to be omitted from the analysis.

We may classify service failure mechanisms following the standard transaction reliability analysis model of Section 8.4.1.

8.4.3.1 Service accessibility

Service accessibility failure mechanisms disrupt the process of setting up or initiating a transaction. Each service accessibility failure mode may be traced to one or more service accessibility failure mechanisms. Here are some examples.

- The US Postal Service's online label printing and payment service: Suppose you are preparing a shipment and wish to pay for the shipment and print a shipping label with postage using this service. The most prominent service accessibility failure mode is inability to access the USPS web page for shipment preparation. Some associated failure mechanisms are:

  - The USPS servers are experiencing a hardware outage, software failure, or a distributed denial of service (DDoS) attack.
  - Some problem occurs in the transport of information from your PC to the USPS servers. This may include packet loss, network congestion, or a local problem such as a malfunctioning network interface.

- Cloud computing: You contract with a cloud computing service provider to back up your local files each night. One applicable service accessibility failure mode is "backup does not start as scheduled." Some of the associated failure mechanisms are:
  - The service provider's servers are unavailable.
  - The mechanism the service provider uses to assign content to different servers has failed.
  - Failure in the communication process between your router and the service provider.
- Self-service gasoline fill-up: Many filling stations require you to swipe a credit card before you can dispense gasoline. A service accessibility failure mode here is pumping does not begin when the pump handle is actuated. Some associated service failure mechanisms are:
  - Failure to register credit card information correctly, due to
    - Failure of the credit card reader,
    - Failure in the communication path between the pump and your bank,
    - Failure of the bank's servers due to hardware outage, software failure, or DDoS attack,
  - Internal gasoline pump failure,
  - External gasoline pump failure (e.g., dispenser handle failure).

Note how, in these examples, more than one service provider is involved in a service failure mechanism: the immediate service provider (the cloud service provider, the filling station, etc.) or the background service provider (a telecom company, your bank, etc.) may be the source of the failure experienced by the user—who is indifferent to the source of the failure. As a business matter, the primary service provider's service reliability planning must include its background partner(s) so that a unified service experience may be presented to users. Users who experience service failures should not be expected to diagnose the source of the problem, and their normal behavior will be to contact the primary service provider if they desire remediation of the problem.

In telecommunications and other services in which the SDI includes shared resources, transaction failures arise from competition for those shared resources. Some transaction requests will be denied because there is not enough capacity in the SDI to accept them all. It is usually not economically feasible to provide enough shared resources to handle every transaction request at all times, even when every element of the SDI is operating normally, so some degree of service inaccessibility is deliberately built into the service. The degree to which this prevails is a decision by the service provider based on their understanding of customer needs and behavior and unavoidable economic trade-offs. When elements of the SDI fail, this congestion, whose symptoms include increasing queuing delays and buffer overflows, increases even when the transaction request intensity stays the same. Any individual customer cannot tell, and should not be expected to tell, whether failure of her transaction request is due to "normal" congestion (the level of service inaccessibility specified in the requirement when all elements of the SDI are operating normally) or to "extra" congestion resulting from the failure of some element(s) of the SDI. Service designers should model service accessibility under normal and disrupted conditions in the SDI so that it is possible to understand the user experience and discern whether improvements (or cutbacks) in the SDI are warranted.

8.4.3.2 Service continuity

Service continuity failure mechanisms are those that disrupt a transaction once it has properly been initiated. Each service continuity failure mode may be traced to one or more service continuity failure mechanisms. The following are some examples:

- The US Postal Service's online label printing and payment service: A prominent service continuity failure mode is an interrupted transaction. Some associated failure mechanisms are
  - Packet loss or congestion in the WAN connecting you to the USPS server,
  - USPS server failure,
  - Failure in the credit card payment network.
- Cloud computing: One applicable service continuity failure mode is "backup does not complete." Some associated failure mechanisms are
  - The mechanism the service provider uses to assign content to different servers fails while the backup is in progress,
  - Failure, occurring while backup is in progress, in the LAN or WAN between you and the service provider.
- Self-service gasoline fill-up: A service continuity failure mode here is pumping stops before the desired amount of gasoline is dispensed. Some associated service failure mechanisms are:
  - Malfunctioning pump handle holding mechanism,
  - Internal pump failure while dispensing is in progress.

8.4.3.3 Service release

Service release failures are any failures that cause inability to dismiss a transaction that has completed its proceeding phase correctly. Each service release failure mode may be traced to one or more service release failure mechanisms. Here are some examples.

- The US Postal Service's online label printing and payment service: A service release failure mode is an incorrect amount billed by USPS to your credit card. Some associated failure mechanisms are
  - Communication failure between the USPS and the credit card company,
  - Interception and altering of the communication by a malicious actor.
- Cloud computing: An applicable service release failure mode is "backup application does not close when the backup is complete." Some associated failure mechanisms are
  - Lack of notification from the service provider that the backup was (or was not) successful,
  - Failure in the communication process between your router and the service provider.
- Self-service gasoline fill-up: A service release failure mode here is "pump does not print receipt." Some associated service failure mechanisms are
  - Printer is out of paper,
  - Printer electronics failure, and
  - Customer keypad failure.

In each case, note how the causes of a user-perceived transaction failure can be traced back to some action or inaction in the SDI. You may say that it is enough to assign reliability requirements for all parts of the SDI, and this will control service transaction failures. However, using this approach ignores the vital information about how users perceive the reliability of the service. In addition, the user perspective allows you to decide how much reliability in the SDI is really needed—it is easy to over- or under-provision the SDI if you are not being guided by the needs and desires of the user community. A reasonable argument could be advanced, for instance, that the old Bell System requirement of switching system availability of at least 0.9999943 (expected downtime of 2 hours in 40 years of operation) was far tighter than necessary to achieve satisfactory POTS[3] reliability given the many redundant paths through the PSTN[4] for any pair of users. Only by understanding the user needs and desires for service reliability can the reliability requirements for the SDI and its components be developed in a rational way. The service reliability

[3] Plain old telephone service.

[4] Public switched telephone network.

requirements need to drive the SDI reliability requirements so that neither overspending nor inadequate provisioning errors are made. See Section 8.7.

8.5 SERVICE RELIABILITY REQUIREMENTS

Service reliability requirements are categorized according to the standard transaction reliability analysis. The service provider uses their understanding of the needs and wants of their service customers, together with their understanding of their SDI's behavior, to devise service reliability requirements that promote user satisfaction and are economically sensible.

8.5.1 Examples of Service Reliability Requirements

Reliability requirements for on-demand services may be organized according to the categories described in Section 8.4.2: accessibility, continuity, and release. Requirements may also be structured to apply to each individual transaction (Section 8.5.1.1) or to some aggregate of transactions, usually over a specified user population (Section 8.5.1.2). In each case, data collection and analysis to verify compliance with requirements are discussed in Section 8.5.2.

8.5.1.1 Per-transaction reliability requirements

Service reliability requirements may be structured to apply to individual transactions by specifying a proportion of transactions that may fail (or will succeed). Frequently, the proportion is expressed as a fraction of unsuccessful transactions per number of opportunities. In telecommunication and other fields where transactions are numerous and of relatively short duration, the proportion is often expressed as "defects per million (DPM) opportunities." The proportion is a reliability effectiveness criterion whose achievement may be demonstrated by modeling (when designing the service) or data analysis (when validating achievement of the requirement after deployment). A related reliability figure of merit is the probability of success (or failure) per transaction.

For example, a service accessibility requirement may look like "the proportion of transactions that fail to initiate properly after a valid customer request shall be no more than 0.005% when all SDI elements are operating normally." When expressed this way, with no conditions on sources of the requests, the requirement applies to any request from any service user. This means that even users accessing the worst (most congested) part of the SDI are to be treated as well as this requirement states. Some (many) users may experience service accessibility (much) better than this under normal conditions, but this formulation does not allow for different classes of service (e.g., better service reliability for a higher price) or for discernment of which parts of the SDI may be improved most efficiently (biggest return on service accessibility per dollar expended on improvement). Some of these objections may be handled by aggregating service users into various groupings.

8.5.1.2 Aggregated reliability requirements

In aggregated service reliability requirements, the requirement structure remains the same except that a requirement is written to pertain only to some specific group of users. For example, requirements may be written for users in a certain city, users purchasing a stated class of service, transactions occurring during a stated period of time, etc. Grouping allows focused improvement based on solid understanding of how service reliability may vary from group to group. It also allows for selling different classes of service—for example, video teleconferencing may be offered with standard definition video, or high-definition video for a higher price. Service accessibility and continuity requirements and service-level agreements may differ across different classes of service.

8.5.2 Interpretation of Service Reliability Requirements

Service reliability requirements are usually written as limits on some percentage of transactions that fail in each of the accessibility, continuity, and release categories. So expressed, the requirement is on a reliability effectiveness criterion. When designing a service, or analyzing data to determine compliance with a reliability requirement in the service, we compare the probability of a successful transaction with the value of the reliability effectiveness criterion. Requirements may also be written as a limit on the overall, or "omnibus," transaction failure proportion (including accessibility, continuity, and release all in one measurement); in that case, the standard service reliability transaction analysis helps the service provider create a plan to meet the omnibus requirement by controlling each of the contributing factors. Many service providers express the percentage as a "DPM" measure, which is a percent multiplied by 10^4, because the number of transaction failures is usually small. Many service providers, especially in the telecom industry, are able to acquire data on every transaction through automated means, and when a census like this is available, comparison of results with requirements needs only a specification of a time period over which the requirement is to hold. However, when a census is not possible, a sample of transactions is taken, and realized service reliability is estimated from the sample. Comparing realized service reliability with requirements then is a problem of estimating a proportion. The statistical procedure for this is given in section 9.1 of Ref. 1. The following example illustrates the ideas in a telecommunications context.

> **Example:** Suppose that a requirement for VoIP telephone service specifies that its reliability shall be no worse than 3.4 DPM. To demonstrate compliance with this requirement, we will test the hypothesis that the probability that a VoIP transaction fails (for any reason) does not exceed $r = 3.4 \times 10^{-6}$. A sample of 100,000 VoIP calls is taken, and the number of failed calls in the sample is 2. It would appear that the requirement is not being met. What is the strength of the evidence for this conclusion?

Solution: Let p_0 denote the omnibus probability of transaction failure (i.e., including all the failure modes that the service provider counts when making the DPM determination) for all transactions in the population of VoIP calls comprehended by the service provider's reliability management plan. We will test the null hypothesis H_0: $\{p_0 \leq r = 3.4 \times 10^{-6}\}$ (the requirement is met) against the alternative H_A: $\{p_0 > r\}$ (performance is worse than the requirement). The appropriate statistical inference procedure is a test for proportions as described in section 10.3 of Ref. 1. The test statistic is the normalized sample proportion $\hat{p} - r / \sqrt{r(1-r)/n}$ (where $\hat{p}$ is the sample proportion), which in this case is 2.8469, yielding a p-value of 0.0022 (this is the probability that, if the null hypothesis were true, you would see the result in the data, that is, 2 or more failed calls in 100,000, by chance). We reject the null hypothesis, and the evidence that the requirement is not being met is very strong (the result is unlikely to be a chance occurrence). If the sample contained 2 failed transactions in 1,000,000, then the test statistic value is –0.7593, yielding a p-value of 0.7762, and strong evidence in favor of the null hypothesis. With 14 failed transactions in 10,000,000, the test statistic value is –5.59, yielding a p-value of 0.9999, very strong evidence in favor of the null hypothesis. The value of large sample sizes in statistical inference for small proportions is clear.

In Chapter 5, we approached demonstrations of this kind by estimating the proportion of (in this case) defective transactions, or, equivalently, estimating the probability that a transaction may fail. Either approach (that of Chapter 5 or that shown here) is suitable. The choice of which to use may depend on which is easier to communicate for the particular audience you face.

If a requirement applies to a stated aggregated population, data collection to verify achievement of that requirement should be limited to members of that population.

8.6 SERVICE-LEVEL AGREEMENTS

A *service-level agreement* is a statement by the service provider that some compensation will be paid to a service user in the event some service reliability requirement is violated by a stated margin. Each customer of the service provider has their own service-level agreement. A service-level agreement is a way for the service provider to make a service more attractive to potential purchasers.

A typical service-level agreement in telecommunications services may read "In the event that the service is unavailable for more than 30 minutes during a single calendar month, the service provider will rebate 5% of the service price[5]

[5] Numbers do not represent any particular service provider or service-level agreement and are for illustrative purposes only.

for that month." The agreement pertains to a specific service. A single service provider and a single service customer may have several service-level agreements in force, one for each service purchased by the customer from that service provider. As the service-level agreement is part of the contract between the service provider and a specific service customer, measurements need to be made for each service and each particular customer with whom the service provider has a service-level agreement. The provisions of the agreement are triggered when the measurements for that particular customer show the provisions are violated.

A key question for the service provider regarding service-level agreements is profitability. Before offering a service-level agreement, the service provider should have some idea of whether it will make or lose money on the agreement. Models based on the ideas in Chapter 4 may be constructed to study profitability of service-level agreements. See Exercise 4 for some ideas.

8.7 SDI RELIABILITY REQUIREMENTS

A basic principle of service reliability engineering is that the reliability of a service, in terms of its accessibility, continuity, and release properties, is determined by actions or omissions taking place in the SDI, so assignment of reliability requirements to elements of the SDI should be set so that when they are met, the service reliability requirements are met. This is a kind of reliability budgeting (Sections 2.8.4 and 4.7.3). It may be accomplished by formal means, as described in Section 4.7.3, or less formally when the precision of the input information available does not justify the expense and time required to complete a formal analysis. Here is an example of the latter case.

Example: In POTS, a dedicated circuit ("talking path") comprising switching and transport elements is set up and held in place during the entire conversation between two parties. The primary cause of transaction interruptions in this service is failure of one (or more) of the elements in the talking path. Suppose a service continuity requirement for interruptions states that no more than 25 calls per 1,000,000 carried are to be interrupted. How should requirements be written for the switching and transport elements of the network so that this service reliability requirement will be met?

Solution: Interrupted transactions ("cutoff calls" in this service) occur when an element in the talking path for that call fails, so the prevalence of cutoff calls is determined by the frequency of failures in switching and transport systems. Let σ denote the number of failures per hour of switching systems and τ denote the number of failures per hour of transport systems (for simplicity, we will take σ and τ to apply to all types of switching and transport systems, respectively). If a talking path contains n switches and $n+1$ transport systems (in which case we say its size is n), then the number of failures per hour in that talking path is $n\sigma+(n+1)\tau$ (why?). Let $r=2.5\times10^{-5}$, and the number of calls per hour be C. Furthermore, let the probability that a talking path contains n switches and $n+1$ transport systems be p_n, $n=1, 2, \ldots, N$, where N is the maximum talking path size allowed by other considerations, such as a loss plan. Then, based on expected values, we require

$$\sum_{n=1}^{N}\left[n\sigma+(n+1)\tau\right]p_n \le rC.$$

If the cost of achieving a failure rate of σ failures per hour in switching is $k_S(\sigma)$ and that for τ failures per hour in transport is $k_T(\tau)$, then a sensible assignment based on minimized expected cost is achieved by solving the optimization problem

$$\text{Minimize} \sum_{n=1}^{N}\left[nk_S(\sigma)+(n+1)k_T(\tau)\right]p_n \text{ subject to} \sum_{n=1}^{N}\left[n\sigma+(n+1)\tau\right]p_n \le rC.$$

While this example pertains to an obsolete technology, and several simplifications were applied, it is intended to illustrate the important idea that service reliability comes from SDI reliability, and these should be considered together when designing a service. More precisely, service reliability requirements should be used to drive reliability requirements for elements of the SDI. The illustration given here is overly simplified but should provide useful guidance for more realistic studies of this kind.

An added feature that sometimes needs to be taken into account when partitioning or assigning service reliability requirements into the SDI is that achievement of a successful transaction may include an element of timing: certain actions need to happen in a specific order to enable a successful transaction (see the service functional decomposition example in Section 3.4.2.3). The reliability models shown in Chapters 3 and 4 cannot incorporate timing, so, for instance, if one wanted to analyze the VoIP service using SIP as shown in Section 3.4.2.3, a richer set of models would be needed. Situations like this are well adapted for modeling using stochastic Petri nets, which do allow for sequencing and timing of events. Description of stochastic Petri net modeling for reliability is beyond the scope of this book. Interested readers are referred to Refs. 2, 4, 12, or Ref. 8 for a good introduction.

8.8 DESIGN FOR RELIABILITY TECHNIQUES FOR SERVICES

The service functional decomposition (Section 8.3) is a good starting point for designing for reliability for services. It contains raw material for service fault tree analysis and service FME(C)A:

- How the SDI is configured to carry out the functions needed to support and deliver the service, and
- Messages that need to be sent and received correctly for a transaction in the service to be successful.

Once configuration of the SDI and appropriate assignment of reliability requirements (Section 8.7) are known, design for reliability techniques for hardware (Chapter 6) and software (Chapter 9) may be brought to bear. Should it be necessary to go into that level of detail, modeling for sequencing and timing of messages or other SDI events is introduced in Section 8.7.

This section discusses modifications or enhancements of fault tree analysis and FME(C)A for use with services. The main idea is that steps are added to the performance of fault tree analysis and FME(C)A that we have seen so far (Chapter 6) to account for the interface between the SDI and the service.

8.8.1 Service Fault Tree Analysis

Top events for fault trees for services can come from the listing of failure modes in the service, which, in turn, come from the service reliability requirements. This is the same pattern we follow for the fault tree analyses studied so far. It promotes a systematic approach for the development of a fault tree for a service. The top event will be some service reliability requirement violation whose causes are sought in the SDI or in the actions of the user. As with the fault trees developed for products and systems in Chapter 6, the Ishikawa, or fishbone, diagram can be a useful aid in developing the fault tree as well as for root cause analysis when diagnosing the SDI events or omissions contributing to service failures.

8.8.2 Service FME(C)A

FME(C)A is a bottom-up analysis that begins with an undesired event in some component of a system and develops the consequences of that event up to the point where a system failure follows. In applying FME(C)A to services, the system is the SDI. So a service FME(C)A begins with some undesirable event in the SDI (e.g., failure of a line card in an edge router). Additional steps are required at the end of the chain of consequences reasoning to determine the consequence(s) for the service. In all other respects, FME(C)A for services is the same as we have used before in Chapter 6.

8.9 CURRENT BEST PRACTICES IN SERVICE RELIABILITY ENGINEERING

8.9.1 Set Reliability Requirements for the Service

If you are a service provider, you will need to understand your customers' experiences on all dimensions of the services you sell, including the reliability dimension. Reliability is a vital part of the value proposition for any service, and understanding and controlling reliability at the service level is best accomplished by assigning reliability requirements at the service level. Furthermore, those reliability requirements should be agnostic with respect to the technologies in the SDI. Most customers don't know or care what technologies are used to provide their service. Their interaction with the service provider is strictly at the service level. Changes in the SDI should be invisible to service users. Service providers will of course advertise and sell the improved service reliability and performance that may result from improved SDI, but the fundamental point is still that what the user sees is the service, and properly managing that interface requires explicit statement of service reliability requirements.

8.9.2 Determine Infrastructure Reliability Requirements from Service Reliability Requirements

Behavior of the SDI determines the reliability of the services it supports. It makes no sense to independently assign reliability requirements for those services and for elements of the SDI because doing so risks conflicts and possible under- or over-provisioning of infrastructure elements. Either of these has economic consequences. Under-provisioning saves in initial capital expenditures at the cost of poorer service and damage to reputation. Over-provisioning causes capital expenditures that may be more than needed to assure adequate service reliability. These risks can be avoided by linking SDI reliability requirements to service reliability requirements as described in Section 8.7.

8.9.3 Monitor Achievement of Service Reliability Requirements

It is no less important to carry out this part of the Deming Cycle for a service as it is for any product or system. Monitoring service transactions does raise the unique issue of privacy, and responsible service providers will employ methods that respect user privacy. Some mathematical models [7, 13] have been developed that aid in this by enabling service-provider-generated transactions (e.g., ping packets used by an ISP to query and categorize network states and user experience) to at least approximately reflect what a user experiences without compromising privacy.

8.10 CHAPTER SUMMARY

This chapter brings to services the engineering principles needed to assure their reliability. Many technological systems are deployed precisely because they provide a service to some community of users. The chapter makes the key point that, in these cases, reliability requirements for the underlying technological systems (the SDI) should be derived from, and be consistent with, the service reliability requirements. Service reliability requirements come to the fore because the service is what the user purchases from the service provider and its characteristics are what will satisfy (or dissatisfy) the user/customer.

Accordingly, the chapter begins with a discussion of on-demand and always-on services, and because reliability of an always-on service is equivalent to the reliability of its delivery infrastructure, the focus moves to on-demand services. The basic unit of on-demand services is the transaction, and we study service reliability requirements for accessibility, continuity, and release of transactions. Examples of service functional decomposition, service failure modes, and failure mechanisms are given, followed by development, interpretation, and verification of service reliability requirements. Techniques akin to reliability budgeting are described for rationally assigning reliability requirements to the SDI so that they are consistent with the reliability requirements for the services it supports. The chapter closes with a discussion of design for reliability for services.

8.11 EXERCISES

1. Identify service failure modes and failure mechanisms for the following:
   a. Facsimile service (sending documents via telephone)
   b. Cloud computing service
   c. A smartphone weather forecasting app
   d. Package delivery service
   Hint: define a transaction in each service first and use the service accessibility, service continuity, and service release formalism. Identify the SDI in each case.
2. Discuss service reliability aspects of a PC application.
3. Suppose each user in a specified population of service users has a random time B for which the user will abandon a transaction setup attempt if the delay in setting up the transaction exceeds B. B may vary from user to user and even from time to time for the same user. Let B have distribution $F(t) = P\{B \leq t\}$ and suppose it is the same for all users. Suppose the service provider's delay in setting up a transaction is a random variable D having distribution function $W(t)$. What is the proportion of users in that specified population who see transaction setups as failed? How should this phenomenon be accounted for in writing a service accessibility requirement?
4. *A high-level model for service-level agreements. Service provider V contracts with a specific group of service customers for whom service accessibility is agreed to be at least 0.9995. The relevant service-level agreement states that if the service is inaccessible for more than 30 minutes in a 30-day month, V will rebate some portion of the price paid for the service for that month. V's SDI suffers outages affecting these customers according to an alternating renewal process whose uptime distribution is exponential with a mean of 3000 hours and whose downtime distribution is exponential with a mean of 1 hour. What is the probability that V will have to pay a rebate in a given month? Assume that the SDI reliability process has been operating for a long time.

REFERENCES

1. Berry DA, Lindgren BW. *Statistics: Theory and Methods*. 2nd ed. Belmont: Duxbury Press (Wadsworth); 1996.
2. Ciardo G, Muppala J, Trivedi K. SPNP: stochastic Petri net package. Proceedings of the Third International Workshop on Petri Nets and Performance Models, 1989. December 11–13, 1989; Piscataway, NJ: IEEE; 1989. p 142–151.
3. Dai YS, Xie M, Poh KL, Liu GQ. A study of service reliability and availability for distributed systems. Reliab Eng Syst Saf 2003;79 (1):103–112.
4. Florin G, Fraize C, Natkin S. Stochastic Petri nets: properties, applications and tools. Microelectron Reliab 1991;31 (4):669–697.
5. Grassi V, Patella S. Reliability prediction for service-oriented computing environments. IEEE Internet Comput 2006;10 (3):43–49.
6. Hoeflin DA, Sherif MH. An integrated defect tracking model for product deployment in telecom services. *Proceedings of the 10th IEEE Symposium on Computers and Communications*. June 27–30, 2005; Piscataway, NJ: IEEE; 2005. p 927–932.
7. Melamed B, Whitt W. On arrivals that see time averages: a martingale approach. J Appl Probab 1990;27 (2):376–384.
8. Sahner RA, Trivedi K, Puliafito A. *Performance and Reliability Analysis of Computer Systems: An Example-Based Approach Using the SHARPE Software Package*. New York: Springer; 2012.
9. Tollefson G, Billinton R, Wacker G. Comprehensive bibliography on reliability worth and electrical service consumer interruption costs: 1980–90. IEEE Trans Power Syst 1991;6 (4):1508–1514.
10. Tortorella M. Service reliability theory and engineering, I: foundations. Qual Technol Quant Manage 2005;2 (1):1–16.
11. Tortorella M. Service reliability theory and engineering, II: models and examples. Qual Technol Quant Manage 2005;2 (1):17–37.
12. Volovoi V. Modeling of system reliability Petri nets with aging tokens. Reliab Eng Syst Saf 2004;84 (2):149–161.
13. Wolff RW. Poisson arrivals see time averages. Oper Res 1982;30 (2):223–231.

9

Reliability Engineering for the Software Component of Systems and Services

9.1 WHAT TO EXPECT FROM THIS CHAPTER

Every technological system or service of any consequence has a significant software component. The share of system failures attributable to failures in the software component is significant. Software has enough unique attributes that it is wise to treat reliability engineering for software as a subject in itself. In this chapter, you will see how software reliability engineering has been handled in the past and be introduced to an approach to software reliability engineering that is consistent with the design for reliability methods discussed in Chapter 6. In keeping with the theme of this book, we give a broad overview aimed at equipping systems engineers to manage software reliability and refer to other sources for details of specific methods.

Reliability, Maintainability, and Supportability: Best Practices for Systems Engineers, First Edition. Michael Tortorella.

9.2 INTRODUCTION

Up to now, we have directed attention to reliability requirements and engineering for entire products or systems on the one hand and intangibles like services on the other. Certainly many, if not all, of the products and services common in a technological society contain a significant software component that is essential to their functioning. Usually, when constructing reliability requirements for these systems and services, it is necessary to pay special attention to the software they contain because

- it is responsible for many of the failure modes uncovered while analyzing the system's or service's attribute requirements, and
- reliability engineering for software presents unique challenges.

We understand a software failure to be either

- a violation of requirements specifically pertaining to the software component of the system, or
- a violation of any system requirement that is caused by misbehavior of the software component of the system.

From here, it is straightforward that the definition of software reliability is the absence of software failures for a stated period of time and under stated operational conditions. This is the same definition we have always used for reliability, except that it is now specifically focused on the software component of the system or service.

Reliability modeling and engineering for the software component of systems or services has traditionally been undertaken using a test, analyze, and fix (TAAF) approach [21, 27, 28, 31]. In fact, time was when "software reliability engineering" was synonymous with the TAAF procedure. This is largely a historical accident, stemming from particular approaches chosen by early practitioners in the field [25]. These emphasized software reliability growth[1] models fed by data gathered during testing. As a rule, TAAF is not favored as a reliability engineering strategy for systems and services because it is costly and time-consuming—impossibly so for the hardware component of a product or system. However, it has had some success in the software arena because

- errors in software are deterministic in the sense that repeated execution of a part of the software that contains an error will always lead to the same failure, and
- correction of errors found in testing can usually be accomplished more quickly in software than in hardware.

[1] A general discussion of reliability growth testing is found in Section 5.7.2.

Even so, TAAF is limited in its remedial abilities. It is most effective for small-scale errors like code errors. Its effectiveness is limited when dealing with more fundamental problems such as ineffective design or inefficient execution of a design where correction might involve extensive and time-consuming rework. Contemporary quality engineering principles indicate that it is always expensive, and usually impossible, to achieve satisfactory product or service quality and reliability solely by testing. These principles emphasize design for reliability and prevention of failures through actions taken at early stages in the design and development process. In this chapter, we review current practices in software reliability modeling and engineering and reexamine the construction of reliability requirements and reliability modeling and engineering for software from this perspective. Indeed, reliability engineering for software has more in common with reliability engineering as we have understood it so far than is usually appreciated. Even though the subject matter, that is, software, looks different in some respects than the systems we have treated so far, our position here is that reliability engineering for software proceeds from the same principles we have used so far: understand the failure modes at play and do what is economically reasonable to prevent these failure modes from appearing through an solid understanding of the failure mechanisms responsible.

9.3 RELIABILITY REQUIREMENTS FOR THE SOFTWARE COMPONENT OF SYSTEMS AND SERVICES

9.3.1 Allocation of System Reliability Requirements to the Software Component

Through our first introduction in Section 3.5 and its continued use thereafter, the approach we advocate to creating reliability requirements should be familiar:

- Catalog the system requirements, paying particular attention to system attribute requirements:
  - Functions the system must perform,
  - Performance goals to be met for each function,
  - Physical characteristics, and
  - Safety.
- Determine the failure modes associated with each of the attribute requirements,
- Determine the customer's tolerance for failure in each of these failure modes,
- Balance the customers' needs and desires regarding reliability with the economics of developing a system meeting the attribute requirements, and
- Document the reliability requirements that result from this analysis.

When system or service attribute requirements involve actions performed by software, those actions may be the source of failures that we would then say are due to the software. For instance, fault tree analysis of the system or service may reveal events that, if the software component does not perform correctly, lead to a system or service failure. For example, in the passenger elevator fault tree analysis example shown in Section 6.6.1.2, event 15 is a software failure in the controller that erroneously turns off power to the elevator motor. Such events and actions are properly the subject of reliability requirements for the software component. We use the system functional decomposition to enumerate essential functions of the software component. Reliability requirements can then be constructed for them.

Example: Consider an oil transportation pipeline network consisting of pipes, terminals, and valves. The essential features of this network are that oil enters the network at certain terminals for transport to other terminals through the pipes in the network, and the valves act as switching elements that control the flow. Automatic control of the flow is implemented by software that opens and closes valves in response to the current demand, the values of the flow throughout the network, and commands from a network control center. The software must be able to read transducers[2] that tell the value of the flow at certain points in the network and respond to the varying values transmitted by all the transducers in the network. The software also needs to perform computations that enable the most efficient flow, satisfying the demands to be realized. Many kinds of failures are possible in such a network because not only is the network required to deliver the demanded oil volume to its destinations (within the engineered capacity of the network), but it is also required to do so safely. That is, in addition to functional requirements, the network also has safety requirements.[3] Violations of any of these requirements are failures. As an example, improper operation of valves may result in excess pressure at some point(s) in the pipeline network; and if this overpressure exceeds the working strength of the pipes, a rupture and spill results. In a fault tree analysis for the top event "rupture and spill," events involving operation of the controlling software are uncovered. Suppose that the service life of the pipeline network is 25 years and the requirement for "rupture and spill" is as follows: The probability of a pipeline rupture leading to a spill at any location in the pipeline network over the 25-year service life of the network shall not exceed 10^{-6}. What are the failure modes associated with this requirement? A systematic approach to an answer is provided by a fault tree analysis based on the top event "Pipeline rupture and spill somewhere in the network." It is not the purpose of this example to develop this fault tree in detail; instead, we focus only on that branch of the tree that includes the event "software failure." What kinds of software failures can lead to a pipeline rupture? At the risk of oversimplifying, we consider

[2] A transducer is a device that converts a physical property (in this case, flow) into a quantitative measurement (e.g., gallons per minute).

[3] Without doubt, it also has performance and physical requirements, but we are not treating those in this example.

only two: improper acquisition of flow transducer value(s) and improper execution of the algorithm that optimizes the flow based on the current transducer readings and the exogenous demands. Some portion of the 10^{-6} probability requirement is allocated to these two software failures, remembering that the fault tree for "pipeline rupture and spill somewhere in the network" will also have some hardware failures (e.g., corrosion and leakage) and externally caused failures (e.g., earthquake) to which part of the probability may be allocated also. Using either a formal optimization method such as that discussed in Section 4.7.3 or an informal cost-benefit analysis, a requirement that looks something like "The probability of a pipeline rupture and spill somewhere in the network due to a failure in the network's controller software shall not exceed 10^{-7} over 25 years"[4] may be developed.

It is worth examining this reasoning in more detail. How can such failures arise if the controller software has been designed and manufactured properly? As noted in Section 9.5.2.1, the software does not deteriorate if unmolested (although it may accumulate faults attendant on changes made to the software through maintenance[5]), so no new causes of failure are being introduced by nature or the passage of time. Some stress–strength interaction failures of the kind described in Section 9.5.2.2 may appear during operation if the particular combination of operational conditions and requirements-legal input conditions (RELICs)[6] causing them were not replicated during testing (for if they had been replicated in testing and caused a failure, then the underlying fault in the software would have been corrected[7]). So at least part of the requirement is to cover incomplete or flawed testing that lets some potential failures slip through. Even if this is the best we can do, it's still uncomfortable because it means that, even for a high-consequence system like the pipeline, our ability to carry out testing that can catch every potential failure is limited. As hard as we may try to ensure that they do not occur, the kinds of manufacturing defects described in Section 9.5.2.3 seem to be an inevitable feature of any software development. Design for reliability (Section 9.6) is intended to ensure that requirements misinterpretations, faulty designs, inefficient execution, and code errors do not occur, but, as is always the as is always the case with processes carried out by humans, imperfections do occur. The result is that development creates faults in the software (compare Section 3.3.6 in which manufacturing is characterized as an opportunity to introduce defects into the product) that may be removed, with greater or lesser success, during testing. These faults are weaknesses in the software. When the proper stress is applied (that particular

[4] The numbers shown in this example are for illustration only and are not intended to be realistic or prescriptive.

[5] This is sometimes known as "rot."

[6] Requirements-legal input conditions: inputs that fall within the boundaries of the permissible region of inputs according to the system's requirements.

[7] Of course, we always need to leave open the possibility that the correction was flawed and did not fix the underlying problem, introduced additional faults into the software, or both.

combination of operational conditions and inputs that executes the part of the code containing the fault), the software fails. To summarize: requirements for software reliability cover failures due to stress–strength interaction failures from defects introduced during manufacturing (faults)—that insert into the software specific weaknesses that cause failures when the particular combination of operational conditions and inputs causes to be executed the part of the software that contains the fault—and are not detected and properly corrected during testing.

9.3.2 Reliability Requirements for Security and Other Novel Areas

It has become apparent that the software that runs the Internet and other communication networks like cellular telephone networks can be misused in ways that lead to security breaches for both users and providers. Security has become a major concern for users and Internet Service Providers (ISPs) alike. Our position is that security, like safety, is (or should be) the subject of system or service requirements, and security problems are violations of those requirements that may be treated by the methods of reliability engineering for the system or service. Consideration of security offers an opportunity to review the system engineer's approach to reliability when confronted by a novel problem. A systems engineer does not need to be an expert in security engineering to create effective security requirements, and a reliability engineer is likely not a security expert. Both need collaboration with security experts to be effective in these aspects of their responsibilities: the systems engineer to learn what's possible for managing security through requirements, and the reliability engineer to develop the catalog of security failure modes and failure mechanisms that will be needed to design for reliability against the security requirements. The results of reliability engineering for security will be

- an improved understanding of the failure modes and failure mechanisms in the system or service that lead to security failures, and
- management of the frequency and duration of security failures and outages.

Note that this same pattern holds for any functional requirements in a new subject matter area.

Users perceive security breaches as failures to meet their expectations (although a cynic might observe that most users don't expect much from security these days). To manage security, systems engineers will encapsulate those expectations in security requirements, and reliability engineers will work to minimize the frequency and duration of violations of those requirements. To design for reliability with respect to security requirements, reliability engineers will need to work with security experts to determine the failure modes and failure mechanisms associated with security so that they may propose suitable countermeasures. As threats continue to evolve, additional understanding is needed to counter them, and design for reliability regarding security requirements is not a stable, finished discipline. Collaboration with security experts will help catalog the relevant failure modes and failure mechanisms needed to begin the design for reliability process to counter the latest threats.

9.3.3 Operational Time and Calendar Time

Much software runs continually, 24 hours a day, 7 days a week. For those that do, it is easy to discern the user experience with failures: their description in terms of calendar time is the same description in terms of operational time. For those that do not, an understanding of how operational time and calendar time are related is needed to be able to understand the user's experience in calendar time. For instance, if a system is in used only 4 hours a day, then the expected number of failures in a (calendar) year is one-sixth the expected number of failures in a year of continual use. Usually, though, customer usage is more irregular, and needs to be treated in a stochastic fashion.

Example: Let t represent calendar time. Suppose that in calendar day n, a customer uses a system for a period of time W_n and that $\{W_1, W_2, \ldots\}$ are independent and identically distributed with a uniform distribution on [2, 8] hours. Suppose that system failures occur according to a homogeneous Poisson process [13] at the rate of 1 per week of use. What is the probability that the user will experience one failure in (calendar) day 65? What is the expected number of failures the user will see in a (calendar) year?

Solution: $W_{65} \sim \mathrm{U}[2,8]$ and failures occur at a rate of 1/168 failures per (running) hour. Let $Z(t)$ denote the number of system failures in $[0, t]$ (operational time, with t measured in hours) and let N_{65} be the number of failures seen by the user in day 65. Then

$$P\{Z(t)=k\} = \frac{1}{k!}\left(\frac{t}{168}\right)^k e^{-t/168}, \quad k = 0, 1, 2, \ldots,$$

and the probability that the user sees one failure in day 65 is

$$P\{N_{65}=1\} = P\{Z(W_{65})=1\} = \frac{1}{6}\int_2^8 P\{Z(W_{65})=1 \mid W_{65}=w\}\,dw$$

$$= \frac{1}{6}\int_2^8 \frac{1}{168} e^{-w/168}\,dw = -\frac{1}{6} e^{-w/168}\Big|_2^8 = 0.0058.$$

In one (calendar) year, the software has accumulated $W_1 + \cdots + W_{365} = H$ hours of use. The distribution of H is the 365-fold convolution of the U[2, 8] distribution, computation of which is straightforward but not easy. Instead, we use the central limit theorem to approximate the distribution of H as a normal distribution with mean 1825 and standard deviation 33.09 to obtain the number of failures in the calendar year as

$$\int_{-\infty}^{\infty} \mathrm{E}[H/168 \mid H = h] dP\{H \le h\} = \frac{1}{168} \int_{-\infty}^{\infty} h\, \varphi_{(1825,1095)}(h)\, dh = \frac{1825}{168} = 10.86.$$

More discussion of operational time and calendar time is found in the Requirements Tip in Sections 2.2.5 and 3.3.7.

9.4 RELIABILITY MODELING FOR SOFTWARE

The field of reliability modeling for software is dominated by the statistical methods underlying TAAF. This models the sequence of times at which failures of the software occur as a nonhomogeneous Poisson process [13] with parameter estimation by classical or Bayesian methods. Other approaches have been proposed but are less commonly used. This section provides a review.

9.4.1 Reliability Growth Modeling for the Sequence of Failure Times

Within the many different styles of software development (waterfall, extreme programming, spiral, etc.), most software development projects alternate between periods of development and periods of testing. That is, after a period of development, a "development release" is declared and testing of that release begins. The purpose of testing is to determine how likely is it that the current development release will meet the software's reliability requirements. During testing, failures occur and root cause analysis of the failures uncovers faults in the software. Each failure initiates a "modification request" (MR) to the development team to repair the fault(s) causing the failure. After a phase of testing is complete, the next period of development includes work to address MRs as well as work on additional functional requirements. At the conclusion of this development period, another development release is issued and testing begins again. Of course, in most cases, development and testing proceed concurrently, but reliability modeling can be accomplished without taking this into account.

From a reliability engineering point of view, this approach to failure management is different from design for reliability. It is rather an example of a test, analyze, and fix (TAAF) procedure. Instead of a design-for-reliability approach that determines the failure modes and failure mechanisms in the software and then arranges the design and construction of the software to avoid their appearance, TAAF proceeds by testing parts of the software as they become available, attempting to correct any faults that may appear, and testing again in a sequence that may continue for some time. While the software industry has been successful in the sense that there is lots of software in the world that seems to work well, there is little doubt that development cost and time-to-market, as well as reliability, could be improved [1, 29]. Some well-known software is constantly being "patched" even as customers are using it, in effect making the customer base (an unpaid) part of the manufacturer's testing forces. In order for TAAF to be effective, it is necessarily time-consuming and expensive. The only reason it is even possible to use TAAF in software is that it is relatively easy[8] (compared to hardware) to make changes in the software so that another round of testing may begin quickly.

A reliability model for the sequence of failure times in a TAAF procedure may be constructed as follows. Postulate a sequence of development releases labeled 0, 1, 2,.... Each development release is followed by a testing interval during which some faults are discovered and MRs for those faults are written. Some faults are addressed in the next development release. If a fault is addressed correctly, it is eliminated. If a fault is not addressed correctly, it may be retained, and it is possible that additional faults may be introduced by a flawed repair. For purposes of modeling, the sequence of failure times, we may consider the length of the development intervals as negligible. So we postulate testing intervals $[t_0, t_1], [t_1, t_2], [t_2, t_3], \ldots$ with $t_0 = 0$; development release i is being tested during $[t_i, t_{i+1}]$ $(i = 0, 1, 2, \ldots)$. During $[t_i, t_{i+1}]$, failures occur according to a (homogeneous) Poisson process with rate λ_i. Let N_i denote the number of failures occurring in $[t_i, t_{i+1}]$. Then we may estimate λ_i by $\hat{\lambda}_i = N_i/(t_{i+1} - t_i)$. If the only activity in the development releases were correcting faults discovered in previous testing intervals, and if all fault correction were perfect, then we could expect that $\lambda_0 \geq \lambda_1 \geq \lambda_2 \geq \cdots$. However, development between releases usually includes new software addressing additional attribute requirements, this new software may contain additional faults, and faults found in testing so far are not always remediated correctly, so monotonicity of $\{\lambda_1, \lambda_2, \ldots\}$ is not guaranteed. The key point with this model, as with all TAAF scenarios, is that each testing interval deals with a different product. During each development interval, the software is altered to include

- new software addressing attribute requirements previously unaddressed and
- corrections to eliminate faults discovered in prior testing.

This is reflected in the model by allowing the λ_i to differ from one testing period to the next.

Many variations on this basic model have been treated in the software reliability engineering literature. Most treatments do not separate the testing intervals but consider them as a single extended testing interval. That is, they consider the entire time T during which testing takes place and postulate that failures appear according to a nonhomogeneous Poisson process[9] over $[0, T]$. If $N(t)$ denotes the (cumulative) number of failures appearing during $[0, t]$, then

[8] Less time-consuming and/or expensive.

[9] This reduces to the previous model if we postulate the intensity function of the NHPP to be a step function with jumps (changes) at the t_i and heights equal to the λ_i.

the cumulative intensity function of the process is estimated by $\hat{\Lambda}(t) = N(t)/t$ for $t \in [0, T]$. If the TAAF process is successful, $\hat{\lambda}(t) = \hat{\Lambda}'(t)$ ought to be decreasing after some reasonable period of time, indicating that most faults have been found and discovery of faults is slowing down. From a statistical point of view, these models have been treated classically [25] and from a Bayesian standpoint [20]. Procedures have been developed to make testing more realistic and effective by using the *operational profile* [26], a catalog of the operations the software is expected to perform together with the proportion of instances in which the operation is used. Criteria have been developed for how much testing is enough, that is, when should testing cease [3]. Those cited are some of the earliest papers concerning software reliability growth, or TAAF, modeling. Definition and application of these models in software reliability modeling may by now be said to be well developed [21, 27, 28, 31]. A more phenomenologically oriented model was introduced in Ref. 6.

When the software is finally released to users, failures occur according to a Poisson process with rate λ which should be equal to $\lambda(T)$ as long as no changes are made to the software after release. If changes are made, reliability modeling must consider that it is now dealing with a different product, and the relevant parameters must be changed accordingly.

9.4.2 Other Approaches

While the TAAF cited earlier is by far the most widely used for software reliability modeling, other approaches have been proposed. Munson [24] and Khoshgoftaar and Munson [14] postulate that the propensity to insert faults into software ought to be proportional to the complexity of the software, so they have investigated correlation between software reliability and complexity. Their idea is that software reliability could be estimated when the values of certain software complexity metrics are known, and computing the values of complexity metrics is relatively easy whereas computing software reliability is not. Structural reliability models using the program flow graph have been proposed by Cheung [2], Littlewood [19], and others. A state diagram approach (Section 4.4.7) has been proposed by Wang et al. [30]. Yet other approaches are described in Ref. 21.

9.5 SOFTWARE FAILURE MODES AND FAILURE MECHANISMS

9.5.1 Software Failure Modes

A failure mode is a description of how a product or service fails, or an overt sign that some requirements violation has occurred. It's an answer to the question, how can you tell that a failure has occurred? In software, a failure occurs when an erroneous output occurs in response to an input that is legal according to the software's requirements (a RELIC). Because software is intangible and has no existence without some hardware to run on, software failure modes can sometimes be obscured by the possibility of failures arising in the platform on which it runs. For example, when an error occurs in a word processing application on a personal computer, it could be caused by a fault in the application or by some anomaly in the PC hardware or operating system. Indeed, the PC user will normally restart the application to see if the anomaly is cleared, and if it is not, will reboot the system. If the anomaly persists after reboot, the usual conclusion is that there is a hardware failure. In addition, when software is part of a larger system (e.g., software that controls switches and signals on a railway), the erroneous output usually has additional ramifications (an incorrect switch may lead to a collision) so that

- the software failure may not be immediately detectable, and
- the fact that the underlying cause of the system failure was a software failure may not be discernible until completion of a root cause analysis.

9.5.2 Software Failure Mechanisms

It is commonly believed that the source of software failures is coding errors. While coding errors are indeed a commonly occurring failure mechanism in software, they are not the only software failure mechanism. Software failure mechanisms may arise in any part of the development process, including

- poorly understood requirements, which leads to not meeting the customer's real needs,
- ineffective design, which again may lead to missing the real problem, or may lead to inefficient solutions to the real problem (e.g., failure to meet a performance requirement in a real-time system), and
- improper execution of a good design, which again may cause inefficiencies if not outright inability to meet system requirements.

Consequently, in addition to treating coding errors, Section 9.6 discusses techniques that help minimize the occurrence of other software failure mechanisms. In particular, fault tree analysis and failure mode, effects, and criticality analysis (FME(C)A) help with this endeavor and are readily adaptable to the software component of systems and services.

Applying the three-phase model for lifetime (Section 3.3.4.4) to software yields some useful insights about software failure mechanisms.

9.5.2.1 Wearout failures

Software does not deteriorate or change with the passage of time. That is, a piece of software does not change by itself. Barring truly bizarre events like the change of a memory location caused by an incident cosmic ray, software only changes by deliberate intervention. In principle, a piece of software that is known to work entirely correctly (that is, every RELIC is known to produce a correct output) will continue to do so as long as it is presented with RELICs and none of the other conditions of its operation have changed (e.g., different hardware, different operating system, etc.). When software is maintained, for example to correct errors that have led to failures, it is changed and

- it is now a different piece of software than before,[10] and
- there is no assurance that it will continue to work as it did before unless deliberate effort is expended to demonstrate this in some fashion, usually through testing ("regression testing").

So there is no wearout failure mode in software. In particular, the notion of *rot*, or the accumulation of defects that eventually may make a software unmanageable [12], is real but it is a characteristic of the sequence of changed versions of the software, not of the original software that no longer exists.

9.5.2.2 Failures during useful life

The stress–strength model explains some software failures (including security failures), at least metaphorically. For instance, a real-time transaction system may include a performance requirement that as long as the number of requests to the system does not exceed 100 per second, the system response time will not exceed 350 milliseconds. The number of requests per second is a stress variable. The system strength is its ability to conjure up a response within the required time. The requirement says that the system strength is to be such that it responds within 350 milliseconds as long as the stress stays below 100 requests per second. Customers receive no assurance about response time whenever the stress exceeds 100 requests per second—some added delay may not even be discernible to a human—but the part of the load-service curve beyond 100 requests per second is unspecified. The system may work perfectly well for some stress exceeding 100 requests per second, or it may "fall off a cliff" and cease operating when the stress reaches 101 requests per second. The lesson from this example is that

- the strength of the software is established by design for reliability and focused testing (this is what operational profile testing [26] is for) and
- as long as RELICs are presented that fall within the strength specified in the requirements, and the software is constructed to have that strength, the software will respond properly.

The stress–strength model for software underscores the importance for reliability of ensuring that inputs to the software are RELICs only. Another way to put this is to say that the software should not be counted on to operate properly when operated outside the environmental conditions specified in its requirements.

[10] Else why trouble to assign a new version number?

9.5.2.3 *Manufacturing defects*

Much of software engineering is concerned with creation of error-free code. This is important because software reliability is deterministic in the following sense: whenever a piece of code containing one or more errors is executed, it will fail. Therefore, a necessary condition for the software to be failure-free is that it contains no code errors (assuming that all parts of the code are essential in the sense that there is no part of the code that is never executed). It is possible to develop a relationship between the proportion of lines of code with errors and the reliability of the software (see Exercise 5 for an example), but these exercises are somewhat beside the point: if you know the software contains code errors, you are going to devote resources to removing them, either during development (as a result of the testing that detects them) or after release to the customer (as a result of customer reports of failures caused by errors undetected by testing). So, at a minimum, good software engineering practices should be followed so that the number of errors introduced in initial development is minimized (because, while it is possible to correct errors found in testing, the correction process creates an opportunity for introducing additional errors; see the discussion of the clumsy repairman problem in Section 4.4.4).

But software manufacturing is more than the generation of code. Creation of a piece of software involves systems engineering and design before code development begins. To promote the creation of a reliable software product, steps must be taken to assure that requirements are interpreted correctly and an effective design that will embody the functional requirements is chosen. These need to be addressed before writing any code.

9.6 DESIGN FOR RELIABILITY IN SOFTWARE

It would be unrealistic to imagine that testing is not going to remain a vital part of software development for the foreseeable future. While we encourage development teams to face reliability considerations as early as possible in the development process and to adopt design for reliability as a preferred reliability enhancement strategy, the facts that

- TAAF is even possible in software and
- software engineering has not yet advanced to the point where procedures for anticipating and eliminating failures are routine and widely known

mean that testing is going to be a part of the software quality engineering and quality control toolset for a long time. Nonetheless, we urge development teams to look forward and use design for reliability techniques from the beginning of development because these offer a new approach to making software reliable in an efficient manner.

Design for reliability in software follows the same reasoning as described in Section 6.4: determine the failure mechanisms that could be active in the product and implement countermeasures that prevent them from becoming active. Fault tree analysis and FME(C)A are usable with software as well as with hardware.

9.6.1 Software Fault Tree Analysis

Fault tree analysis is readily adapted for use with software entities. Leveson and Harvey [16] and Leveson et al. [18] used fault tree analysis as a reliability improvement tool for safety-related failures.[11] Lyu [21] devotes a chapter to fault tree analysis.

Example: Consider event 12 "the controller erroneously turns off power to the motor" in the fault tree for the passenger elevator example shown in Section 6.6.1.2. In the illustration in Chapter 6, event 12 "controller erroneously turns power to the motor" is shown as the "or" of events 14, "hardware failure," and 15, "software failure." In the example in Chapter 6, event 15 is shown as an elementary event in that, for the purposes of that example, is not analyzed further into other causing events. Let's do that analysis here to illustrate a software fault tree analysis. That amounts to listing the possible errors in the software that would cause it to erroneously turn off power to the motor. These errors include

- The software does not read one or more of the box position sensors correctly, or
- The software contains one or more errors in assigning actions in response to box position sensor readings, or
- The software reads the box position sensors correctly but sends an erroneous signal to the gate controlling the motor's power relay.

If the root causes of these errors are code errors that slipped through the design reviews and other quality management processes, they should be caught during test. If these errors happen during operation, either testing missed them or operational conditions in the elevator changed in a way the software is unprepared for. For example, the software may not have been designed to respond correctly to a position sensor that becomes dirty or intermittent. A more detailed elaboration of the fault tree should include this, and other similar possibilities, so that the software design can be properly created to handle them. Again, this is only a toy illustration, not a fault tree analysis detailed enough to properly treat real passenger elevator system software, but it does illustrate the reasoning process that goes into creating the fault tree in a software entity.

[11] Both of these represent a further illustration of the idea that systems may be designed to avoid safety failures the same way they are designed to avoid other types of requirements violations. See Section 6.7.

9.6.2 Software FME(C)A

The same is true for FME(C)A. While Chapter 6 discusses FME(C)A from a hardware perspective, it works equally well in software entities. For example, consider an intelligent home smoke alarm that sends an SMS to a designated telephone number when it detects an event (smoke). This functionality is accomplished through software. The FME(C)A procedure examines what happens when each part of the software misbehaves. In this example, the software contains routines to accept a signal from the smoke detector hardware, close a relay or switch to turn on the audible alarm, read the telephone number from memory, and send its digits to a dialer (hardware that will generate the correct DTMF tones for dialing), and send the dialed number to the landline or the cellular network. As usual, we ask for the consequences of improper operation of any of these elements, both in failing to act when required and in acting when not required. This is the same reasoning we used with hardware elements in Chapter 6, and the information so uncovered is used to develop countermeasures for the major events on the Pareto chart of FME(C)A outputs. In the case of software elements, the countermeasures may include design features that it might not have been apparent were needed before the FME(C)A results were seen.

Finally, the use of data flow and control flow diagrams [4] can help expose causality paths that can be useful when carrying out a software FME(C)A.

9.6.3 Some Software Failure Prevention Strategies

9.6.3.1 Software design patterns

Design patterns are established solutions to commonly occurring problems. They are not necessarily blocks of code, but rather are abstract prescriptions about the best way to solve certain problems. For example, there may be many ways to write a calendar function in different languages. A design pattern for a calendar function tells what actions the calendar function needs to accomplish and the best way to accomplish them. Design patterns are one level of abstraction removed from software reuse (Section 9.7.3) in that reuse concerns blocks of code carried over wholesale or minimally adapted to a new system from one already known to be working as desired. Design patterns help enhance reliability because they are proven procedures from which errors have been removed by repeated refinement. Further information about design patterns may be found in Refs. 11, 22, and others.

9.6.3.2 Exception handling

Exception handling is a form of fault tolerance. It is the provision of built-in routines that respond to unusual conditions ("exceptions" such as floating point division by zero) occurring during the execution of a program. If provision for responding to exceptions is not included, exceptions may go on to cause failures. An exception handler will typically save the state of the program when the exception occurs, transfer control to an exception handling routine which may attempt to rectify the condition, and return to the main program when it satisfactorily resolves the problem. For more information on exception handling, see Ref. 5.

9.6.3.3 *Choice of language*

Some software languages are more adapted to creating error-free routines. For example, C and C++ both require the programmer to release resources upon completion of their use. Failure to do so results in buffer overflows, "memory leaks," and other undesirable conditions which may even be exploited by an attacker for nefarious purposes. By contrast, Java has built-in "garbage collection" so that the programmer does not have to remember to explicitly release resources, and memory leak failures are less common in applications written in Java.

9.7 CURRENT BEST PRACTICES IN RELIABILITY ENGINEERING FOR SOFTWARE

9.7.1 Follow Good Software Engineering Practices

Software engineering is still an actively studied field, largely because of the recognition that software development remains a hard problem. Many maintain that developing software is still too expensive, too time-consuming, and the results are too unreliable. Debate continues as to which software engineering practices may best overcome these problems [9, 10, 23]. It is not within the scope of this book to judge the merits of any side in this debate. For reliability engineering purposes, it is more important that some systematic, documented software engineering body of practice be followed. This book cannot recommend which software engineering style may be appropriate in any particular case because all developments have unique properties that may argue for choice of one or another and the author is not a software engineering expert. But it is more important that some software engineering discipline be followed rather than none at all. At a minimum, this will help cut down the number of code errors, and should help with the more fundamental design and development choices as well.

9.7.2 Conduct Design Reviews Focused on Reliability

Design reviews serve many useful purposes in software development. Effective development processes use design reviews to knit together the perspectives of all team members to promote better outcomes. For software reliability engineering, design reviews are most useful for

- selecting a design that is most likely to have better reliability,
- suitably arranging internal communication paths (data flows and control flows) so that they are not a source of failures,

- confirming detailed design and initial development as reliable, and
- helping eliminate code errors before testing.

9.7.3 Reuse Known Good Software

Many functions are common to different software applications. For instance, an application may need to compute an elapsed time between two events. It can do this using a calendar function. Every computing language has a calendar function already included. There are few, if any, good reasons for a development to create a new calendar function rather than re-using the one provided in the language. From a reliability perspective, reuse of a known good function can improve the reliability of the using element provided

- all the inputs and outputs are checked for compatibility between the reused function and the new code,
- it can be verified that all operational conditions to be encountered by the reused function have been previously certified as having been handled correctly by the function.

These cautions come directly from the definition of a good reliability requirement that includes the operational or environmental conditions prevailing in the system operation. Inattention to this concern has been the source of disastrous failures in the past. Notably, improper reuse of software in the medical device Therac-25 [17] caused the death of several patients. Software reuse continues to be actively studied [7, 15], and the last word on reuse and reliability has yet to be written [8]. Nonetheless, reuse has advantages that go beyond improved reliability, so there may be pressure in some developments to adopt a reuse strategy. Systems engineers should be prepared to examine the reliability implications of reuse in their particular developments and promote reuse when reliability advantages can be discerned.

9.7.4 Encourage a Prevention Mindset

As usual, systems engineers are not likely to be involved in the day-to-day execution of specific software engineering tasks. To promote effective software design for reliability, systems engineers should urge the software development team to adopt a design for reliability approach as recommended here (see also Ref. 6) in which proactive methods to prevent software failure mechanisms from becoming active are a useful complement to TAAF as a reliability assurance strategy.

9.8 CHAPTER SUMMARY

Our aim in this chapter is to prepare systems engineers to work with reliability requirements and design for reliability in the software component of systems and services. The chapter reviews the definition of reliability for software, reliability requirements for software, and allocation of system reliability requirements to the software component. We also touch on the TAAF approach that has been commonly used as a software reliability engineering technique. To prepare for design for reliability for software, we review software failure modes and failure mechanisms. We review fault tree analysis and FME(C)A as design for reliability tools that can be applied equally well to software as to systems and services as a whole. The reader should leave this chapter with a sense that, despite obvious differences in the attributes of software and hardware, reliability assurance for both is best served by a prevention approach that emphasizes design for reliability in addition to TAAF as usually practiced.

9.9 EXERCISES

1. Critically examine the requirement given in the pipeline network example in Section 9.3. Does the requirement contain a quantitative expression of a suitable reliability effectiveness criterion? Does the requirement clearly state a time period over which it is to apply? How does the requirement treat (or not treat) the conditions over which it is to apply?
2. The expected number of failures of a software program is 3 failures per hour. The system operates for 3 hours per day. What is the expected number of failures per day of the program?
3. The number of failures of a software program has a Poisson distribution with mean of three failures per hour. The system operates for 3 hours per day. What is the expected number of failures per day of this program?
4. The number of failures of a software program has a Poisson distribution with mean of three failures per hour. The number of hours per day that the program is in use has a uniform distribution over [2, 21]. What is the distribution of the number of failures of the program per day? What is the expected number of failures of the program per day?
5. Define a *blot* as a line of code that contains one or more errors. Suppose blots occur in n lines of code according to a Poisson process with rate βn. Suppose that a module containing 1076 lines of code is executed X times per day, where X has a uniform distribution on [7, 36], and all other modules in the software contain no errors. What is the expected number of failures per week of this software?

REFERENCES

1. Brooks FP Jr. *The Mythical Man-Month, Anniversary Edition: Essays on Software Engineering*. Hoboken: Pearson Education; 1995.
2. Cheung RC. A user-oriented software reliability model. IEEE Trans Softw Eng 1980;2:118–125.

3. Dalal SR, Mallows CL. When should one stop testing software? J Am Stat Assoc 1988;83:872–879.
4. DeMarco T. *Concise Notes on Software Engineering*. Volume 1133, Englewood Cliffs: Yourdon Press; 1979.
5. Dony C, Knudsen JL, Romanovsky AB, Tripathi A. *Advanced Topics in Exception Handling Techniques*. New York: Springer-Verlag; 2006.
6. Everett WW, Tortorella M. Stretching the paradigm for software reliability assurance. Softw Qual J 1994;3 (1):1–26.
7. Frakes WB, Kang K. Software reuse research: status and future. IEEE Trans Softw Eng 2005;31 (7):529–536.
8. Frakes WB, Tortorella M. Foundational issues in software reuse and reliability. Department of Industrial and Systems Engineering, Rutgers University; 2004. Working Paper 04-002.
9. Frakes WB, Fox CJ, Nejmeh BA. *Software Engineering in the UNIX/C Environment*. Englewood Cliffs: Prentice Hall; 1991.
10. Ghezzi C, Jazayeri M, Mandrioli D. *Fundamentals of Software Engineering*. Englewood Cliffs: Prentice Hall; 2002.
11. Hanmer R. *Patterns for Fault-Tolerant Software*. New York: John Wiley and Sons, Inc.; 2007.
12. Izurieta C, Bieman JM. A multiple case study of design pattern decay, grime, and rot in evolving software systems. Softw Qual J 2013;21 (2):289–323.
13. Karlin S, Taylor HM. *A First Course in Stochastic Processes*. 2nd ed. New York: Academic Press; 1975.
14. Khoshgoftaar TM, Munson JC. Predicting software development errors using software complexity metrics. IEEE J Sel Areas Commun 1990;8 (2):253–261.
15. Krueger CW. Software reuse. ACM Comput Surv 1992;24 (2):131–183.
16. Leveson NG, Harvey PR. Software fault tree analysis. J Syst Softw 1983;3 (2):173–181.
17. Leveson NG, Turner CS. An investigation of the Therac-25 accidents. Computer 1993;26 (7):18–41.
18. Leveson NG, Cha SS, Shimeall TJ. Safety verification of Ada programs using software fault trees. IEEE Softw 1991;8 (4):48–59.
19. Littlewood B. Software reliability model for modular program structure. IEEE Trans Reliab 1979;28 (3):241–246.
20. Littlewood B, Verrall JL. A Bayesian reliability growth model for computer software. J R Stat Soc Series C 1973;22 (3):332–346.
21. Lyu MR. *Handbook of Software Reliability Engineering*. New York: McGraw-Hill; 1996.
22. Martin RC. 2000. Design principles and design patterns. Available at www.objectmentor.com. Accessed November 10, 2014.
23. Mills HD, Linger RC. *Cleanroom Software Engineering: Developing Software Under Statistical Quality Control*. New York: John Wiley & Sons, Inc.; 1991.
24. Munson JC. Software faults, software failures and software reliability modeling. Inf Softw Technol 1996;38 (11):687–699.
25. Musa JD. A theory of software reliability and its application. IEEE Trans Softw Eng 1975;3:312–327.
26. Musa JD. The operational profile in software reliability engineering: an overview. Proceedings of the Third International Symposium on Software Reliability Engineering. October 7–10, 1992; Piscataway, NJ: IEEE; 1992. p 140–154.

27. Musa JD, Iannino A, Okumoto K. *Software Reliability*. New York: McGraw-Hill; 1987.
28. Rook P. *Software Reliability Handbook*. New York: Elsevier Science; 1990.
29. Verner JM, Overmyer SP, McCain KW. In the 25 years since *The Mythical Man-Month* what have we learned about project management? Inf Softw Technol 1999;41 (14):1021–1026.
30. Wang WL, Pan D, Chen MH. Architecture-based software reliability modeling. J Syst Softw 2006;79 (1):132–146.
31. Xie M. *Software Reliability Modelling*. Volume 1, Singapore: World Scientific; 1991.

Part II

Maintainability Engineering

10

Maintainability Requirements

10.1 WHAT TO EXPECT FROM THIS CHAPTER

We pass now to the second major division of this book. Maintainability is the second of the three sustainability disciplines that form a major part of the system engineer's responsibilities and skills regarding continued satisfactory operation of a system or service beyond its initial installation. These disciplines pertain to the actions that need to be taken during design and development to ensure that a system or service will continue to operate properly and profitably throughout its intended life. This chapter begins the study of maintainability by first achieving an understanding of maintainability as a system property and then devising maintainability effectiveness criteria and figures of merit that are consistent with this understanding. Both corrective and preventive maintenance are covered. We are then in a position to deal with maintainability requirements. Examples of maintainability requirements and their interpretation are discussed. A review of contemporary best practices in developing maintainability requirements and a summary of the chapter bring the chapter to a close and prepare for the design for maintainability material in Chapter 11.

Reliability, Maintainability, and Supportability: Best Practices for Systems Engineers,
First Edition. Michael Tortorella.

10.2 MAINTAINABILITY FOR SYSTEMS ENGINEERS

10.2.1 Definitions

We discussed reliability at length in Part I of this book. If a system or service is designed for reliability, then the number of system or service failures (requirements violations) should be smaller than it would be if the system or service had received no design for reliability attention. But we live in an imperfect world. Even with the best of intentions and technology, failures will occur. This is inevitable. So when a failure occurs, it is of interest to restore the system or service to normal operation as quickly as possible. This is consistent with keeping the system achieved availability (Section 10.6.4) as high as possible. Maintainability is related to this goal in that it refers to how readily the system can be repaired when it fails. The reasoning is that the more easily the system may be repaired, the more quickly such repairs may be completed, the duration of any outage associated with the failure will be smaller, and system availability will increase.

A commonly accepted definition of maintainability is *the ability of a system to be repaired and restored to service when maintenance is conducted by personnel using specified skill levels and prescribed procedures and resources* [13]. As with reliability, we note that this is an *ability*, an abstract property of a system (or service) that, in this case, tells how adapted the system is to repair and restoration to usefulness. It is fruitful to think of maintainability as that collection of system properties or characteristics that promote speedier, lower cost, and less error-prone repair. If system A is more maintainable than system B, then it is likely that the repair times for system A will mostly be shorter than the repair times for system B.[1] A system whose maintainability is poor will suffer extended outage durations because it is difficult to repair for some of the reasons we explore when we discuss design for maintainability in Chapter 11. In other words, some of the factors that promote shorter repair times may not have been properly considered, controlled, or monitored when the system maintenance concept (Section 10.2.2) was created.

Some of the factors that influence maintainability are

- the basic maintenance and support policies applied to the system, also known as the system maintenance concept,
- appropriate use of maintainability effectiveness criteria and requirements,
- appropriate use of preventive maintenance,
- the locations where maintenance will be performed and the types of maintenance to be performed at each location (referred to as "levels of maintenance"),
- organizational responsibilities,

[1] One may wish to make this a formal definition by saying that system A is defined to be more maintainable than system B if the repair times for system A are stochastically less than those for system B, but this is not yet common practice.

- design features associated with maintenance elements,
- the anticipated maintenance environment, and
- warranties.

Design for maintainability (Chapter 11) comprehends choosing appropriate values for each of these factors to achieve the customers' desired maintainability.

10.2.2 System Maintenance Concept

One of the first steps in system development is to create a catalog of the functions the system is supposed to perform. Once these functions are known, planning should begin for actions that are to be taken to maintain the performance of those functions, or restore the performance of those functions, in the event of system or subsystem failure(s). This planning forms the start of the system maintenance concept, a comprehensive plan for how the system will be repaired when it fails. The system maintenance concept comprehends

- maintainability requirements,
- an overall repair strategy to meet the requirements, including a program for specific repair procedures (remove-and-replace, immediate repair on site, etc.),
- designating which parts of the system are replaceable,
- what preventive maintenance, if any, is to be applied,
- where different kinds of repairs are to take place,
- who is responsible for repairs,
- what features will be incorporated into the design that bear on maintainability,
- whether a warranty will be offered, and
- any other system design and operation factors that influence maintainability.

As the system design concept is refined, so should the system maintenance concept be along with it. This chapter helps flesh out the system maintenance concept by providing information relevant to each of these factors. Chapter 11 discusses specific design for maintainability procedures you can use to create specific maintenance concept sections to promote satisfaction of the system's maintainability requirements. We begin here with a discussion of the most basic decision in maintainability: do we repair at all?

10.2.2.1 Repair, immediately, or restore, then repair?

The definition quoted in Section 10.2.1 says "repaired and restored to service." It is important to consider that these need not always be inseparable. Sometimes, it may be desirable to take some quick action to at least partially restore service and defer complete repair until a reasonable later time. For instance, a server failure in a large group of servers may not need to be attended to right away because the failure may cause only a minimal, perhaps barely noticeable, amount of service degradation. In an extreme example, some telecommunications service providers maintain a fleet of central offices on semitrailers so that in the event of a major disaster, such as a fire, at a central office, a temporary replacement can be trucked in quickly to restore some level of service before a complete repair is undertaken. The temporary replacement may not have the same capabilities or capacity as the permanent installation, but it does restore some level of service quickly while the permanent repair is under way. When we talk about corrective maintenance actions, it will be important to distinguish whether the corrective action is focused on quick restoration of some level of service, or if it is intended to be a complete repair of the failure that has taken place.

Many of the systems we consider in this book are elements of very large and complex service delivery infrastructures supporting sophisticated services like Internet service, postal services, etc. At any time, it is almost certain that some number of service delivery infrastructure elements are failed, yet service continues with little or no discernible degradation. We have previously discussed this property of network robustness, or network resiliency, in connection with design for service reliability (Chapter 8). Here we focus instead on maintainability implications. In large and complex service delivery infrastructures, it may not be necessary to attend to every element failure as soon as it occurs. If the service delivery infrastructure is sufficiently resilient, or if traffic loads are low, a failure of a small number of elements may not be noticeable to service users. For instance, a failure of 10 servers in a farm of 10,000 servers probably does not affect service very much unless the farm is very heavily loaded and utilization is very high. So repairs need not always begin immediately. Different repair scenarios are possible for these situations: repair technicians may be dispatched only after the number of failures reaches some predetermined threshold, or there may be a schedule instituted by which technicians are dispatched on a periodic basis to carry out repairs on whatever failures may have accumulated since their last visit. Should it be desirable to follow one of these protocols, a mathematical optimization can be developed to define the operation that minimizes cost.

Henceforward, when it is necessary to make a distinction, we will refer to the earlier two cases as restoration with delayed repair in the first case and full corrective maintenance in the second.

10.2.2.2 What gets repaired and when

The fundamental decision concerning repair is whether something is to be repaired at all, or is to be discarded when it fails, and whether replacement after discard is warranted. As discussed in Section 2.2.2, economics plays an important role in determining these policies. In most large military, telecommunications, or other complex technological systems, replacement of the entire system with a new one when there is a system failure is impractical and uneconomical. Consequently, arrangements to repair the part of the system that has failed are usually made. The first iteration of the system's maintenance concept, or maintenance and support policy, produces an initial designation of the maintainable sections and parts of the system. The concept is refined as systems engineering and design proceeds and more detailed knowledge about the system's structure is developed. The level of repair analysis (LoRA) covered in Chapter 11 is one of the last steps in the development of the maintenance concept. The LoRA is the final division of the system into maintainable units and tells where and by whom each unit will be repaired or replaced. This information is also needed for the system reliability model and for design for supportability.

10.2.3 Use of Maintainability Effectiveness Criteria and Requirements

As with reliability and supportability, the systems engineering principle "if you don't measure it, no one will pay attention to it" applies in maintainability too. Maintainability requirements focus attention on those aspects of system or service design and operation that pertain to the ability to repair a system when it fails. Ideal maintenance promotes repair that is

- rapid,
- as inexpensive as possible, and
- error-free.

Choose requirements that bring to the fore those aspects of speed, cost, and effectiveness that are most relevant to the design and operation of the system or service. The use of requirements in this way creates a rational framework for maintainability management by fact. Data gathered to verify achievement of maintainability requirements are analyzed to discern properties of maintenance planning and operations and point out areas where improvement may be needed. Recurring measurements (e.g., percentage of jobs in a facility that exceed some duration threshold per week) may be tracked over time using a control chart to help distinguish between changes due to normal statistical fluctuations and changes due to special causes that should be investigated for possible corrective action. In all cases, the systems engineering approach should be to encourage

- incorporation of maintainability at the early stages of system or service design,
- use of design for maintainability (Chapter 11) to refine and update the system maintenance concept as the design progresses,
- creation of a rational framework for managing maintainability by fact, and
- institution of the systems and processes that will promote success by making this systematic and repeatable.

Chapter 2 defined effectiveness criteria, figures of merit, and metrics as applied to reliability. The same general framework of effectiveness criteria, figures of merit, and metrics is underlies maintainability requirements. A maintainability effectiveness criterion is a quantitative description of a frequency of occurrence, duration, or other aspect of a maintenance factor. Maintainability effectiveness criteria are usually conceptualized as random variables because the particular values they may take in any given situation usually cannot be precisely known because of the influence of numerous noise factors. A maintainability figure of merit is some abbreviation or summary of a maintainability effectiveness criterion, such as a mean, variance, percentile, etc. A maintainability metric is a statistic, or a quantity derived from data, pertaining to a maintainability effectiveness criterion or figure of merit. Maintainability effectiveness criteria, figures of merit, and metrics are used to focus attention on the maintainability features that systems engineering marks as important based on their understanding of customer needs. Several maintainability effectiveness criteria are reviewed in Section 10.3.

Maintainability effectiveness criteria and figures of merit are used in defining maintainability requirements (Section 10.4). Maintainability requirements link to system reliability requirements through any system requirements pertaining to availability (Section 10.6.4). That is, availability is determined by the characteristics of the operating time (up-time) random variables and the outage time (downtime) random variables, so requirements for availability and for repair time cannot be imposed independently.

Example: Suppose the mean operating time for a system is determined by reliability modeling to be 10,500 hours, and we wish to attain inherent availability (Section 10.6.4) of at least 0.9999 in the long run. If we assume renewal repair (Section 4.4.2), we may solve the inequality

$$\frac{10,500}{10,500+\nu} \geq 0.9999$$

for the largest mean downtime ν that will permit this desired long-term availability to be attained. The solution is $\nu \leq 10.5$ hours. That is, attaining the desired long-term achieved availability will be impossible unless the mean outage time is limited to no more than 10.5 hours. In case of the existence of a standard limiting the mean outage time to some smaller value (i.e., 4 hours in Telcordia's GR-284-CORE [16] for telecommunications switching systems), this inequality indicates that reliability will have to be improved:

$$\frac{\mu}{\mu+4} \geq 0.9999$$

from which we obtain $\mu \geq 39,996$ hours. Even though the mathematics of this example applies only when there is renewal repair, it shows that if the system has an availability requirement, then the reliability (operating times) and maintainability (outage times) may not be prescribed independently.

Maintainability metrics are computed to verify compliance with maintainability requirements (Section 10.6).

10.2.4 Use of Preventive Maintenance

Preventive maintenance improves maintainability by the application of actions intended to forestall possible system failures. For example, a schedule of periodic lubrication of bearings is used to prevent undue wear so that the known wearout failure mode of mechanical wear is postponed and does not cause an unscheduled outage. Preventive maintenance may be performed

- on a fixed schedule based on system age (using whatever variable, such as time, mileage, number of operations, etc., is being used to track system age),
- in response to a reading from some sensor(s) in the system, or
- according to any of a number of predictive schemes.

Predictive maintenance uses what is known about the forward recurrence time to the next system failure to apply a maintenance action aimed at postponing or eliminating the predicted failure. Predictive maintenance is a form of reliability-centered maintenance [4, 11]. It is also related to condition-based maintenance [9]. See Section 11.5.3.

10.2.5 Levels of Maintenance

Levels of maintenance, or levels of repair, are the locations at which repair can be performed and the types of repair that are supported at each location. The specification of levels of repair is an important component of design for maintainability, which we cover in depth in Chapter 11 where we introduce level of repair analysis (LoRA).

Depending on the type of system, its uses, and the types of failures it may undergo, various levels of repair are possible. For simpler systems, or systems whose operators may be unsophisticated, it is possible that the only reasonable choice is to have the original manufacturer perform repairs. This can be a common (or a forced) strategy for certain kinds of consumer electronics: for example, battery replacement in many Apple™ products is normally performed only by Apple or its dealers. Home appliances and automobiles usually may be repaired by the user, by the original manufacturer (or its representative), or by an independent repair provider such as a nonaffiliated mechanic shop. Some repairs may be simple enough that the user may elect to perform them herself (replacing a burned-out headlight bulb, for instance), and others may require expertise that can only be obtained from trained specialists at the dealer or mechanic shop. Large technological systems such as defense systems and telecommunications systems typically employ a more formal maintenance strategy. Certain kinds of repairs may be performed by the user in the field, while other kinds of repairs may only be capable of being performed at a depot (a centralized location, off-line and away from the area of use) or at a manufacturer's location. We explore these issues in more detail in Chapter 11.

10.2.5.1 Maintenance performed by the customer

After any warranty expires, the system customer may choose to perform maintenance using their own personnel. This is common in defense systems. For instance, in the United States, many military occupational specialty (MOS) categories involve deep knowledge of weapons systems, communications systems, aircraft, etc., so that military personnel can perform preventive maintenance and repairs. The notion of "levels of repair" has evolved from the way defense systems are traditionally maintained, where maintenance may be done on-site, at a remote central repair location (depot), at some higher level of aggregation, or by the manufacturer.

10.2.5.2 Maintenance performed by the supplier under contract

As an alternative, the customer may contract with the system supplier for maintenance. This is common for software products like operating systems, enterprise management software, enterprise telecommunications systems, etc. In this case, the supplier has the opportunity to optimize their repair facilities and processes for minimum cost, or the facility may run as a profit center. Design for supportability now becomes important for the supplier because this bears on efficiency in repair operations and costs to the repair provider. LoRA also pertains because a repair contract may also specify on-site visits for certain preventive maintenance and repair activities and repair at a supplier location for other preventive maintenance and repair activities.

10.2.5.3 Other maintenance providers

Maintenance may be also carried out by other providers such as independent repair shops. This is quite common in consumer products, automobiles, etc. We are not going to consider this case further in this book other than to note that the further removed a facility is from the supplier, the more difficult it is for the supplier to manage maintainability. The supplier has less control over documentation, tools, spare parts, etc., so maintainability requirements have less force in these situations. Nevertheless, many customers find this an attractive option.

10.2.6 Organizational Responsibilities

In selecting appropriate levels of maintenance, it is important to also specify who will be assigned to each level and each type of repair to be performed. Different options here entail different costs and different requirements for training and support (test equipment, documentation, tools, etc., covered extensively in Chapter 11). A large part of design for maintainability consists of creating mathematical models for costs of the various options and seeking minimum cost solutions using these models. Typically, assignment of organizational responsibilities becomes part of the LoRA through its cost and effectiveness variables.

10.2.7 Design Features

Maintainability can be improved or damaged by design choices. In particular, Chapter 11 is devoted to discussion of the positive aspects of design for maintainability. As an introduction, here are a few examples of how a design feature had a positive or negative on the maintainability of the product.

In the 1990s, Lexmark made widespread use of using fastening technology that did not need screws or rivets that required time and effort to assemble and remove. They used snaps, tab-and-slot alignment aids, and other "fastenerless" technologies to speed up both assembly (during manufacturing) and disassembly[2] (for maintenance). This design feature has a positive impact on maintainability by decreasing the time needed to disassemble the product to reach internal parts that may need to be replaced or refurbished.

Conversely, it is possible to make maintenance more difficult by failing to pay attention to maintainability. The 1975–1980 Chevrolet Monza and related General Motors (GM) vehicles had a V6 engine that was large relative to the engine compartment available space, and to change the rearmost two spark plugs required unbolting the engine mounts and lifting the engine slightly so that enough space could be acquired to reach these two spark plugs. It is not clear whether this was an oversight or a deliberate choice. In either case, when owners realized how much tuneups cost, it is likely that word spread and GM probably sold fewer of these vehicles than they might have otherwise.

A lesson for systems engineering is, as with reliability, if attention is not paid to maintainability during design, then the product or system is completed with a haphazard collection of maintainability-related features, some of which may be good and some bad, but you get what you get essentially at random. Focused attention to product or system design features relating to maintainability is essential to a successful development.

10.2.8 Maintenance Environment

Part of planning for maintainability must include a consideration of the environment in which maintenance is to take place. Different design features are necessary if maintenance is to take place in a desert environment, a marine environment, or an indoor, climate-controlled facility. Again, this is an important maintainability-affecting factor that must be planned for in advance. More detail is found in Chapter 11.

[2] Provided procedures were well documented. Anyone who has tried to disassemble a fastenerless design without instructions will readily appreciate this.

10.2.9 Warranties

In some cases, a product or system supplier will offer a warranty. This is hardly a new concept. An example of a written warranty statement dating back to approximately B.C. 429 is cited on page 2.9 of *Juran's Quality Handbook* [10]:

> An early example was on a clay tablet found amid the ruins of Nippur in ancient Babylon. It involved a gold ring set with an emerald. The seller guaranteed that for twenty years the emerald would not fall out of the gold ring. If it did fall out of the gold ring before the end of twenty years, the seller agreed to pay to the buyer an indemnity of ten mana of silver.

Most warranties cover product repair or replacement in case of failure during some period of time or usage beginning when the product was sold. Recently, some auto manufacturers have begun to supplement this with a promise to also pay for scheduled maintenance up to some mileage limit.

Warranties are a tool used by suppliers to make their products more attractive to buyers. Some warranties can create a substantial cost burden for the supplier. Others, like the 30-day warranty common in consumer electronics products, are merely nugatory. Warranty costs are external failure costs [19] borne by the supplier of the product or system. Inadequate attention to design for maintainability increases these costs, and, while they do vanish after the expiration of the warranty period for the last product sold, some manufacturers have been held liable many years later for product defects affecting safety. As noted in Section 4.7.2.1, full treatment of warranty modeling is beyond the scope of this book. See Ref. 3 for a good introduction to this topic.

10.2.10 Preventive Maintenance and Corrective Maintenance

Preventive maintenance refers to any actions that are taken in order to forestall failures. Lubrication is a simple example of preventive maintenance. Metal-to-metal contact in bearings would cause rapid wear. Lubrication is specified as a preventive measure that prevents direct metal-to-metal contact and dramatically increases the service life of the bearing. The time spent on preventive maintenance for a system or subsystem is recorded as "preventive maintenance time" associated with that system or subsystem. Preventive maintenance increases system inherent availability (Section 10.6.4.1) because some outages that may have occurred absent the preventive maintenance now do not occur and therefore contribute zero to the overall system downtime.

Corrective maintenance refers to those actions that are taken to restore a system to service after a failure has occurred. Corrective maintenance time records the amount of time taken to complete a repair on a system or subsystem, and is associated with that system or subsystem. Many concepts and ideas of maintainability pertain to both these classes of maintenance, and are so discussed in this chapter. Preventive maintenance requires additional consideration of relevant properties such as scheduling, failure prediction, and optimization, and these will be treated separately where appropriate. See also Section 10.2.2.1.

10.2.11 Maintainability for Services

10.2.11.1 Introduction

Up to this point, our maintainability discussion has focused only on products and systems. Many products and systems exist to perform certain actions for, or provide services to, a user base. We have seen in Chapter 8 how products and systems used in service delivery infrastructures determine the quality and reliability of the services seen by service users/purchasers. When there is a service outage, restoration of service is accomplished by rectifying failures that have taken place in the service delivery infrastructure (see also Section 10.2.2.1). Maintainability requirements for services are related to maintainability requirements for elements of the service delivery infrastructure in the same way that reliability requirements were related. Service outages may be decomposed into two parts: the first part for support times and the second part for repair durations as shown in Figure 10.1 for system outages. Supportability for services is discussed in Section 12.3.

10.2.11.2 Service maintenance concept

The service maintenance concept is a plan that tells how a service will be restored to proper functioning when a service outage occurs. Service outages are caused by failures in the elements of the service delivery infrastructure, so actions taken to restore the service to proper functioning are usually taken on elements of the service delivery infrastructure. This creates the need to understand how changes to elements of the service delivery infrastructure are reflected in behavior of the service. For example, when repairs are made to an element of the service delivery infrastructure, we need to know

- what proportion of normal service functioning is restored by the repair and
- how rapidly that happens.

This is similar to the reasoning we used in relating service delivery infrastructure element failures to service failures (Chapter 8). Here, we relate service delivery infrastructure element repairs to service restorations (partial or complete). As noted in Section 10.2.2.1, immediate complete service restoration may not be necessary or desirable, and the service maintenance concept will address this is specific terms for each service and service delivery infrastructure.

Elements of the service maintenance concept include

- service maintainability requirements,
- development of specific service repair procedures (whether the service will be restored gradually, in stages, or in full immediately),

- responsibility for repairs,
- service features that bear on maintainability,
- whether a service-level agreement (see Section 8.6) will be offered, and
- any other service design and operation factors that influence maintainability.

Service maintainability requirements will place limits on the duration of service outages and related variables (i.e., the number of service outages per month exceeding a given threshold). They may also set targets for the amount of time it may take to restore given proportions of the service capability (e.g., service response time shall return to 150% of nominal within 10 minutes of the start of the outage). See Section 10.2.2.1 for some factors to consider when planning for full or partial service restoration. Note that in cases in which large sums are at stake, it would be wise to formulate and solve optimization models for the operations involved in the service restoration. For example, how many server failures in a large server farm should be permitted to accumulate before technician(s) are dispatched to begin performing repairs? This can be formulated as an optimal delay and minimum cost problem where the costs accrue from

- loss of revenue due to delays or lost transactions, service-level agreement penalties and
- costs for technician dispatch and repair actions.

The first continues to increase as the outage persists. The second cost remains zero for some period of time (waiting to dispatch), includes a fixed cost component (travel and related expenses), and a portion that increases proportionally to the number of servers requiring repair or replacement once dispatch takes place. The details of the solution vary with the type of service, number of servers in the farm, etc., but the illustration serves to point out that optimization models help make better decisions. In other words, an initial service maintenance concept might include a vague plan of this nature, and as the service definition is further refined, information becomes available upon which a quantitative optimization study may be based.

As with systems and products, responsibility for the activities required to bring a service back to normal functioning after an outage must be clearly delineated. Given the distributed nature of many service delivery infrastructures, this point cannot be ignored at the risk of extending outages because of disorganized response by uncoordinated organizations. As always, the initial idea for assignment of responsibility need not be very precise but should always be refined as the service definition becomes refined.

Finally, service outage times are not independent of outage times of the elements of the service delivery infrastructure. We saw in Chapter 8 how failures in elements of the service delivery infrastructure are failure mechanisms for the service those elements support. Similarly, the duration of maintenance activities on the elements of the service delivery infrastructure determines how long service outages will last. The point is that maintainability requirements for elements of a service delivery infrastructure should not be developed independently of requirements for service failure frequency and outage duration. As with reliability, a model connecting outage times for elements of a service delivery infrastructure to outage times for the service it supports is indispensable.

Example: Suppose a service is provided on a service delivery infrastructure consisting of two elements, A and B, in a series configuration (Section 3.4.4). Failures take place in units A and B according to Poisson processes with rates λ_A and λ_B, respectively, and outages are independent and identically distributed with means $1/\mu_A$ and $1/\mu_B$, respectively. μ_A, μ_B, λ_A, and λ_B all have the same units of 1 per hours. If λ_A and λ_B are small and μ_A and μ_B are large, then we may ignore the possibility of overlapping outages and service failures will occur approximately according to a Poisson process with rate $\lambda_A + \lambda_B$ and their mean duration will be approximately

$$\frac{\lambda_A \mu_A + \lambda_B \mu_B}{\lambda_A + \lambda_B}$$

hours. So if a service outage requirement for a mean outage time of no more than 2 hours is required, then it is necessary to find values of μ_A, μ_B, λ_A, and λ_B that satisfy

$$\frac{\lambda_A \mu_A + \lambda_B \mu_B}{\lambda_A + \lambda_B} \leq 2$$

in order that the service outage requirement may be met.

Obviously, this example is oversimplified and realistic requirements assignments of this kind may need solutions of much more complex network problems. The point, which might have been obscured by complicated computations, is that maintainability requirements for elements of a service delivery infrastructure should derive from maintainability requirements for the service supported by that infrastructure so that the service customer is the ultimate driver of related maintainability requirements. In other words, maintainability requirements for elements of a service delivery infrastructure are not arbitrary but are rationally defensible as supporting the service maintainability requirements.

10.3 MAINTAINABILITY EFFECTIVENESS CRITERIA AND FIGURES OF MERIT

10.3.1 Products and Systems

We have seen effectiveness criteria used before in Part I where reliability, availability, number of failures, etc., were explained as ways to summarize system characteristics pertaining to failure-free operation that are important to customers. It bears repeating that effectiveness criteria serve to direct systems engineering, design, and development attention to those system performance variables that customers feel are important and/or desirable. As maintainability deals with the time required to carry out repairs on a failed system, or the time needed for failure prevention actions, many maintainability effectiveness criteria concern frequency and duration of maintenance- and repair-related actions. Other maintainability effectiveness factors concern cost and labor hours. Commonly used maintainability effectiveness criteria include

- Preventive maintenance time: the time required to complete a preventive maintenance operation or operations. It may refer to a single preventive maintenance operation, or to an aggregate of preventive maintenance times over a stated time interval (e.g., a year). If the latter, it should be so indicated by a suitable modifier (e.g., total preventive maintenance time per year). While increasing preventive maintenance time causes achieved and operational availability to decrease (Section 10.6.4), this needs to be traded off against the potential improvement in inherent availability because less corrective maintenance may be required. Depending on the customer's needs, more or less preventive maintenance may be optimal. These considerations may be rationalized by mathematical optimization. Some examples include Refs. 7, 14.
- Corrective maintenance time: the time required to complete a corrective maintenance operation or operations. It may refer to a single corrective maintenance operation, or to an aggregate of corrective maintenance times over a stated time interval (e.g., a year). If the latter, it should be so indicated by a suitable modifier (e.g., total corrective maintenance time per year). Decreasing corrective maintenance time leads to increased system availability. In turn, decreased corrective maintenance time is promoted by a maintenance concept based on removing and replacing subassemblies with known good ones (in preference to, say, removing a subassembly, repairing it, and putting it back into the system). Corrective maintenance time may refer to restoration with deferred repair or to complete corrective action (Section 10.2.2.1); use an appropriate modifier to make the distinction clear when necessary.
- Maintenance time: the sum of preventive maintenance time and corrective maintenance time.
- Active preventive maintenance time: some sources add the modifier "active" to preventive maintenance time if they wish to emphatically distinguish maintenance time from any associated support time. In the terminology used in this book, this is not necessary because we clearly distinguish supportability from maintainability.
- Active corrective maintenance time: see earlier text.
- Active maintenance time: the sum of active preventive maintenance time and active corrective maintenance time.

- Times between preventive maintenance actions: conceptualizing the times at which preventive maintenance actions occur (are initiated) as a point process (Section 4.3.3.1), the times between preventive maintenance actions are the intervals in this point process. If defined in this fashion, note that the times between preventive maintenance actions include the preventive maintenance times themselves. This is like the distinction between times between failures and operating times. See Section 4.3.3.
- Times between corrective maintenance actions: conceptualizing the times at which corrective maintenance actions occur (are initiated) as a point process, the times between corrective maintenance actions are the intervals in this point process. If defined in this fashion, note that the times between corrective maintenance actions include the corrective maintenance times themselves.
- Preventive maintenance actions per (hour, day week, month, year, other): self-explanatory.
- Corrective maintenance actions per (hour, day week, month, year, other): self-explanatory.
- Times between replacements: considering a replacement as a specific type of preventive or corrective maintenance, and conceptualizing the times at which replacements occur as a point process, the times between replacements are the intervals in this point process.
- Replacements per (hour, day week, month, year, other): self-explanatory.

All of these are effectiveness criteria, that is, random variables. Figures of merit based on these, using means, medians, etc., are in common use.

It is apparent that there are many possibilities for maintainability effectiveness criteria; the above list is certainly not exhaustive. The number of maintainability effectiveness criteria and the depth of detail the systems engineer chooses to implement depend on the type of system in question and the customer's needs and desires. It may be tempting to institute every possible maintainability effectiveness criterion in every development. While one can see the "heart in the right place" nature of this, beware of dilution of focus. Use as many effectiveness criteria as needed to focus on the important system characteristics that matter to customers. Think twice about adding more than this necessary number to avoid cynicism and despair among the development team. If the team feels that they are being overwhelmed by the need to predict and track a large number of effectiveness criteria, clear traceability of the chosen effectiveness factors to the customer needs analysis should be widely disseminated and explained. If such explanations are not possible, consider eliminating those effectiveness criteria that are thus shown to be superfluous. A good general principle is that you can measure anything you like, as long as

- there is an identified need for that measurement,
- it is given a descriptive name so that its use is unambiguous, and
- it is used consistently throughout the development.

10.3.2 Services

The maintainability effectiveness criteria and figures of merit used for services that carry over from the list of product and system maintainability effectiveness criteria and figures of merit are

- corrective maintenance time,
- times between corrective maintenance actions, and
- corrective maintenance actions per (hour, day week, month, year, other).

An added detail that is useful for services maintainability comes from the possibility of partial restoration of service when a service outage occurs (Section 10.2.2.1). A useful way of incorporating partial service restoration into maintainability effectiveness criteria and figures of merit is to use a scale of restoration. For example, we may use the time to restore $p\%$ of service capability as a maintainability effectiveness criterion, where p is chosen to be meaningful in the service context (i.e., bandwidth, pumping capacity, line voltage, etc.).

10.4 EXAMPLES OF MAINTAINABILITY REQUIREMENTS

Return to the system history diagram we first encountered in Figure 3.1. Figure 10.1 is an extract from the system history diagram focusing only on a single outage. In this diagram, time increases to the right along the horizontal direction. As before, the upper horizontal lines (at $y=1$ when using the 0–1 model described in Section 4.3.2) represent time periods during which the system is operating normally. The lower horizontal line (at $y=0$) represents a time interval during which the system is out of service; a failure occurs at the time instant represented by the beginning of this horizontal line. Now we add more detail: the outage interval is divided into two parts. The left-hand part is the time during which preparations for repair are being made; this is the support interval, discussed at length in Part III. The right-hand part is the time during which repairs are actually being performed. This is the maintenance interval: the period of time during which the repair is in progress (and the right-hand endpoint of this line represents the time instant at which the repair is completed and normal functioning of the system begins anew). It is this time

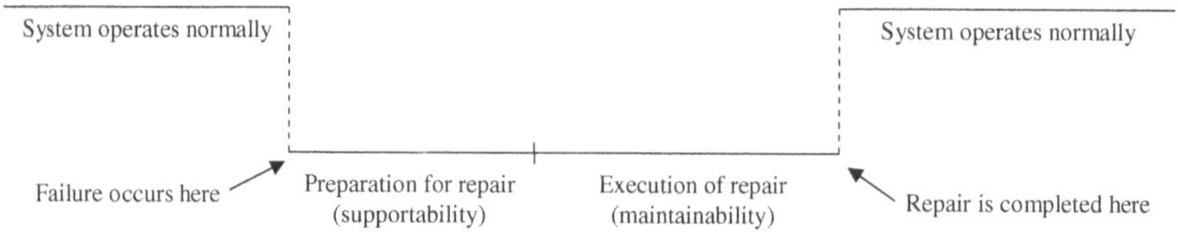

Figure 10.1 Extract from a system history diagram.

duration of the maintenance interval, and the factors that go into making it what it is, that we refer to when we speak of maintainability as promoting speedy repairs.

Maintainability requirements are written to control

- the length of time needed to carry out repairs,
- the factors that contribute to this time duration,
- the effectiveness of preventive and corrective maintenance actions, and
- the costs and labor hours associated with maintenance activity.

The number of repairs that will be needed is determined by the number of failures the system undergoes, and that is determined by the system reliability (Chapter 4).

Any maintainability time, cost, labor hours, etc., variable may be the subject of a requirement, depending on the needs of the customer and the warranty servicing and post-warranty servicing strategies adopted. For customers like airlines, defense, telecommunication, etc., who do their own servicing after a warranty has expired, limiting the difficulty and time required for maintenance is a benefit for customers and its value to the supplier is in greater product attractiveness. For warranty service and post-warranty service that is performed by the supplier or a subcontractor, ease of operations and shorter maintenance times are a direct benefit through decreased internal costs, especially important if the repair operation is run as a profit center.

Some examples of maintainability requirements include

- 98% of the times required to carry out repair procedure [designator] using the instructions [designator] and tools specified shall be no greater than 4 hours.
- the mean time required to complete a repair on the system, when work is carried out according to the specified instructions and using the specified tools, shall not exceed 8 hours.
- the proportion of repairs that are erroneous and require rework shall not exceed 0.01.
- the system achieved availability (Section 10.6.4.1) shall not be less than 0.99 after the first month of operation.

These examples are built on maintainability effectiveness criteria and figures of merit: proportions in the first and third example, and means in the second and fourth (availability is the expected value of the system reliability process). Maintainability requirements should state

- the system and subsystems to which it applies,
- the conditions under which it applies, and
- any other qualifiers that may be needed to avoid ambiguity or misunderstanding.

Any repair times specified in maintainability requirements pertain to the second (right-most) part of the system outage time represented in Figure 10.1. Requirements for support times (the first or left-most part of the system outage time represented in Figure 10.1) are discussed in Chapter 12.

10.5 MAINTAINABILITY MODELING

As with reliability modeling as discussed in Section 4.7.1, maintainability modeling should be guided by an understanding of what are you going to see when you collect maintainability data to verify compliance with maintainability requirements. For instance, if a requirement specifies mean task duration, data will be collected on task durations and analyzed to estimate the mean task duration so it can be compared with the requirement. A model consistent with this should produce the mean task duration as its output.

10.5.1 Duration and Labor-Hour Effectiveness Criteria and Figures of Merit

Maintainability variables come in continuum and discrete flavors. Continuum variables include event durations, task durations, labor hours, etc. Labor hours may be different from task durations because there may be several people's labor to account for in carrying out a task, so labor hours will be the sum of the hours spent by each person on the task (an individual person may not spend the entire duration of the task working on it: he/she may be called on to carry out some step(s) of the task but not necessarily be involved for the entire task, so labor hours need to be calculated from an accounting perspective, not simply by multiplying the task duration by the number of persons that worked on it). This section discusses modeling for continuum maintainability variables. Discrete (count) variables are discussed in Section 10.5.2.

Duration effectiveness criteria are continuum variables (may take on any real value, need not be integer-valued only) and, in the maintainability context, usually have the units of time. The precedence diagram or activity network representation [8] is a useful model for the duration of, and labor hours consumed in, maintenance operations. These tools are commonly used in project management and scheduling of scarce resources and are well adapted to this application. They are a special case of stochastic flow network [5] in which maintenance tasks are accomplished at one or more workstations and pass from workstation to workstation according to rules reflecting the task needs. Network models used as planning and optimization tools for maintenance facilities are treated in detail in Section 13.4.1. For maintenance task duration modeling, the quantity of interest is the total time a job spends in the facility; this is the duration of the maintenance task that would be compared with a duration requirement. The expected value of this duration is given in the sixth list item in Section 13.4.1. Labor hours are computed from the sojourn times at the individual nodes in the activity network.

Example: Consider the maintenance facility described in Section 13.4.1. The expected total time a job spends in the facility is given by the following equation:

$$(I-R)^{-1}(S\#R)(I-R)^{-1}$$

where I is the identity matrix, R is the routing matrix for the network (Figure 13.2), and S contains information about the individual workstation sojourn times and the inter-workstation transit times. This reduces to a simple addition if the workstations are arranged in a sequential (series) order. Let T_i denote the time spent by a job in workstation i and let n_i denote the number of persons staffing workstation i, $i=1,\ldots,7$. Then, regardless of the overall facility's network structure, and assuming that all operators at a workstation are involved in servicing every job visiting that workstation, the total labor hours attributable to a job is

$$\sum_{i=1}^{7} n_i T_i$$

with expected value (mean)

$$\sum_{i=1}^{7} n_i \mathrm{E} T_i .$$

This expression results from adding up all the labor hours consumed at each workstation in the facility. This is, in general, different from the total time consumed by a job in the facility because a job may not visit every workstation (in this example, because some workstations are duplicated to handle additional jobs).

The figures of merit for duration of maintenance events in most common use are the mean and percentiles. As always, when writing a requirement for the mean of some maintainability effectiveness criterion, it helps to have some understanding of the amount of variability that is possible in the underlying effectiveness criterion. For example, suppose X has a normal distribution with mean 60 and variance 1 and Y has a normal distribution with mean 60 and variance 625. While both populations have a mean of 60, $P\{X \le 30\} < 10^{-10}$ while $P\{Y \le 30\} \approx 0.11$. In a sense, we know much more about a population of durations described by X than we do about one described by Y. In the X case, most of the duration values cluster around 60 ± 3, whereas in the Y case, almost a quarter of the durations fall <u>outside</u> the interval [30, 90]. There is much more spread, or dispersion, in the Y-population than in the X-population, and, in the language of quality engineering, we say that the quality of our information about the X-population is greater than the quality of our information about the Y-population. You can see how this matters when setting requirements. Imagine writing a requirement like "The mean duration of maintenance operation 14.7 shall not exceed 65 minutes." If you knew that the population of durations of operation 14.7 was like the X-population, you could have a lot more confidence that most of the results you would get would comply with the requirement. If it were like the Y-population, the results would be much more spread out and not as many would comply with the requirement. The point is the same as we made in Section 2.7.2.1: controlling the mean alone can open the door to unwanted small or large values of the effectiveness criterion unless you have some understanding of the variability in the population being controlled.

10.5.2 Count Effectiveness Criteria and Figures of Merit

Count effectiveness criteria are discrete (integer-valued) variables. Examples of discrete maintainability effectiveness criteria include number of preventive maintenance actions per month, number of maintenance task durations exceeding a given threshold duration per week, etc. The latter offers an example of how a continuum variable may be studied with counting concepts.

Caution: the expected value or mean of a count-based effectiveness criterion need not be an integer.

10.6 INTERPRETING AND VERIFYING MAINTAINABILITY REQUIREMENTS

10.6.1 Duration Effectiveness Criteria and Figures of Merit

We use the same interpretation framework for maintainability variables that we did with reliability variables in Chapter 5. When maintainability requirements are built using effectiveness criteria, modeling and verification can only address the probability that the requirement will be or is being met (unless a census of the installed population is available, in which case a yes-or-no decision is possible). When maintainability requirements are built using figures of merit, again a census permits yes-or-no decisions, while sampling allows a variety of approaches including hypothesis testing, confidence interval estimation for the figure of merit, Bayesian methods, etc. Though the language in Chapter 5 is that of reliability, the statistical ideas introduced there apply to maintainability variables as well, so the following treatment is less extensive than that in Chapters 3 and 5.

10.6.1.1 Duration effectiveness criteria

As with reliability requirements based on effectiveness criteria, modeling and verification for requirements based on maintainability effectiveness criteria can only speak to the probability that the requirement will be or is being met unless a census of the installed population is available.

Example: The maintainability requirement is "Maintenance task 14.7, when carried out according to published instructions and using the provided tools, shall be completed in 65 minutes or less." The following data, in minutes, are collected on the durations of task 14.7: 45, 81.2, 58, 69, 52.1, 71, 60.8, 64, 47.5, 58.7, 62, 68.5, 63, and 64.5. Was the requirement met in this set of tasks? Is the requirement being met in the maintenance shop from which these data were collected?

Solution: If these data represent a census of all the tasks from a particular day, say, then the requirement is not met on that day because there are four durations in this set of tasks exceeding 65 minutes. To settle the second question, we will estimate the probability that the requirement is being met based on treating the data shown as a random sample (not necessarily from a single day) from the shop's typical operations. The point estimate of the probability that the requirement is being met in the shop is the sample proportion $\hat{p} = 10/14 = 0.71$. The standard error of $\hat{p}$ is $0.12/\sqrt{14} = 0.032$, so a 90% confidence interval for the probability that the requirement is being met is

$$\left[0.71 - 1.645 \times 0.032, 0.71 + 1.645 \times 0.032\right] = \left[0.66, 0.76\right]$$

so the evidence is quite strong that the requirement is not being met in this shop. The same problem could be solved using a hypothesis test approach, similar to that shown in Section 8.5.2. We may also approach this problem from a Bayesian point of view which would be appropriate if this shop has been operating for a while and has accumulated a history of instances of task 14.7. From this history, we might assume that the proportion of durations of task 14.7 that do not exceed 65 minutes has a beta distribution with parameters $r = 7$ and $s = 3$ (this distribution has a mean of 0.75). The posterior distribution, given the data above, is also beta and its parameters are $r = 17$ and $s = 8$, which has mean $17/24 = 0.71$. This is the Bayes estimate of the proportion of durations of task 14.7 that do not exceed 65 minutes. See example 9.11a of Ref. 2. Note that, if the facility is meeting the requirement only about 3/4 of the time, management will most likely have directed corrective attention to the facility and its performance would have been improved to the point where our estimate of the prior probability of meeting the requirement would be much larger than 0.75.

10.6.1.2 Duration figures of merit

If a duration requirement is built on a figure of merit, such as the mean, median, or percentile, that figure of merit may be estimated from data to determine whether the requirement is being met. In the example of Section 10.6.1.1, suppose the requirement had been "The median duration of maintenance task 14.7, when carried out according to published instructions and using the provided tools, shall be 65 minutes or less." The median of the 14 observations in the data set of Section 10.6.1.1 is 62.5, so if these data represent a census of the durations of interest, then the requirement is met for those durations. If the data represent a sample from some larger population of task 14.7 durations, we estimate the median of that population by the sample median, which is 62.5. The standard deviation of the sample median is approximately 1.253 times the sample standard deviation [2]. The sample standard deviation for these data is 9.52, so the standard deviation of the sample median is approximately 11.92, and the corresponding standard error of the median is approximately $11.92/\sqrt{14} = 3.19$. This yields a 90% confidence interval $[62.5 - 1.645 \times 3.19, 62.5 + 1.645 \times 3.19] = [57.25, 67.65]$ for the median of the population. It is possible that the requirement is being met, but no degree of confidence can be attached to that statement from this analysis. When this happens, one strategy is to find the largest confidence interval not containing the requirement; then the probability that the requirement is not being met is the confidence level for that interval. Here is an example: solve the equation $62.5 + \alpha(3.19) = 65$ for $\alpha = 0.78$. Then find the confidence level corresponding to the (two-sided) confidence coefficient 0.78, or how much of the standard normal distribution is between −0.78 and +0.78, which in this case is approximately 0.565. The probability that the requirement is not being met is about 56.5%.

Requirements tip: See how the different requirements (in the first case, a requirement on a maintainability effectiveness criterion and the second on a particular figure of merit) lead to the same behavior's being unacceptable in one case and acceptable in the other. So when setting a requirement, decide first what behavior you are trying to promote and then write the requirement accordingly.

10.6.2 Count Effectiveness Criteria and Figures of Merit

10.6.2.1 Count effectiveness criteria

Count effectiveness criteria are integer-valued. These are used for requirements on the number of certain events related to maintenance actions. For example, maintainability effectiveness criteria in count form include the number of preventive maintenance actions per month over a specified population, the number of maintenance tasks exceeding 4 hours in a given week at a given maintenance facility, etc. Count effectiveness criteria are used in frequency and rate contexts by referencing a number of counts to a given time interval.

Example: A maintainability requirement is instituted at a certain maintenance facility and states "No more than one job per week in this facility shall require more than 10 hours to complete." The following data on job durations (in hours) are collected during 1 week of the facility's operation: 3, 6.2, 8, 11.8, 8, 9.1, 7.5, 5.7, 8.1, 6, 9.8, 3.5, 4, 7.1, 8, and 9.5. Is the requirement being met this week?

Solution: If these are census data, then the requirement is met because there is only one job whose duration is greater than 10 hours. If the data are a sample from all the jobs processed through that facility in this week, we can use these data to estimate the probability that there are two or more jobs in the facility that take more than 10 hours to complete, that is, the probability that the requirement is not met. The sample proportion of jobs whose durations are more than 10 hours is 1/16 = 0.0625 with a standard error of 0.0605. An approximate 95% one-sided confidence interval (based on the approximate normal distribution of the sample proportion, acceptable because there are 16 observations) for the probability that a job takes more than 10 hours to complete is

$$[0.0625, 0.0625 + 1.645 \times 0.0605] = [0.0625, 0.162].$$

The requirement is not met in a given week if there are two or more jobs in that week taking more than 10 hours to complete. Based on these data, we can be approximately 95% confident that the probability that a job takes more than 10 hours to complete is less than or equal to 0.162. If the facility processes w jobs in a week, the probability the requirement is not met is the probability that a binomial random variable with parameters w and 0.162 is 2 or more. Let N denote the number of jobs per week taking 10 or more hours to complete. Using the conservative value 0.162, the probability that the requirement is not met in various cases of jobs per week is shown in Table 10.1.

See also Exercises 7 and 8.

10.6.2.2 Count figures of merit

Count-based maintainability requirements may also be expressed in terms of figures of merit, just as was the case for reliability requirements. We will illustrate this with an example.

Example: Suppose the requirement shown in Section 10.6.2.1 reads instead "No more than 5% of jobs in this facility shall exceed 10 hours duration.[3]" With the data as shown in Section 10.6.2.1, is this requirement met?

Solution: If these data form a census of all jobs worked during some time period, then the requirement is not met for that time period (one job out of 16, or 6.25% of the jobs in the census, had a duration exceeding 10 hours). If the data shown are a sample from all jobs in the facility over some time period, we will test the hypothesis H_0 that q, the proportion of jobs in the

TABLE 10.1 *P*{Requirement Not Met}

Jobs per Week	$P\{N \geq 2\}$
10	0.211
20	0.651
50	0.991
100	0.999

[3] Another way to read this requirement is that the 95th percentile job duration in the facility should be 10 hours or less.

facility that do not exceed 10 hours in duration, is at least 0.95 against the alternative H_1 that $q<0.95$. The sample proportion of jobs that do not exceed 10 hours in duration is $15/16=0.9375$. Under the null hypothesis, the probability that either 15 out of 16 jobs do not exceed 10 hours or all 16 jobs do not exceed 10 hours is 0.46:

$$\binom{16}{15}(0.95)^{15}(0.05)^{1}+\binom{16}{16}(0.95)^{16}(0.05)^{0}=0.46$$

Thus we fail to reject the null hypothesis that less than 95% of the jobs take 10 hours or less to complete, so the requirement is not being met. Oddly enough, with only 16 observations it is not possible to reject the null hypothesis even if all 16 jobs took less than 10 hours; the p-value in this case would be 0.44. Collection and analysis of additional data would help reach a stronger conclusion.

As an alternative, we may compute a confidence interval for q. An approximate 95% confidence interval for q is $[0.84, 1]$ which contains the requirement value 0.95, so no conclusion about meeting the requirement can be drawn at the 95% level of confidence: the interval contains many unacceptable values.

10.6.3 Cost and Labor-Hour Effectiveness Criteria and Figures of Merit

Cost and labor hours are continuum variables, like duration, and the same kinds of statistical analyses as were used with duration variables can be used in these cases. See Sections 10.6.1.1 and 10.6.1.2.

10.6.4 Three Availability Figures of Merit

Current practice distinguishes three related availability figures of merit:

1. Inherent availability,
2. Achieved availability, and
3. Operational availability.

All three concern the proportion of time that a system is operating. They differ in what is counted against downtime.

10.6.4.1 Inherent availability

The inherent availability, $A_I(t)$, is defined as the probability that a system or piece of equipment, when used under stated conditions in an ideal support environment, will operate satisfactorily at the point t in time. The key words here are "ideal support environment." They imply that only corrective maintenance time is considered when aggregating system downtime. That is, periods of preventive maintenance and periods of support are excluded from the computation when modeling or estimating inherent availability.

TABLE 10.2 Downtimes Included in Availability

Availability	Symbol	Downtime Included
Inherent	$A_I(t)$	Corrective maintenance only
Achieved	$A_A(t)$	Preventive and corrective maintenance
Operational	$A_O(t)$	All, including support times

10.6.4.2 Achieved availability

Achieved availability, $A_A(t)$, is defined as the probability that a system or piece of equipment, when used under stated conditions in an environment including preventive maintenance, will operate satisfactorily at the point t in time. Achieved availability differs from inherent availability only in that preventive maintenance time is included in the system downtimes. Periods of time consumed by support activities are not counted in modeling or estimating achieved availability.

10.6.4.3 Operational availability

The operational availability, $A_O(t)$, is defined as the probability that a system or piece of equipment, when used under stated conditions in an actual operational environment, will operate satisfactorily at the point t in time. The key words here are "actual operational environment." All contributors to downtime, whether preventive or corrective maintenance or support activities, are counted when modeling or estimating operational availability.

Table 10.2 summarizes these three cases.

When using the models in Sections 4.4.2 and 4.4.3 for one of these types of availability, arrange the system reliability process so that the $D_1, D_2, \ldots$ downtimes denote only the portions of outage times appropriate to the availability type.

10.7 MAINTAINABILITY ENGINEERING FOR HIGH-CONSEQUENCE SYSTEMS

Chapter 6 recommends additional design for reliability methods to make high-consequence systems as reliable as possible. Even when those are implemented, though, unforeseen failure modes or combinations of failure mechanisms that may have been dismissed as too rare for serious consideration[4] may occasionally cause a failure. When a failure does occur in a high-consequence system, there is a premium on speedy restoration of service. So the priorities for maintainability for high-consequence systems are

- speed,
- accuracy, and
- low cost.

[4] A significant source of these kinds of failures is the use of stochastic independence in modeling events that may not, in fact, be physically independent.

in that order. In addition to the standard design for maintainability practices discussed in Chapter 11, maintainability planning for high-consequence systems can also benefit from business continuity and disaster recovery plans developed for IT systems [1, 6, 15], where short-duration downtimes are desired. Some principles and practices learned from this field include

- Emphasize preventive maintenance, both scheduled and predicted, to minimize the need for corrective maintenance,
- Incorporate system design features to isolate failures not only to improve reliability and minimize effects on customers and the environment but also to minimize the time needed for corrective maintenance,
- Run drills and exercises periodically to ensure that maintenance personnel are up to date with skills and system properties,
- Carry out root cause analysis of past failures and outages not only to improve reliability but also to improve maintenance processes,
- Automate recovery to the greatest degree possible but always include manual overrides for special circumstances, and
- When system changes are made, train maintenance personnel immediately so that they are as familiar with the new conditions as they were before.

Other high-consequence systems from which transferrable maintainability engineering practices may be learned include nuclear power plants, oil refineries, and electric power distribution.

- The US Nuclear Regulatory Commission has developed inspection procedures to verify that power plant operators use standard, approved practices in all relevant areas, including maintenance. Inspection Procedure 42451B [17] aims to confirm that plant maintenance procedures are prepared to adequately control maintenance of safety-related systems within applicable regulatory requirements. Inspection Procedure 62700 [18] looks for specific plant maintenance practices promoting reliable and safe plant operation. These practices include traceability of spare parts, post-maintenance testing, documented procedures for corrective maintenance for frequently occurring failures, and provisions for control of equipment even when it is out of service and being maintained.
- From refinery operations, we learn the importance of documenting maintenance that has been performed, taking steps to minimize maintenance staff turnover, and focusing on safety and personal protection for maintenance workers.
- Electric power distribution has suffered some highly publicized outages. From the root cause analyses of these outages, principles for improved maintenance were developed [12]. Many outages have been caused by damage to outside plant from falling trees and other debris during storms,

so more frequent tree trimming has been implemented by many power distribution companies. Some outages affected large geographical areas, so technology to improve failure isolation is useful.

10.8 CURRENT BEST PRACTICES IN MAINTAINABILITY REQUIREMENTS DEVELOPMENT

The purpose of this section is to offer some suggestions for developing maintainability requirements using contemporary quality engineering principles and quantitative reasoning. Customer needs for speedy, low-cost, and error-free repair are the starting point. Once maintainability requirements are in place, use the design for maintainability techniques in Chapter 11 to arrange the system to meet the requirements. If the supplier contracts to perform repairs, the requirements also drive design for supportability so that a successful and profitable maintenance operation may be developed.

10.8.1 Determine Customer Needs for Maintainability

As with reliability, a clear understanding of customer needs for maintainability in quantitative terms is needed to craft successful maintainability requirements. Customers will be concerned not only with the time needed to accomplish repairs but also with the support they expect to receive if performing preventive maintenance and repairs themselves. The frequency of maintenance actions is controlled through design for reliability even though the maintenance action rate (e.g., number of maintenance actions per week) may be tracked as part of maintainability management. So reliability and maintainability need to be considered together to fully understand the customer's experience with outages. In particular, reliability and maintainability are linked through availability requirements: it is not possible to arbitrarily specify reliability, maintainability, and availability requirements because availability is determined by the times between outages (reliability) and the outage durations (maintainability and supportability). See Section 10.2.3 for an example.

We discussed in general terms in Chapter 1 the techniques systems engineers can use to acquire information about customer needs and desires as a first step in developing requirements. Consideration should be given to use of those tools for maintainability needs also so that clear, unambiguous, quantitative statements may be understood in the same way by all parties on the supplier and customer teams.

10.8.2 Balance Maintenance with Economics

It may seem that only in the case of maintenance performed under contract by a supplier should cost and labor-hour factors be significant. However, even in the case where maintenance is to be performed by the customer, it is in the supplier's interest to understand and factor into maintainability requirements the cost and labor-hour burdens created by maintenance procedures. Customers want to see that the supplier understands their needs and is acting to help them succeed. When possible, the supplier should work with customers to understand their processes and arrange the system to better align with those processes to improve maintainability.

10.8.3 Use Quantitative Maintainability Modeling to Ensure Support for Maintainability Requirements

It is difficult to foresee the effects of proposed requirements or changes to requirements without some quantitative maintainability model. The frequency of preventive maintenance actions may be readily determined because they are scheduled or deducible from predictive maintenance schemes (Section 11.5.3). The distribution of the duration of each specific preventive maintenance action should be relatively tight (have small variance) because these actions are predetermined and predefined; again, this may be determined from historical data or from a time-and-motion study.[5] The frequency of corrective maintenance actions is driven by the reliability of the replaceable subassemblies of the system. A model for this is found in Section 11.4.2.1. The durations of corrective maintenance actions may be somewhat more variable than the durations of preventive maintenance actions if only because failures are sometimes accompanied by confusion, may occur at odd times so may not always be attended by the most trained personnel, etc. Historical data on the same or similar corrective maintenance actions, as well as time-and-motion studies, can again provide insight into this distribution.

10.8.4 Manage Maintainability by Fact

We have recommended writing requirements in quantitative form so that it is possible to collect data and verify whether they are being met. Routine verification using a systematic, repeatable process approach is recommended so that a realistic understanding of realized maintainability may be acquired.

An important part of this process is the ability to discern when results you may be seeing from data indicate a real signal or just statistical noise. See Exercise 8 for an illustration using data from a maintenance shop. Measurements from any process that is influenced by noise variables as well as control variables will exhibit some degree of statistical fluctuation. This is certainly true of maintainability where the vagaries of human behavior and other noise factors play a prominent role. All stakeholders are best served when managers respond to signals that indicate a real change in the conditions of the process and ignore signals that represent simple statistical fluctuation. Control charting [19] offers

[5] In the maintainability context, these studies are grouped under the title "maintenance task analysis."

a systematic approach to distinguishing statistical noise ("common cause" fluctuations) from signals that indicate a real change in process conditions ("special causes"). Tracking maintainability requirements achievement over periods of time will produce a sequence of measurements to which these ideas readily apply. Use the results of tracking to improve

- maintainability when special causes are noted and
- the tracking process overall.

In particular, you need to know what your maintenance operation is capable of now so that you can decide whether to accept this performance or set a requirement that incorporates an improvement goal. Process capability should be part of the information you consider when creating a requirement. It's acceptable to incorporate an improvement goal in a requirement, but it should be based on an understanding of what the process is currently capable of doing and whether the proposed improvement makes economic and customer sense. A desired improvement must be accompanied by provision of the tools needed to make that improvement, be they hardware and software, or training, or new facility layouts, etc. Requiring a major improvement without concomitant investment in equipment and training is asking for failure.

It may be that the current process capability is not adequate to meet customer needs or, in a maintenance contract situation, it may be that the current process capability entails excess cost. When these prevail, a requirement incorporating an improvement goal may be warranted. For example, it may be desirable to reduce the mean sojourn time of a job at a particular workstation from 4 to 3 hours based on the economics of the operation. Changing the requirement this way should be accompanied by the changes in procedures, tools, training, etc., that make such an improvement possible.

10.9 CHAPTER SUMMARY

This chapter has dealt with maintainability both in general terms, as a system property, and in specific terms, listing several commonly used maintainability effectiveness criteria and figures of merit. It emphasized the importance of creating an initial system maintenance and support policy as soon as the system functionality is defined, and of refining this plan as the system becomes better defined. Consider whether it is always necessary to immediately start a corrective maintenance repair, or whether it may be possible to make some reduced effort to restore some degree of system functioning (or service delivery) and postpone full repair until a later time. Maintainability requirements are constructed based on maintainability effectiveness criteria and figures of merit, and verification of requirements

- during design, using maintainability modeling and
- during operation, using maintainability metrics

is discussed with examples of relevant statistical techniques. Maintainability and reliability are connected through three types of availability: inherent, realized, and operational.

10.10 EXERCISES

1. A server farm contains 10,000 servers, and each server contains 1,000 Web pages. The time required for a server to handle a request to serve up a Web page has an exponential distribution with mean 10 milliseconds. The server farm receives requests for the pages it contains according to a Poisson process with rate 1,000,000 requests per second. A request that cannot enter the farm because all servers are busy is blocked and lost. What is the probability that a request is blocked? Suppose 10 servers are failed. Now what is the probability that a request is blocked?
2. Give an example of preventive maintenance in a software product (e.g., a server operating system).
3. Critique the maintainability requirements examples shown in Section 10.4 for completeness, ambiguity, appropriateness, etc.
4. Carry out a hypothesis test on the data shown in the example in Section 10.6.1.1. Discuss how your conclusion and the conclusion shown in the example are related.
5. Draw two graphs of the data in the example in Section 10.6.1.1 as follows: Label the horizontal axis form 0–100. For the first graph, to each x on the horizontal axis, draw the number of data points greater than x on the vertical axis. On the second graph, draw instead the proportion of data points greater than x on the vertical axis. The second graph is an example of an empirical survivor function for these data.
6. Carry out a hypothesis test for the example in Section 10.6.1.2.
7. Discuss the requirement shown in the example of Section 10.6.2.1.
   a. What data would you collect to determine whether the requirement is being met over a long period of time, say 1 year? Is it feasible to collect census data?
   b. Is the requirement complete? Is it clear and unambiguous? Discuss what might be added or stated differently if you find the given statement wanting.
   c. Would it be more useful or appropriate to write the requirement in terms of the mean number of jobs per week exceeding 10 hours? What are the business implications of the requirement as it is stated and as it would be revised to use the mean?
8. What are the implications of a requirement not being met? Consider again the requirement in Exercise 6. Suppose you collect (census) data over a period of 25 weeks on the number of jobs per week exceeding 10 hours in duration, with the following results: 0, 0, 0, 1, 0, 0, 5, 1, 0, 0, 2, 0, 0, 0, 0, 4, 3, 0,

0, 0, 1, 0, 0, 1, 0. What should you conclude about the operation of the facility? (Hint: consider using a control chart to determine when a search for special causes is warranted.)

REFERENCES

1. Arnell A. *Handbook of Effective Disaster Recovery Planning*. New York: McGraw-Hill; 1990.
2. Berry DA, Lindgren BW. *Statistics: Theory and Methods*. 2nd ed. Belmont: Duxbury Press (Wadsworth); 1996.
3. Blischke WR, Murthy DNP. *Warranty Cost Analysis*. New York: Marcel Dekker; 1994.
4. Bloom NB. *Reliability Centered Maintenance*. New York: McGraw-Hill; 2006.
5. Buzacott JA, Shanthikumar JG. *Stochastic Models of Manufacturing Systems*. Volume 4, Englewood Cliffs: Prentice Hall; 1993.
6. Cimasi JL. *Disaster Recovery & Continuity of Business: A Project Management Guide and Workbook for Network Computing Environments*. CreateSpace Independent Publishing Platform; 2010.
7. Dekker R. Applications of maintenance optimization models: a review and analysis. Reliab Eng Syst Saf 1996;51 (3):229–240.
8. Freivalds A. *Niebel's Methods, Standards, and Work Design*. Volume 700, Boston: McGraw-Hill Higher Education; 2009.
9. Jardine AK, Lin D, Banjevic D. A review on machinery diagnostics and prognostics implementing condition-based maintenance. Mech Syst Signal Process 2006;20 (7):1483–1510.
10. Juran JM, Godfrey AB. *Juran's Quality Handbook*. 5th ed. New York: McGraw-Hill; 1999.
11. Nowlan FS, Heap HF. Reliability-centered maintenance. 1978. Defense Technical Information Center document no. AD-A066579.
12. Pflasterer R. Maintenance work management—best practices guidelines. 1998. Electric Power Research Institute technical report no. TR-109968.
13. Schaeffer M. *Designing and Assessing Supportability in DoD Weapons Systems: A Guide to Increased Reliability and Reduced Logistics Footprint*. Washington, DC: Defense Acquisition University Guidebook; 2003.
14. Sherif YS, Smith ML. Optimal maintenance models for systems subject to failure—a review. Naval Res Logist Q 1981;28 (1):47–74.
15. Snedaker S. *Business Continuity and Disaster Recovery Planning for IT Professionals*. 2nd ed. New York: Syngress (Elsevier); 2013.
16. Telcordia Technologies. Reliability and quality switching systems generic requirements, Issue 1. 2003. Telcordia document no. GR-284. Piscataway, NJ: Telcordia Technologies.
17. US Nuclear Regulatory Commission. Maintenance procedures. 1975. Inspection Procedure no. 42451B.
18. US Nuclear Regulatory Commission. Maintenance implementation. 2000. Inspection Procedure no. 62700.
19. Wadsworth HM, Stephens KS, Godfrey AB. *Modern Methods for Quality Control and Improvement*. New York: John Wiley & Sons, Inc.; 2002.

11

Design for Maintainability

11.1 WHAT TO EXPECT FROM THIS CHAPTER

Once a system's maintainability requirements are known, properties of the system need to be arranged so that the requirements will be satisfied. Deliberate actions must be taken to guide the system to a state in which it is likely that fulfillment of its maintainability requirements will become more than a fervent hope. This chapter reviews design for maintainability techniques, including

- Quantitative maintainability modeling,
- Level of repair analysis (LoRA),
- Preventive maintenance,
- Reliability-centered maintenance (RCM).

Each of these is intended to add features, properties, and characteristics to the system's design that will enhance its ability to be repaired quickly, inexpensively, and with few errors.

11.2 SYSTEM OR SERVICE MAINTENANCE CONCEPT

Try as we may to design for reliability to prevent failures, it is rare that we are completely successful. So when planning a new system, product, or service, it is a good idea to pay attention to how the system, product, or service will be

Reliability, Maintainability, and Supportability: Best Practices for Systems Engineers, First Edition. Michael Tortorella.

repaired and restored to operation when it fails (corrective maintenance) and to procedures needed for preventing failures after the system is in operation (preventive maintenance). In more formal terms, when we begin to design a system, we also create the beginnings of a plan for how that system will be maintained. This plan is called the *system maintenance concept*.

As noted in Section 10.2.2, the system maintenance concept addresses

- what parts of the system will be maintained, and how will this maintenance be accomplished,
- how many levels of maintenance are anticipated before any formal planning is carried out (see Section 11.4.2),
- what types of repairs and other functions are anticipated to be performed at each level,
- what maintainability requirements (Section 10.3) will be instituted to meet the particular needs of this system,
- what design features should be incorporated to simplify system repair, speed it up, and make it less error-prone,
- preliminary ideas on other maintenance elements such as type of testing and diagnostic procedures to be employed, staff skills needed, etc., and
- relevant environmental requirements (e.g., special precautions to be taken in case maintenance is performed in deleterious environments such as in a desert, on shipboard, in polluted atmospheres).

At the very early stages of design, when the maintenance concept is first explored, one should not expect firm answers for all these issues. However, it is in keeping with the spirit of prevention and quality engineering to begin thinking about these issues as soon as is practical. The maintenance concept should be continually updated and become more precisely specified as greater understanding and specificity of the design are attained.

Some parts of the maintenance concept shade over into the system support concept, and it could be argued that some relevant activities could be reasonably placed in either category. Some of these include spares inventory planning, logistics planning for the transport of failed units, spares, and repaired units, provision of online or off-line test procedures and equipment, planning (layout, staff sizing, etc.) of a maintenance facility, etc. Rather than laboring over semantics, it is best to make sure that important activities are covered. A good way to do this is to integrate maintenance staff and support staff into the design team so that should one or the other side inadvertently omit a needed activity, the chances that this omission will be caught and rectified are increased.

The maintenance concept for a service requires an understanding of the service delivery infrastructure and how the maintainability of each of its elements contributes to the maintainability of the service. Chapter 8 showed some examples of how service reliability is driven by the reliability of elements of the service delivery infrastructure [20]. The same reasoning applies to maintainability: the duration of service outages is influenced by the durations of outages in the elements of the service delivery infrastructure (as well as other factors, including the service delivery infrastructure architecture and backup provisions). Most often, at least a monotone relationship can be asserted: the longer outages persist in the service delivery infrastructure, the longer the service outages will be. Quantitative modeling for relating the duration of service outages to the duration of outages in the service delivery infrastructure is needed so that outage duration requirements (i.e., maintainability and support requirements) for elements of the service delivery infrastructure are developed on a rational basis.

11.3 MAINTAINABILITY ASSESSMENT

11.3.1 Maintenance Functional Decomposition and Maintainability Block Diagram

Section 3.4.1.2 introduced the system functional decomposition, a systematic description of how the various elements of the system work together to carry out each system function. The system reliability block diagram is an important by-product of the system functional decomposition. The reliability block diagram indicates how a system failure (violation of one or more requirements) is caused by the failure of an element of the diagram. We have called this the "reliability logic" of the system. Thought of in this way, the reliability block diagram is like the inverse of the system functional decomposition: while the system functional decomposition tells how an element of the system contributes to the system's functioning, the reliability block diagram tells how the failure of an element of the system contributes to failure of the system.

At times, failure of an element of the system does not cause a system failure. This is the case, for instance, when a system element is backed up by a redundant element ("spare") so that when the system element fails, the spare element takes over the function(s) of that element, and the system continues to function without (or with only a brief[1]) interruption. Maintenance planning requires that these events be taken into account because

- some cost is incurred each time this happens,
- an action to replace the failed unit may be needed,
- if not attended to, some such events could leave the system in an undesirable brink-of-failure state.[2]

[1] A brief outage that may occur when a spare element is switched into service is called "failover time."

[2] A system is said to be in a brink-of-failure state if the next failure of any system element causes the system to fail. For example, after the second unit failure in a three-unit hot standby redundant ensemble, the ensemble is in a brink-of-failure state because there are no more spares to take over when the third unit fails.

The maintenance functional decomposition facilitates this accounting. The maintenance functional decomposition is a systematic description of whether the failure of a system element requires that a maintenance action be performed (e.g., replacement of the failed element). We may derive a *maintainability block diagram* from the maintenance functional decomposition in the same way the reliability block diagram is derived from the system functional description. The maintainability block diagram expresses in pictorial form the way in which a maintenance action may follow from the failure of an element of the diagram. The simplest maintainability block diagram is a series system in which each element of the series arrangement causes a maintenance action when it fails.

If an element that is a single point of failure fails, then both a system failure and a maintenance action to replace the failed unit need to be recorded. The reliability block diagram scores a system failure, and the maintainability block diagram scores a maintenance action. If an element fails that is backed up by a redundant unit, a maintenance action may or may not be called for. For instance, if the element that fails has a hot standby spare, it may be decided, as part of the system maintenance concept, to leave the failed unit in place until the second unit also fails, at which time both are replaced in a single maintenance action. Also, when the system maintenance concept calls for the repair of certain subsystem failures to be deferred until some later time (Section 10.2.2.1), the maintainability block diagram does not include this subsystem. Whenever a unit or ensemble failure entails a maintenance action, we place that unit or ensemble in a series configuration in the maintainability block diagram, even if that unit in the reliability block diagram may have redundant backup. When a unit does not require a maintenance action when it fails, that unit enters the maintainability block diagram in a parallel configuration with the number of "spare" or "backup" units determined by the number of unit failures that have to take place before a maintenance action is required. For instance, a two-unit hot standby ensemble that is not maintained until the second unit fails (so the entire ensemble fails at that time) enters the maintainability block diagram as a two-unit parallel system. If the system maintenance plan calls for each unit to be replaced when it fails, even though a spare unit is in place and enabling system operation to continue, then the two-unit hot standby system enters the reliability block diagram as a two-unit parallel system but enters the maintainability block diagram as a series system of two units (because the failure of each unit increases the maintenance action counter by one).

Example: Consider the single server rack in a server farm example shown in Section 4.4.5. All units in the rack are single points of failure except for the redundant power supply. There are two choices concerning the replacement of the power supply when a failure occurs: we may either replace each power supply unit whenever it fails, regardless of the status of the other supply unit, or we may wait until both power supply units have failed and then replace the ensemble of the two units together. In the first case, a maintenance action is called for every time a power supply unit fails, in effect treating the two power supplies as nonredundant from the maintenance action point of view. The maintainability block diagram for this case is a series system of all 16 units comprising the rack. In the second case, no maintenance action is called for until both power supply units have failed, and replacement of the ensemble of the two redundant units requires only one maintenance action. The maintainability block diagram for this case is a series system of 15 units, 1–12, 15, and 16, and another single unit representing the parallel ensemble of units 13 and 14.

11.3.2 Quantitative Maintainability Modeling

11.3.2.1 Frequency of maintenance actions

Once a maintainability block diagram is in place, projections about the number of maintenance actions, the times between maintenance actions, etc., may be made using the same techniques that were used in Chapters 3 and 4 for reliability block diagrams. The key is to prepare a block diagram that reflects the number of maintenance actions instead of the reliability of the system. As a rule, a system will undergo more maintenance actions than it will undergo failures, mainly because redundant units, while useful for preventing outages, may require attention when they fail in order not to leave the system in a brink-of-failure state.

This section discusses the use of the separate maintenance model (Section 4.4.5) as a model for the number of maintenance actions over a given time period. To implement the separate maintenance model in this application, begin with a maintenance functional decomposition (Section 11.3.1) in which every replaceable unit in the system is individually identified as an element of the decomposition. A maintenance functional decomposition is like a system functional decomposition, but it is constructed so that every maintenance action is recorded—even if it does not cause, or result from, a system failure. There may be other elements in the maintenance functional decomposition, but every subassembly or LRU that is designated as replaceable in the system maintenance plan should appear in the decomposition. A maintainability block diagram is a reliability block diagram based on the maintenance functional decomposition. Separate the diagram into two parts so that one part contains all the replaceable units. Let $\varphi_M(X_1,\ldots, X_n)$ denote the structure function of that part of the diagram containing the replaceable units (numbered $1,\ldots, n$).[3] Finally, denote by $Z_1(t),\ldots, Z_n(t)$ the reliability processes (Section 4.3.2) of the n replaceable units.[4] The separate maintenance model is the reliability process $Z_M(t) = \varphi_M(Z_1(t),\ldots, Z_n(t))$ of the ensemble of the replaceable units. We use $Z_M(t)$ to obtain the number of maintenance actions for the system.

[3] When it is necessary to make the distinction, we call this the "maintainability structure function" to distinguish it from the reliability structure function introduced in Section 4.6.

[4] In fact, this is a description of the operating and outage times of the socket containing the replaceable unit.

Example: Continue the server rack example from Section 11.3.1. Let $N_1(t),\ldots,$ $N_{16}(t)$ denote the number of unit failures in the time period $[0, t]$ for units 1,…, 16, respectively. We consider two cases:

1. Each power supply module is replaced when it fails. In this case, the number of maintenance actions is $N_1(t)+\cdots+N_{16}(t)$ because every unit failure, including each power module, causes a maintenance action. When the operating time and outage time distributions for each of the system's components are as shown in Table 4.1, the expected number of maintenance actions over 5 years is 49.363 if the units are replaced by new ones when they fail [21, 23]. If each unit is revived (Section 4.4.3.1) when it fails, the expected number of maintenance actions over 5 years is 79.579 (Exercise 2). The number of failures with revival is greater than the number of failures with renewal because of the increasing hazard rate nature of the life distributions for the server and the power supply.
2. When the power supply module in service fails, the backup power supply (if it is not already failed) is put into service; the ensemble of two hot standby power supply modules is replaced at the time of the second power module failure. In this case, the number of maintenance actions is $N_1(t)+\cdots+N_{12}(t)+N_{15}(t)+N_{16}(t)+J(t)$, where $J(t)$ denotes the number of replacements of the ensemble of two hot standby power modules. The expected value of $J(t)$ may be determined from equation (4.10) of Ref. 22 if an alternating renewal model for the modules is acceptable and the on- and off-time distributions in that model are assigned.

11.3.2.2 Duration of maintenance actions

Maintenance action durations of interest include

- duration of an individual operation at a single workstation and
- total time needed for a single maintenance job to transit a facility.

These may pertain to preventive or corrective maintenance.

The sojourn time of a job at a single workstation may be estimated from historical data or may be measured using a time-and-motion study [9]. In a maintainability context, these studies are also known as maintenance task analysis [3].

Information about the total time a job spends in a maintenance facility may be gained from a precedence diagram (critical path method) or activity network model [9] for the facility. For purposes of this analysis, a maintenance facility may be conceptualized as a network of workstations with jobs flowing around the network in a pattern determined by the type of equipment being serviced, its service needs, and the types of tasks that can be performed at each of the workstations in the facility. Discussion of this variable is postponed until Section 13.4.1 in which we describe a stochastic network flow model for performance analysis and optimization of a maintenance facility.

11.4 DESIGN FOR MAINTAINABILITY TECHNIQUES

11.4.1 System Maintenance Concept

The system maintenance concept (Section 11.2) serves as a foundation for maintenance planning. While it begins at the early stages of system design and so at that time is necessarily incomplete and lacking in detail, good practice encompasses continual updating of the maintenance concept as the design progresses. There is a natural link between the maintenance concept and system reliability modeling: maintainable system reliability modeling undertaken (e.g., as in Section 4.4) is driven by the system maintenance concept because the locations and types of maintenance performed are the raw material for system reliability models such as a separate maintenance model (Section 4.4.5) or a state diagram reliability model (Section 4.4.7).

A strategy for assigning certain maintenance actions to different locations is a vital part of the maintenance concept. The locations at which repairs are performed are referred to as "levels," a terminology deriving from early use of this procedure in defense systems. Options for maintenance levels include

- online maintenance at the site where the system is being used,
- offline maintenance at a location near the site where the system is being used,
- offline maintenance at a location distant from the site where the system is being used, and/or
- offline maintenance at a manufacturer's plant or supplier's facility.

A system maintenance concept need not include all these levels. The choice of which levels to employ is accomplished by a LoRA (Section 11.4.2), an economic exploration and optimization of maintenance operation through implementation of options from this menu. Some additional factors that influence design of maintenance and assignment of specific maintenance procedures to each level include

- what repairs need to be done, including an assessment of how complicated each repair type may be,
- what spare parts, tools, documentation, and other materials need to stored to enable maintenance at that level,
- what skills will be required of the repair staff, including an assessment of how much repair can be accomplished by the system operators who may not necessarily be trained to carry out maintenance tasks, and
- how often maintenance (corrective and preventive) may be needed.

Details pertaining to each level of maintenance are considered in Section 11.4.2.

11.4.2 Level of Repair Analysis

A LoRA is an economic optimization that determines the least expensive assignment of repair operations to one or more of the four levels of maintenance commonly considered. Division of maintenance activities into levels comes originally from the defense industry in which systems may be deployed in far-flung locations and rapid repairs are usually required, so that the option of repairing a failed system on-site was developed.

11.4.2.1 Online maintenance

Online maintenance refers to preventive or corrective maintenance actions that take place where the system is deployed. Usually, online maintenance comprehends simpler, shorter-duration, or less frequently–occurring tasks, such as cleaning, minor adjustments, periodic condition monitoring, and the like, that may be accomplished by system operators without taking the system out of service. Online maintenance may also be preferred for rapid replacement of critical items to minimize system outage time.

Maintenance training for system operators incorporates

- ability to initiate and interpret diagnostic routines for fault location,
- procedures for the maintenance tasks (preventive or corrective) that are determined by the LoRA to be performed on-site, and
- ability to discern when a repair is outside the scope of on-site maintenance but needs to be referred to a higher level for completion.

Planning for environmental influences on repair procedures is more important at the online level because systems may be deployed in a variety of differing environments: shipboard, aircraft, automotive, poor air quality, arctic or equatorial, etc.

11.4.2.2 Off-line maintenance on-site or at a nearby site

Certain preventive or corrective maintenance actions may require that a system be taken out of service before they can be performed. The system is said to be off line, and the maintenance may then be undertaken on site or at a nearby fixed or mobile location. This, and off-line maintenance at a remote location (Section 11.4.2.3), is referred to as "intermediate maintenance." For example, some line-replaceable units (LRUs) may require the system to be powered down before they can be safely removed or replaced.

11.4.2.3 Off-line maintenance at a remote location

This is the second type of intermediate maintenance, sometimes called depot-level maintenance. It should be considered for repairs that

- may be more complicated,
- may require specialized test equipment and/or tools,

- may require more specialized expertise to accomplish, or
- may occur less frequently (so that on-site stocking of spare part(s) required for this repair may be costly).

Turnaround times will be greater than for online maintenance or for off-line maintenance on site or nearby, and transportation costs are also incurred.

This is also the first level of maintenance where it is reasonable to consider repair of LRUs. Many LRUs are valuable enough that they are not discarded when they fail but are instead repaired and placed into a spares inventory for use in future system corrective maintenance. Repair of an LRU usually entails replacement of some component(s) on the unit, so solder rework stations and other specialized tools may be required. This is also usually exacting work, and it is not reasonable to expect that it could be performed under field conditions (even an environmentally controlled location such as a telephone central office, while offering a benign (stable temperature and humidity, low vibration, etc.) environment, would not be equipped with the workspaces and workstations needed for these repairs). When a system contains valuable LRUs that are repaired and not discarded when they fail, depot- and/or manufacturer-level maintenance is almost mandatory. Figure 5.1 shows an example of the flows of material and information supporting a repair scheme in which LRUs are repaired at an independent facility under contract to the system manufacturer. That is, the system manufacturer has outsourced the repair of LRUs that it might have performed itself. Note that Figure 5.1 does not allude in any way to the potential political difficulties that may arise in seeking to share reliability data across unrelated organizations. While this is an important consideration, it is beyond the scope of this book.

11.4.2.4 Off-line maintenance at a manufacturer's or supplier's facility

For systems or products using a multilevel maintenance concept, this is the last resort for repairs. Consider reserving this level of maintenance for

- particularly intractable failures that are difficult to diagnose,
- situations where long turnaround times can be tolerated, or
- tasks that are beyond the capability of on-site or intermediate maintenance staff.

A multilevel maintenance scheme also offers the possibility of spilling over to the next higher level repairs that have been attempted but were unsuccessful. It is reasonable to expect (but should be verified) that the manufacturer of the system has the specialized expertise, tools, and diagnostic systems to handle almost every type of failure. For some types of products (e.g., consumer entertainment products), this may be the only option offered by the manufacturer (even though the owner may be able to perform repairs himself or may contract repair to be conducted by an independent shop).

As with intermediate maintenance, costs will be incurred for transportation of materials both to and from the facility.

11.4.2.5 Analysis and optimization

The LoRA described in this section helps choose the least cost repair scheme to fit the particular needs of your system. From the four levels of repair and the possibility of discarding the item, there are at most 31 combinations (ranging from using on-site repair only up to using all four levels plus discard). Some combinations may be ruled out by other conditions prevailing in system use or operation, so the number of choices is usually limited to a small number. Choice of the least cost option is readily accomplished through use of a spreadsheet-based accounting procedure. LoRA is described in detail in MIL-STD-1390D [24], which, while no longer supported by the Department of Defense, contains a wealth of information and procedures that help in practical LoRA studies. LoRA is also used in the automotive and aerospace industries [19]. Software to allow rapid completion of LoRA has been described [7, 11]. As with all off-the-shelf software, users should verify that the assumptions used by the software developers are appropriate for the study being undertaken before relying on the answers generated by the software.

To begin a LoRA, choose a time horizon for the decision process. Use a time horizon that reasonably reflects the time over which you expect that the system will be supported by the maintenance-level scheme. Use one spreadsheet worksheet for each item type to be maintained in the multilevel scheme. Use one column for each level-of-repair option (e.g., a three-level maintenance scheme uses four columns, one for each level and one for the "discard" option). Use one row for each of the following costs:

- Acquisition cost: the first cost of purchasing the item,
- Expected maintenance labor cost: labor cost per hour multiplied by the expected total duration of all maintenance tasks on that item over the chosen time horizon,
- Maintenance staff training cost allocated to the item,
- Maintenance facility costs (rent, utilities, janitorial costs, capital costs, etc.) allocated to the item (if necessary, separate capital costs from expenses),
- Inventory acquisition cost for the item,
- Inventory carrying cost for the item,
- Repair parts inventory acquisition cost for the item,
- Repair parts inventory carrying cost for the item,
- Cost of test equipment, tools, documentation, software, etc., for the item,
- Transportation costs for the item,
- Recycling and disposal costs allocated to the item.

Each cell is populated with the cost associated with its row for the level associated with its column. Use the spreadsheet to add up the costs down each column. The LoRA procedure chooses the option corresponding to the column with the lowest total. If you were making a decision for that one item only, you could do it now based on the spreadsheet results. If there is more than one item in the plan and all items are to be repaired using the same multilevel plan, make a weighted average of all the total costs, weighted by the proportion of the total population of items represented by each individual item. For instance, if there are two items A and B, and item A represents 30% of the total number of both items and item B represents 70% of the total, weight the total costs of items A and B by 0.3 and 0.7, respectively. Make separate worksheets for the two items and average the total costs over the two worksheets for A and B using these weights. The lowest weighted average total cost over the options considered is the LoRA solution. If there is more than one item in the plan but each item may have a separate repair scheme, the items may be treated individually with an optimal level of repair selected for each, and the weighted-average procedure is not needed.

The remainder of this section is devoted to a small example of a LoRA, not so much as a generic example to be followed to the letter in particular applications but more as an illustration of the procedure and the reasoning process that makes LoRA useful.

Example: This example concerns two LRUs, A and B, that are themselves repairable. The options considered include intermediate-level repair, depot-level repair of the LRUs (a failed LRU is replaced by a working one from a spares inventory so that the system is restored to service; the failed LRU undergoes repair itself using the scheme to be decided by the LoRA), and discard (when an LRU fails, it is not repaired but is discarded or recycled). The units are not repairable on-site. We will illustrate a LoRA for these units using a 10-year time horizon. To fill in the spreadsheet, some facts about units A and B are required.

1. A system containing units A and B is installed in 10 submarines. Each submarine contains two systems. Each system contains three A units and seven B units. The systems are in service 12 hours a day, and the submarines run on a 6-months-on, 2-months-off schedule, so over the 10-year study horizon, each system accrues a total of 32,400 hours of operation.[5] The type A units accrue a total of 1,944,000 unit-hours, and the type B units accrue a total of 4,536,000 unit-hours.
2. Type A units cost $18,000 each if designed to be repairable and $15,000 each if designed to be discarded. The corresponding costs for unit B are $3500 and $2750, respectively.
3. The labor rate, including all overhead, is $35 per hour at the intermediate level and $55 per hour at the depot level. Unit A takes an average of 4 hours to repair, while unit B takes an average of 3 hours to repair.

[5] Using a 30-day month, which is a common simplifying assumption in such studies.

4. The estimated failure intensity of unit A is 6×10^{-5} failures per hour and that of unit B is 2×10^{-5} failures per hour.[6]
5. Training repair personnel costs \$60 per hour for intermediate-level staff and \$80 per hour for depot-level staff.
6. Facility costs allocated to units A and B are \$1.50 per maintenance hour at the intermediate repair level and \$2.50 per maintenance hour at the depot repair level.
7. The spares inventory size is 2 type A spares and 5 type B spares per system at the intermediate level and 10 type A spares and 25 type B spares (covering all systems) at the depot level when the units are repaired. If the units are discarded, one spare is needed for every unit in the field. Inventory carrying costs are approximately 7% of the inventory value per year.
8. Parts consumed in the repair of unit A amount to \$78 per repair and for unit B amount to \$24 per repair.
9. Costs of test equipment, tools, etc., allocated to units A and B together are \$12,000 per installation for intermediate-level repair and \$50,000 per installation for depot-level repair.
10. Transportation costs \$500 for any number of units from the submarine to either the intermediate or depot repair facility, regardless of the number of units being shipped.
11. Disposal costs \$25 for unit A and \$15 for unit B. Fifty percent of those disposed are recycled, and recycling unit A (B, respectively) brings in \$80 (\$10, respectively) in revenue.

The spreadsheet for unit A is shown in Table 11.1.

The spreadsheet for unit B is shown in Table 11.2.

In both cases A and B, the analysis indicates that intermediate repair is preferred. Note that training costs are the same for both units A and B, so these could have been left out of the analysis, and the conclusion would not change. If we needed to consider both units A and B together (i.e., only a single level of repair strategy was to be selected for both units), the additional step of taking weighted averages of the results for A and B would be required if the two analyses pointed to two different choices for A and for B. In this example, this is not needed because the conclusion is the same for A and B: use intermediate-level repair.

This example is oversimplified and not intended to provide a template for any particular LoRA. Rather, it is intended to show in general terms how LoRA is carried out and what the underlying reasoning process is. Practical LoRA is facilitated by standards such as Refs. 19, 24 and off-the-shelf software such as Refs. 7, 11.

[6] Note that because the failure intensities are stated as constants with no time dependence, this assumes that failures will appear in the field according to homogeneous Poisson processes with the stated rates. See Section 4.3.3.1.

TABLE 11.1 Unit A LoRA Spreadsheet

Cost Description	Intermediate Repair ($)	Depot Repair ($)	Discard Option	Notes
Acquisition	1,080,000	1,080,000	900,000	20 systems, 3 type A units per system
Labor	16,330	25,661	0	Expected number of failures over 10 years is 116.64
Training	11,520	2,880	0	Eight students for 3 days at intermediate, two students for 3 days at depot
Facility	700	1,166	0	
Spares inventory acquisition	720,000	180,000	900,000	
Spares inventory carrying	504,000	126,000	630,000	
Repair parts consumption	9,098	9,098	0	
Test equipment, tools, etc.	3,600	4,500	0	
Transportation	0	0	0	Not included because it is the same in all scenarios
Recycling/disposal	0	0	−3,208	
Total	2,345,248	1,429,305	2,426,792	

TABLE 11.2 Unit B LoRA Spreadsheet

Cost Description	Intermediate Repair ($)	Depot Repair ($)	Discard Option	Notes
Labor	22,226	34,927	0	Expected number of failures over 10 years is 90.72
Training	11,520	2,880	0	Eight students for 3 days at intermediate, two students for 3 days at depot
Facility	408	680	0	
Spares inventory acquisition	350,000	87,500	385,000	
Spares inventory carrying	245,000	61,250	269,500	
Repair parts consumption	2,177	2,177	0	
Test equipment, tools, etc.	8,400	10,500	0	
Transportation	0	0	0	Not included because it is the same regardless of the number of units
Recycling/disposal	0	0	227	
Total	639,731	199,914	654,727	

11.4.3 Preventive Maintenance

Preventive maintenance is the application of occasional interventions intended to forestall possible system failures. Preventive maintenance is most effective in cases in which there are one or more wearout failure modes present in the system for which suitable countermeasures were not implemented, either because none could be identified or because identified countermeasures were considered too expensive, and measures are known that may be applied to forestall the wearout failure mode. An example that clearly illustrates preventive maintenance, because the wearout failure mode is readily discernible, is lubrication of bearings. In the absence of lubrication, metal-to-metal contact in rolling or sliding bearings would cause rapid wear and failure of the system of which they are a part. Therefore, lubrication is specified as a part of most, if not all, bearing applications to postpone the time at which this wearout failure mode may activate. An additional preventive maintenance aspect of this example is that, in some cases, such as an internal combustion engine, the lubricant may itself wear and need to be replaced from time to time to maintain its effectiveness. So a schedule of lubricant replacement is recommended: change the engine oil every 7500 miles or once a year, whichever comes first. This is an example of a fixed schedule in which "age" is measured both by elapsed time and by elapsed mileage.

Fixed preventive maintenance schedules may not be optimal. For instance, in the internal combustion engine example, lubricant wear also depends on other factors, such as style of driving and environmental conditions. Highway driving at a more-or-less constant speed is less wearing on lubricants than stop-and-go city driving. Driving in dusty or sandy environments causes faster lubricant wear. But a fixed schedule of lubricant replacement does not account for these variables and may cause

- premature replacement of lubricant that may have many miles of safe use remaining, or
- tardy replacement of lubricant that may have already worn past the point of effectiveness.

As industries began to recognize the economic implications of these (we might call them) type-1 and type-2 errors, a search for better preventive maintenance schemes began. One result was RCM.

11.4.4 Reliability-Centered Maintenance (RCM)

When initially conceived, preventive maintenance was envisioned as a regularly scheduled activity that would take place regardless of other conditions prevailing in the system. It soon became apparent that a fixed schedule of preventive maintenance was not optimal. You could be carrying out preventive maintenance long before it was necessary, or the failure mode it was intended to forestall occurred before the preventive maintenance could be applied. So it became sensible to look for ways to incorporate knowledge of the system's current condition and past failure behavior into a preventive maintenance scheme. Techniques of this type fall under the category of RCM [5, 17]. We will describe two types of RCM, predictive maintenance and condition-based maintenance.

> **Language tip:** "Predictive maintenance," "condition-based maintenance," and "RCM" are not used consistently in the community. We have chosen in this book a usage that we hope aligns the term with the process so that it will be easier to remember which term applies to which concept. Thus, we use RCM as the general term for all preventive maintenance schemes that involve using information about the reliability of the system to plan the next, or the next series of, preventive maintenance intervention(s). We reserve predictive maintenance for plans based on the knowledge of stochastic properties of times to failure or operating times of the system, and condition-based maintenance for plans based on following the progress of some degradation process active in the system or on the results of some specific testing applied to the system. Be aware that usage varies and take pains to verify which is being used for which when clarity is important.

11.4.4.1 Predictive maintenance

There are two broad classes of RCM schemes: ones based on the knowledge of the pattern of system failures in time (or other age-measuring variable), and ones based on an understanding of some degradation process at play in the system. We will describe the first class in this section on predictive maintenance and the second class as condition-based maintenance in Section 11.4.4.2.

We have previously conceptualized the times at which failures of a maintainable system occur as a point process (Section 4.3.3). If it is possible to characterize the operating times in this point process thoroughly, what we know about the distributions of the operating times in the process may be used to construct a preventive maintenance schedule. For example, suppose $n-1$ failures have occurred in the system so far. At the end of the current outage, the system is repaired and returned to service, and the next operating time U_n begins. If we know the distribution of U_n, we may choose a time at which we are, say, 90% certain that the next failure will occur beyond this time (this would be the 10th percentile of the distribution of U_n). Of course, this choice is not going to be arbitrary (although even an arbitrary choice has a chance of improving on the fixed schedule scheme) but will be determined through an optimization balancing the cost of carrying out the preventive maintenance too early against the costs of a failure. Some examples of optimization of predictive maintenance schemes can be found in Refs. 6, 14.

11.4.4.2 Condition-based maintenance

As an alternative to acquiring stochastic characterization of the system's operating times, some physical characteristic(s) of the system may be measured and tracked over time[7] in an attempt to predict when a failure may be imminent. The measurement may be passive (i.e., no special stimulus is applied to the system but rather some existing operating characteristic or sensor reading is followed) or active (i.e., some stimulus is applied to the system, and a response is measured).

In the first case, the measured characteristic is usually conceptualized as a stochastic process $\{X(t): t \geq 0\}$ ($X(t)$ is the value of the measurement at time t), and the connection with maintenance is that the system fails at the first time τ that this process crosses some stated threshold x_0. That is,

$$\tau = \inf\{t \geq 0 : X(t) \geq x_0\}$$

if we think of the process $X(t)$ as nondecreasing. For instance, $X(t)$ may represent the percentage of oxidation in a steel structure; when that percentage reaches a predetermined threshold x_0, some remedial action is taken to forestall collapse of the structure. $X(t)$ may represent the level of vibration measured in some rotating machinery; when the measured vibration is too great, preventive action is taken to forestall a failure that may take place if bearing wear (or some other failure mechanism) were to be allowed to increase unchecked. A schedule of inspections is needed so that maintenance personnel will know when the system should be monitored. A static schedule calls for measurements at fixed, predetermined times. If the system is continuously monitored, this procedure is called *condition monitoring* [8]. A dynamic schedule may be developed based on an understanding of how rapidly $X(t)$ changes, that is, slowly changing phenomena need not be inspected as often. A dynamic schedule may be created by estimating $X'(t)$ at the same time $X(t)$ is measured so that the time of the next inspection will be longer if $X'(t)$ is small and shorter if $X'(t)$ is large.

The measurements (longitudinal data) obtained from these inspections are values of $X(t)$ at the inspection times $t_1, t_2, \ldots$. Statistical treatment of these data in an engineering context was introduced by Carey and Tortorella [4] and developed to a high standard by Lu and Meeker [13] and others. Degradation analysis is now a standard part of statistical analysis of reliability data [15] and is widely used in many condition-based maintenance analyses [2, 12, 13, 14, 25].

In the second case, inspection entails measuring the response of the system to a defined stimulus. For example, inverse acoustic scattering may be used to detect cracks in structural materials [1], enabling early detection of deterioration that, if left unchecked, could lead to catastrophic failure [10]. In other respects, this procedure is like passive condition-based maintenance with the added feature that periodic inspection of elements not ordinarily visible may be accomplished.

[7] Statisticians would call the data that result from this scheme *longitudinal data*.

11.5 CURRENT BEST PRACTICES IN DESIGN FOR MAINTAINABILITY

11.5.1 Make a Deliberate Maintainability Plan

The degree of maintainability to be achieved by a system or service is determined by customer needs for rapid, low-cost, and error-free repair or restoration, and the business case for the system or service. It may be possible to develop an optimization model to guide the proper balance or determine just how much maintainability customers will be willing to pay for. Even if this is done only informally, through discussions between marketing and systems engineering, greater understanding of the trade-offs involved forms a more rational basis for action. Some deliberate action must be taken, or the result will be a system or service that has some degree of maintainability that is achieved essentially at random. That is, without deliberate planning and attention to design for maintainability, the system or service maintainability is what it is because of actions or omissions that were unguided, and a good outcome would be the result of good luck rather than a solid plan. So the first best practice is to determine just how much maintainability is appropriate for the system or service. This should be undertaken before a maintenance concept (Section 11.2) is considered.

11.5.2 Determine Which Design for Maintainability Techniques to Use

Not every system or service will require a high degree of maintainability, but every system or service should have a maintenance concept, for it is here that the fundamental decisions about maintenance are made and documented. From these decisions, maintainability requirements should be constructed. Once the requirements are known, you can decide how much effort in design for maintainability will be needed. This chapter discusses three relevant designs for maintainability techniques: the system or service maintenance concept, LoRA, and preventive maintenance.

Every system or service needs a maintenance concept, even if the concept is "do no maintenance" (and if this is the decision, it must be deliberate). LoRA is not needed for systems that don't use the replace-and-reuse concept for subassemblies. If a system is deployed in widely separated geographic locations, and is repaired using a replace-and-reuse concept, a LoRA is needed so that costs of the repair operation can be identified and a rational, minimum-cost repair scheme is selected. Even legacy repair infrastructures should be considered for cost-saving opportunities that may arise from batching strategies, repair-on-demand possibilities, etc.

Investigate possible preventive maintenance schemes. Determine whether the system harbors any wearout failure mechanisms and whether these are likely to activate before the end of the system's intended service life. If so, consider developing preventive maintenance, either scheduled or predictive, as appropriate, to forestall the large number of failures that may occur because all deployed systems have the same wearout failure mechanism(s). This is analogous to the treatment of design flaws in reliability engineering: if a failure mode is due to a design flaw, then every copy of the system contains the same flaw, and (depending on the environment in which the system operates) the flaw will cause a failure sooner or later in every copy of the system. As always, the preventive maintenance decision is an economic one. It may not be economically sensible to spend large sums on prevention for systems having short useful lives, of low value, or that do not generate large external failure costs.

For high-consequence systems, the same principle used in design for reliability (Chapter 7) applies here too: require justification for any elimination of design for maintainability interventions. The contrast is with "ordinary" (not high-consequence) systems in which economic considerations usually require justification of inclusion of design for maintainability interventions. The balance between prevention costs and external failure costs should always be considered.

11.5.3 Integration

Maintainability (and supportability) and reliability are connected through availability requirements (see the example in Section 10.2.3). Therefore, reliability modeling and maintainability modeling (and supportability modeling, Chapter 12) should be linked so that

- availability implications of maintenance and support policies can be discerned, and
- maintainability requirements can be set on a rational basis.

While the frequency of failures is determined by reliability, the duration of outages is driven by supportability and maintainability. If there are downstream effects of failure frequency,[8] these should be considered when developing requirements for reliability. Downstream effects of outages in elements of a service delivery infrastructure should factor into the development of maintainability (and supportability) requirements.

11.5.4 Organizational Factors

Integrate maintenance team members with design team members as soon as possible. We have seen examples of how design features may promote or inhibit maintainability. It is wise to begin interdisciplinary communication

[8] For instance, circuit-switched telephone networks suffer stable call cutoffs at the moment a failure occurs in a network element in the talking path of the call. This phenomenon is directly linked to the onset of the failure, and so its prevalence is determined by the frequency, not the duration, of failures in the network [20].

as early as possible in the design process so that decisions that affect maintainability are reviewed and rationalized to avoid adverse effects. Design reviews are a natural forum for these discussions.

11.6 CHAPTER SUMMARY

As usual in this book for systems engineers, this chapter offers many recommendations for design for maintainability, but only some of these are covered in detail. Most often, these actions will be carried out not by the systems engineer but by other maintainability specialist engineers on the design team. Some of the resources that can be used to fill in the details of these procedures include Refs. 3 and 16, and, for insight into early thinking on the subject [18].

11.7 EXERCISES

1. Continue the server rack example from Sections 11.3.1 to 11.3.2. Suppose that replacement units are new and that unit i has an exponential life distribution with parameter $\lambda_i, i=1,\ldots,16$, where $\lambda_1=\cdots=\lambda_{12}=2, \lambda_{13}=\lambda_{14}=8, \lambda_{15}=1$, and $\lambda_{16}=0.01$ failures per year. Suppose that the downtime for unit i (including support and replacement time) has a uniform distribution on [1, 4] for every $i=1,\ldots,16$.
   a. In case every unit is replaced when it fails, what is the expected number of unit replacements in the rack over the first year of operation?
   b. In the case where the power supply ensemble is replaced only after both power supplies have failed, what is the expected number of ensemble replacements in the first year of operation? (Hint: see section 4.2 of Ref. 22.)
2. Complete the example in Section 11.3.2.1 using the method found in Section 4.4.3.1.
3. List the 31 possible combinations of repair levels noted in the beginning of Section 11.4.2.5. Do all 31 combinations make sense? Are there any situations in which it makes sense not to have an on-site repair option?
4. Cite and discuss two examples of preventive maintenance. Identify the wearout failure mode(s) the preventive maintenance is intended to forestall. How is "age" measured in your examples?

REFERENCES

1. Baltazar A, Wang L, Xie B, Rokhlin SI. Inverse ultrasonic determination of imperfect interfaces and bulk properties of a layer between two solids. J Acoust Soc Am 2003;114 (3):1424–1434.
2. Besnard F, Bertling L. An approach for condition-based maintenance optimization applied to wind turbine blades. IEEE Trans Sustain Energy 2010;1 (2):77–83.

3. Blanchard BS, Verma D, Peterson EL. *Maintainability: A Key to Effective Serviceability and Maintenance Management*. Volume 13, New York: John Wiley & Sons, Inc; 1995.
4. Carey MB, Tortorella M. Analysis of degradation data applied to MOS devices. Sixth International Conference on Reliability and Maintainability; Strasbourg, France. 1988.
5. Carter AB. Reliability centered maintenance. 2011. *United States Department of Defense Manual 4151.22-M*. Washington, DC: US Department of Defense.
6. Chu C, Proth JM, Wolff P. Predictive maintenance: the one-unit replacement model. Int J Prod Econ 1998;54 (3):285–295.
7. Elliott-Brown JA, McPherson SW. NAVSEA level of repair analysis (LORA) software. Naval Eng J 1995;107:59–66.
8. Elsayed EA. *Reliability Engineering*. 2nd ed. Hoboken: John Wiley & Sons, Inc; 2012.
9. Freivalds A. *Niebel's Methods, Standards, and Work Design*. Volume 700, Boston: McGraw-Hill Higher Education; 2009.
10. https://en.wikipedia.org/wiki/Mianus_River_Bridge
11. Ituarte-Villarreal CM, Espiritu JF. A decision support system for the level of repair analysis problem. Proceedings of the 41st International Conference on Computers & Industrial Engineering. October 23–25; Los Angeles, CA; 2011. p 666–671.
12. Jardine AKS, Banjevic D, Makis V. Optimal replacement policy and the structure of software for condition-based maintenance. J Qual Maint Eng 1997;3 (2):109–119.
13. Lu CJ, Meeker WQ. Using degradation measures to estimate a time-to-failure distribution. Technometrics 1993;35 (2):161–174.
14. Lu S, Tu YC, Lu H. Predictive condition–based maintenance for continuously deteriorating systems. Qual Reliab Eng Int 2007;23 (1):71–81.
15. Meeker WQ, Escobar LA. *Statistical Methods for Reliability Data*. New York: John Wiley & Sons, Inc; 1998.
16. Okogbaa OG, Otieno W. Design for maintainability. In: Kutz M, editor. *Environmentally Conscious Mechanical Design*. New York: John Wiley & Sons, Inc.; 2007. p 185–248.
17. Rausand M. Reliability centered maintenance. Reliab Eng Syst Saf 1998;60: 121–132.
18. Rigby LV, Cooper JI, Spickard WA. Guide to integrated system design for maintainability. Defense Technical Information Center document AD-0271477. 1961.
19. Society of Automotive Engineers. Level of Repair Analysis standard AS-1390. 2014.
20. Tortorella M. Cutoff Calls and Telephone Equipment Reliability. Bell Syst Tech J 1981;60 (8):1861–1889.
21. Tortorella M. Numerical solutions of renewal-type integral equations. INFORMS J Comput 2005;17 (1):66–74.
22. Tortorella M. On cumulative jump random variables. Annal Oper Res 2013;206 (1):485–500.
23. Tortorella M, Frakes WB. A computer implementation of the separate maintenance model for complex-system reliability. Qual Reliab Eng Int 2006;22 (7):757–770.
24. US Department of Defense. *Military Standardization Document 1390D*. Washington, DC: US Department of Defense; 1993.
25. Williams JH, Davies A, Drake PR, editors. *Condition-Based Maintenance and Machine Diagnostics*. New York: Springer-Verlag; 1994.

Part III

Supportability Engineering

12

Support Requirements

12.1 WHAT TO EXPECT FROM THIS CHAPTER

We have completed two-thirds of our journey through sustainability engineering for systems engineers, and our final push is through the landscape of supportability engineering. While we strive to avoid failures through design for reliability, failures are nonetheless inevitable at least because it is at least possible, if not indeed likely, that we may not be able to anticipate all the possible failure modes in a system. So a well-designed system, anticipating that outages will occur, incorporates features that help minimize their duration. In the extract from a system history diagram shown in Figure 10.1, the outage period is divided into two parts: an initial period of time devoted to preparation for repair and a subsequent period of time devoted to execution of repair. This distinction is made so that each duration may be minimized separately by the application of techniques particular to the needs of the two activities of preparation and execution. Maintainability engineering, the body of knowledge connected with execution of repairs, was the subject of Part II of this book. We now turn to supportability engineering, the body of knowledge connected with preparation for repair, in Part III.

This chapter begins the study of supportability engineering by examining support requirements. To do this, we need to know

- what constitutes system or service support,
- what activities are included in supportability engineering,

Reliability, Maintainability, and Supportability: Best Practices for Systems Engineers, First Edition. Michael Tortorella.

- how supportability may be measured and monitored, and
- how supportability requirements are developed and interpreted.

We begin with a discussion of supportability as a system attribute and how systems engineers influence supportability by creation of appropriate requirements and undertaking or commissioning quantitative studies for system features and attributes that directly affect supportability. As we did previously with reliability and maintainability, we list and discuss various supportability effectiveness criteria and figures of merit that are routinely used in the creation of support requirements. These help us understand what makes a good support requirement and form the basis for interpreting support requirements and comparing support performance against requirements. The chapter closes with a review of current best practices in support requirements development.

12.2 SUPPORTABILITY FOR SYSTEMS ENGINEERS

12.2.1 Supportability as a System Property

Time was when support was considered synonymous with logistics. When, for example, multilevel repair schemes were first implemented, logistics constituted a clearly visible large cost. However, continued operation of more and more sophisticated repair operations made it clear that there were many other factors involved in ensuring that repairs could be executed quickly. These other factors are grouped under the heading of supportability.

Supportability is a system property comprehending preparation for effective and efficient repair. We may say that supportability is the degree to which a system contains features and procedures that enable rapid, inexpensive, and error-free repair to be carried out. In other words, supportability addresses the degree to which the system is prepared to execute maintenance. Supportability comprehends the properties and operations that make a system more or less ready to have maintenance performed on it in a speedy, low-cost, and error-free manner. Sometimes, the synonym *serviceability* is used.

When a system failure occurs, users are vitally concerned with rapid restoration of system functioning. In addition, while an outage persists, external failure costs mount up:

- The system operator loses revenue while the system is not able to serve customers.
- Backlogs of material and incomplete jobs grow while the system is not able to work on them.
- The system operator's reputation suffers while their customers are unable to access services provided by the system.
- Expenses of repair and service restoration increase.

Through study of work, industrial engineers have determined that the activity of repair is fruitfully divided into two parts: one part concerning preparation for repair and the other part concerning execution of repair. This division promotes more effective and efficient service restoration because it allows for the study of each part separately and application of techniques and tools adapted to each part separately. This chapter concerns the first part, preparation for repair. Chapters 10 and 11 concerned the second part, execution of repair.

Language tip: Refer to Figure 10.1. The figure shows a typical outage interval divided into two parts: a part involving preparation for repair and a subsequent part involving execution of repair. The first part, the preparation for repair, is a *support interval*, and its duration is a *support time* or *supportability time*. This interval is normally considered to include activities directly connected with preparation for the particular repair to be proximately undertaken. Such activities include collecting the tools, instructions, and spare parts required for this repair, clearing the workstation of clutter from a previous repair, assembling the staff necessary for the repair, etc., but as pertaining to only the individual repair that is about to be begun. We will describe these as "online" supportability activities. On the other hand, *supportability* as a system property may include other factors that promote more efficient, speedier repair as a general property of the repair facility and processes. Such factors may involve the design of repair workstations and their layout in the repair facility, creation of broadly applicable documentation and workflow management policies and supporting tools, spares inventory provisioning and management, and other more general design features beneficial to repair execution overall. Supportability actions taken for the general improvement of the repair facility, without respect to any particular repair, will be called "off-line" supportability and are discussed in Chapter 13, Design for Supportability. In the literature, *supportability* is used to describe both of these types of activities, usually without further distinction.

Language tip: *Support* and *supportability* are not the same thing. We use *support* to refer to specific activities undertaken to prepare for efficient and effective repair. Support includes things like spares inventory size determination, provision of tools and documentation, repair workstation ergonomics, and other specific processes and procedures needed to make repairs go smoothly. Supportability, on the other hand, is the result of support procedures and processes. It is the degree to which the system is prepared to be repaired quickly, inexpensively, and in an error-free manner. Effective support procedures and processes make for good supportability. Haphazard, unplanned, or otherwise inadequate support procedures and processes mean that supportability will not be as good. In this book, we sometimes use "degree of supportability" and "supportability" as the same thing even though "degree of supportability" is redundant. Finally, not every document or discussion makes a distinction between "support" and "supportability." It is a contention of this book that being careful about language is a necessary condition for systems engineers to be better able to do their jobs effectively.

12.2.2 Factors Promoting Supportability

Certain factors under the control of the system development team can improve supportability. These should be considered as part of the team's efforts to produce a system that meets customer needs for short outages. Relevant factors include

- incorporation of system features to allow for rapid diagnosis of the cause of a failure and location of the faulty subassembly or subassemblies,
- design to use standard fasteners and tools, wherever possible, and to use fastenerless assembly (such as snap fastening) where appropriate,
- provision of appropriate tools, equipment, documentation, and system ergonomics and interfaces so that replacement of faulty subassemblies is as simple and foolproof as possible,
- provision of appropriate documentation and workflow management so that information needed by repair staff is available with minimal delay,
- availability of regularly scheduled training for repair staff,
- layout of the repair facility and design of its individual workstations,
- design of maintenance and repair procedures that are easily completed while minimizing errors,
- provision of adequate numbers of spare subassemblies, repair parts, and consumables so that they are available to repair staff with minimal delay,
- design of transportation and logistics to support the level of repair analysis (LoRA) created in the design for maintainability,
- management of spare parts aging, obsolescence, and replacement,
- management of vendors who provide materials, temporary staffing, or other necessities of support, and
- costs of supportability activities and equipment.

This list contains only off-line factors in the sense introduced in Section 12.2.1. Appropriate attention paid to these factors can help improve the supportability of the design. Section 13.3 describes implementation issues, including some quantitative models that would be useful in the optimal allocation of resources to these tasks.

12.2.3 Activities Included in Supportability Engineering

The scope of supportability engineering includes any activities that take place before execution of repair begins that promote faster, more efficient repair. In keeping with the factors cited in Section 12.2.2, these may include

- design and provision of built-in test, online test, and off-line test procedures and facilities,
- design of system ergonomics and interfaces to minimize the chance for error during preventive and corrective maintenance,
- specification and acquisition of test equipment required for off-line test and other maintenance and repair activities,

- design and layout of repair facilities, including human factors such as lighting, HVAC, and ergonomics, as well as staff sizing and sourcing,
- design and creation of documentation to support repair execution, including design and implementation of workflow management systems as appropriate,
- design and provision of computing facilities supporting documentation, work execution, etc.,
- quantitative modeling and optimization for transportation, logistics, spares and consumables inventories, ordering procedures for out-of-stock situations, and other relevant repair preparation issues,
- design and implementation of appropriate staff training, and
- data collection to support monitoring of current performance and to enable continual improvement.

12.2.4 Measuring and Monitoring Supportability

As usual, the task of measuring and monitoring supportability falls to the creation of suitable support effectiveness criteria and figures of merit, developing requirements for these, and collecting data from repair and maintenance support operations to enable determination of the degree to which support requirements are helpful and are being met. We discuss support effectiveness criteria and figures of merit in Section 12.4.

12.2.5 Developing and Interpreting Support Requirements

Once effectiveness criteria and figures of merit have been defined for the support factors that are important to the system or service operation, requirements for these may be developed. Requirements form part of the overall support concept for the system or service. Not all support concepts need be alike. Support requirements focus development and management attention on those particular supportability issues that are important for the design of the product, system, or service. We introduce the notion of the support concept in Section 12.3. Sections 12.5 and 12.6 discuss design and interpretation of support requirements.

12.3 SYSTEM OR SERVICE SUPPORT CONCEPT

We refer to the overall plan to support a system or service as the system (or service) support concept. This plays the same role in system or service design for supportability as the system (or service) maintenance concept does for maintainability. The system support concept comprehends

- assignment of preventive and corrective maintenance to appropriate maintenance levels,
- design and management of facilities to be used in preventive and corrective maintenance operations,

- life cycle cost estimation,
- inventory sizing and management for spare parts and subassemblies, consumables, and other items needed in maintenance,
- logistics of preventive and corrective maintenance,
- diagnostic and fault location procedures and equipment,
- documentation, tools, workflow managers, and other supplies needed to carry out preventive and corrective maintenance,
- policies for maintenance staff training, and
- data collection, analysis, and archiving for verifying conformance with reliability and maintainability requirements and other maintenance-related items.

At the very early stages of design, when the support concept is first defined, one should not expect firm answers for all these issues. However, it is in keeping with the spirit of prevention and quality engineering to begin thinking about these issues as soon as is practical. The support concept should be continually updated and become more precisely specified as greater understanding and specificity of the design and how it is to be repaired are attained. When the system development is complete, the support concept, or plan to support the system, should contain specific elements addressing each of the items listed earlier.

While services are intangible, it is still necessary to prepare for service restoration when a service outage occurs. In addition to system support issues that need to be addressed for elements of the service delivery infrastructure, support for a service also comprehends

- data collection, analysis, and archiving for verifying conformance with service reliability and maintainability requirements and for any relevant regulatory requirements and
- Design of procedures to support the service restoration process decided as part of the service maintenance concept (see Section 10.2.2.1).

12.4 SUPPORT EFFECTIVENESS CRITERIA AND FIGURES OF MERIT

The most basic support effectiveness criterion is the overall amount of time required to complete all tasks preliminary to, or in preparation for, a particular repair. Because it is related to a particular repair, and not the processes connected with repairs in general, this is an on-line supportability effectiveness criterion. Call this the *support time* and denote it by W. This time is represented by the first part of the outage interval pictured in Figure 10.1. As usual, we treat W as a random variable because of many potential noise factors influencing this duration. A figure of merit reflecting the complete supportability picture is the probability that W does not exceed x ($x \geq 0$ is a discretionary variable).

This is the distribution of W and as such is often difficult to estimate. As usual, abbreviations are employed: the mean support time is the most common, and the median support time is also useful if a figure of merit is needed that is less influenced by extremely large or extremely small values.

Support time comprises other individual activity durations, and when it is desirable to focus attention on these other durations, individual effectiveness criteria may be defined for them. These include

- logistic delay time: time spent waiting for the delivery of spare parts, repair parts, tools, consumables, or any other items needed to begin a repair. Logistic delay time begins to accrue once a repair is committed and it is found that one or more physical objects (tools, spare parts, etc.) that are needed to execute the repair are not in place at the workstation at the beginning of the repair.
- administrative delay time: time spent waiting for the completion of any relevant administrative tasks, such as securing approvals for the use of certain facilities or tools, arrival of repair personnel from remote locations, etc.
- inefficiency delay time: time wasted in overcoming errors such as locating misplaced tools or documentation, correction of withdrawal of an incorrect repair part from inventory, dealing with injuries to personnel while preparing for repair, etc.

Creation of support effectiveness criteria for these contributory components of support time is in keeping with the key concept in the systems engineering that if it is necessary to focus attention on a particular supportability issue, then one or more quantitative measurements related to that issue should be devised, monitored, and publicly tracked. The variables covered by this reasoning typically are arrived at during the "check" portion of the Deming Cycle when procedures have been established but unsatisfactory performance is taking place. When unsatisfactory performance is noted, it is necessary to determine

- whether the unsatisfactory performance is due to common causes or to one or more special causes and/or
- whether the process may be improved by redesigning it to eliminate the problematic step(s).

Should a special cause be identified as negatively affecting an operation, Table 12.1 may be helpful in diagnosing underlying ailments. That is, all the symptoms listed in column 1 of the table happen from time to time. It is up to the process manager to determine whether their occurrence is part of normal statistical fluctuation within the process capability or is an extraordinary occurrence pointing to a change in underlying process conditions that needs to be tracked down and remedied. A control chart [2] is a good, simple approach to this discrimination task.

TABLE 12.1 Supportability Tracking Variables

Symptom	Possibly Indicating
Repair workstation buffer overflows	Inadequate workstation staffing or facilities, failure to adequately characterize the time needed for a repair, or other workstation deficiency
Frequent stockouts[a]	Improper inventory sizing
Replenishment order delays	Inadequate management of suppliers and/or logistics functions
Excess time consumed in a particular repair operation	Inadequate documentation, training, tools, etc., and inadequate characterization of activities needed for a repair
Failure to meet a fault location time requirement	Improper design of, or error(s) in, fault location routines
Lost shipment(s)	Lack of robustness in the supply network

[a] A stockout is an instance in which the inventory of a spare part is depleted (empty) at a time the spare part is needed.

Support effectiveness criteria may be developed for each symptom listed. For instance, if frequent stockouts are noted, a support effectiveness criterion "number of stockouts per week" may be implemented. If finer control were needed, this effectiveness criterion could be defined for each spare part type in the facility. Successive values of effectiveness criteria may be tracked on control charts to ensure that design and/or process changes in the repair facility are undertaken only in response to true signals (special causes) rather than to normally expected statistical fluctuation within the process capability. While a good deal of benefit may be obtained from even the simplest control charts, more complicated situations may benefit from some of the more advanced control chart techniques [2].

In addition to time- or duration-related support effectiveness criteria, effectiveness criteria relating to other important supportability factors may be found useful. Some examples include

- number of times (per week, month, quarter, year) that a stockout takes place,
- (weekly, monthly, quarterly, annual) cost of required transportation of failed and repaired material,
- utilization for each workstation in a repair facility,
- number of repair jobs (per day, week, month, quarter, year) experiencing buffer delays of greater than a specified amount,
- number of instances in which tools, documentation, or other capital equipment is missing or misplaced,
- injury rate in the maintenance facility, and
- diagnostics error rate.

Some of these are discrete (count) effectiveness criteria. As is true with maintainability, there are a great many supportability variables that can potentially be tracked. A good principle is to choose a minimal set of useful variables that will give an adequate picture of supportability given the particular needs of the system or service. As noted previously, you can measure anything you like as long as there is a demonstrated need for and benefit from the measurement (outweighing its cost), you give the measurement a descriptive name, and it is used consistently throughout the development.

12.5 EXAMPLES OF SUPPORT REQUIREMENTS

As with reliability requirements and maintenance requirements, support requirements may be written in terms of either relevant effectiveness criteria or in terms of related figures of merit. The choice determines how the requirement will be interpreted and how supportability data will be analyzed to verify conformance with the requirement. In addition, guiding principles for support requirements are similar to those for reliability and maintenance requirements. Requirements should be written

- in quantitative terms so that it will be possible to unambiguously determine whether the requirement has been met over a defined period of time,
- so that the population of items to which the requirement applies can be unambiguously determined, and
- for a minimal set of useful supportability variables to ensure that
  - The variables that are important for the system or service are tracked, and
  - Systems engineers and the development team are not distracted by an excessive number of requirements that cannot be directly traced to an explicitly identified customer need or profitability concern.

In addition, it is beneficial to be able to make some judgment about whether a requirement is being met without having to wait for an extended period of time. Neither the customer's nor the supplier's interest is served if it takes 20 years to determine whether a requirement has been met or not. This is sometimes a problem for reliability requirements, where durations of operating times tend to be (and it is better if they are) long. It is easier to accomplish this objective in maintenance and support requirements because the durations of maintenance and support events tend to be shorter.

12.5.1 Support Elapsed Time (Duration) Requirements

The most basic online support elapsed time requirement places a limit on the support time itself, as, for example, "the support time shall not exceed 1 hour for a repair of type A, 2 hours for a repair of type B, and 3 hours for all other repair types, when repair is undertaken using the specified procedures." In this example, the requirement is written in terms of a support effectiveness criterion, and as we have done previously for reliability and maintenance requirements, we interpret a requirement written for an effectiveness criterion as applying to every relevant instance (in this case, every repair commenced).

Elapsed time requirements may also be written in terms of appropriate figures of merit, such as the mean support time, median support time, 90th percentile support time, etc. In that case, the requirement is interpreted as pertaining to a population of relevant instances (in this case, repairs commenced). As before, if a census of the population is available, simple arithmetic allows one to see whether the requirement is being met. Otherwise, standard statistical inference procedures, similar to those considered in Chapters 2, 5, and 10, are needed to estimate the probability that the requirement is being met.

Other important online support elapsed time effectiveness criteria include

- the amount of time a repair job spends in each workstation buffer waiting for repair activity to start,
- the total logistic (or administrative or inefficiency) delay time applicable to a repair that has been committed but not commenced, and
- performance management and fault localization elapsed times.

The total time elapsed between when spare parts are ordered and the delivery of those parts is an important example of an important off-line support duration effectiveness criterion. Other off-line support effectiveness criteria that are continuum variables but are not necessarily duration-based include

- facility expenses, including energy, janitorial services, facility maintenance and repairs, etc.,
- facility capital costs, and
- cost of compliance with government regulations.

Here are some additional examples of support duration requirements:

- Severity-1 faults (100%) will be detected and located to the responsible line-replaceable unit within 15 seconds; 100% of severity-2 faults will be detected and located to the responsible line-replaceable unit within 2 minutes; 90% of severity-3 faults will be detected and located to the responsible line-replaceable unit within 15 minutes.
- The median transportation time for the return of repaired material to the spares inventory location shall not exceed 24 hours.
- 90% of workstation buffer delays shall not exceed 1 hour.

12.5.2 Support Count Requirements

In addition to monitoring important elapsed times as indicators of process efficiency, support effectiveness criteria involving counting discrete items may be the subject of requirements. Some of these include

- number of stockouts (per week, month, etc.) for each spare part type,
- number of spare parts of each type in inventory at the end of each week, month, etc.,

- number of repair jobs (per week, month, etc.) whose support time component exceeds stated time, and
- number of fault localizations consuming more than (stated number of) minutes to carry out.

12.6 INTERPRETING AND VERIFYING SUPPORT REQUIREMENTS

Our framework for interpreting support requirements is the same as we have used for reliability and maintenance requirements. If a support requirement is written in terms of an effectiveness criterion, then

- it is interpreted as applying to every instance of the event in the requirement,
- modeling can only address the probability that the requirement may be met,
- supportability data analysis has two aspects:
    - If the data collected are a census of all the events of interest in some stated time–space region,[1] then each observation is compared to the requirement, and a yes-or-no answer is possible regarding whether the requirement is met in that time–space region.
    - If the data collected represent a sample of the events of interest during some stated time–space region, statistical techniques designed to cope with sampling variability are needed to reach conclusions about whether the requirement is met in that region. The conclusion is in the form of an estimate of the probability that the requirement is met in that time–space region.

If a support requirement is written in terms of a supportability figure of merit, then

- it is interpreted as applying to some aggregate or population of events in the requirement over a stated time–space region;
- modeling may compute the same figure of merit over that time–space region;
    - if the model computation is accompanied by a dispersion estimate (this is rare in current practice), then a probability statement may be made about whether the requirement is likely to be met, or
    - if the model computation is not accompanied by a dispersion estimate (this is usual in current practice), then the model computation is compared to the requirement for a yes-or-no result;

[1] For example, a time–space region might be all the jobs in a stated repair facility during a stated month.

- Supportability data analysis has two aspects:
  - if the data collected are a census of all the events of interest in some stated time–space region, then the figure of merit is computed from these data, and a yes-or-no answer is possible regarding whether the requirement is met in that region, or
  - if the data collected represent a sample of the events of interest during some stated time–space region, statistical techniques designed to cope with sampling variability are needed to reach conclusions about whether the requirement is met in that region. The conclusion is in the form of an estimate of the probability that the requirement is met in that time–space region.

These considerations apply whether the variable in the requirement is continuum (e.g., an elapsed time, duration, or cost variable) or discrete (e.g., a count variable). Here are two examples.

Example: The requirement is "there shall be at most one stockout in the facility, over all inventoried part types, in each month." The following data were collected over a period of 2 years: 0, 0, 0, 3, 0, 1, 1, 0, 0, 0, 0, 0, 2, 0, 5, 0, 0, 0, 0, 0, 1, 0, 1, 0. Each observation is the number of stockouts in the facility over 1 month. The requirement is written on a supportability effectiveness criterion. The data form a census of the facility's stockouts over a period of 2 years. The number of months in which the requirement is not met is 3. If no changes are made to the inventory operation, we may estimate the probability that the requirement will be met in a given month by treating these observations as a sample from some (currently invisible) future stream of data. The sample proportion of months in which the requirement is met is 21/24 = 0.88 with a standard error of 0.05. A 95% confidence interval for the probability that the requirement will be met in a future month is

$$[0.88 - 1.96 \times 0.05, 0.88 + 1.96 \times 0.05] = [0.79, 0.96].$$

Example: The requirement is "the mean logistics delay time for repair jobs in the facility shall not exceed 1 hour." The following data were collected in the facility on logistics delays (in hours): 0, 0, 4, 0.5, 0, 0, 0, 0.3, 14, 0, 0, 1, 0, 0, 0, 0, 2.1, 0, 0, 0.7. If these 20 observations represent a census of the jobs in the facility over some period of time, the fact that the sample mean of these data is 1.13 shows that the requirement is not met for these 20 jobs. If instead the data represent a sample from some period of operation of the facility (not all logistics delays were measured over that period of time), we may estimate the probability that the requirement is met over that period of time. The sample standard deviation is 3.18, so the probability that the requirement is met over this period of time is 0.48, the probability that a normal random variable having mean 1.13 and standard deviation 3.18 is less than or equal to 1. In this example, we may wish to examine observation

14 to determine whether it is an unusually large value, or whether the process commonly throws out values that large or larger.[2] If we eliminate this observation from the data, we now have a sample mean of 0.45 and a sample standard deviation of 0.98. The probability that the requirement is met over this time period is now 0.71, quite a bit higher than before. The decision about whether observation 14 is routine (within the process capability) or due to a special cause has important consequences.

12.7 SUPPORTABILITY ENGINEERING FOR HIGH-CONSEQUENCE SYSTEMS

After significant efforts to design for reliability and assure that failures and outage in high-consequence systems are infrequent, it is still of interest to complete as speedily as possible corrective actions on any failures that do occur. Section 10.7 offered some suggestions for minimizing the time required for repair execution in high-consequence systems; a capsule summary of those recommendations is essentially to consider implementing all design for maintainability practices in a high-consequence system. In other words, while profitability may still be an important factor in high-consequence systems, the serious consequences of failures and outages make it reasonable to assert that systems engineers need to justify exclusions of these practices, rather than having to justify inclusions as would be the case for ordinary (not high-consequence) systems. The same holds true for supportability. Chapter 13 discusses many online and off-line designs for supportability practices. In most systems that are not high consequence, only a few of these practices will be implemented, depending on the particular needs of the system and its business case. Because external failure costs in a high-consequence system are large, consider requiring that the omission of any design for supportability practices be justified in the system's business case—where, in a system that is not high consequence, the inclusion of design for supportability practices may need to be justified.

12.8 CURRENT BEST PRACTICES IN SUPPORT REQUIREMENTS DEVELOPMENT

As was the case with maintenance, recommended practices for developing support requirements are based on contemporary quality engineering principles and quantitative reasoning. Maintainability is concerned with speedy, low-cost, and error-free repair. Supportability engineering prepares the way so that is

[2] A formal way to do this would involve making a control chart of the successive observations, but there is not enough information in this example to do that—in particular, there is no information given about the process capability and the $\pm 3\sigma$ limits. More history with the process would be required to incorporate this additional formalism. Otherwise, statistical techniques for identifying outliers [1] could apply.

becomes possible to execute speedy, low-cost, and error-free repair. Depending on whether maintenance is performed by the customer, the supplier, or a third party, various opportunities for optimizing parts of the support environment are created.

Once support requirements are in place, use the design for supportability techniques in Chapter 13 to arrange the system and its supporting infrastructures to meet the requirements. Success here creates a stronger value proposition for the customer when the customer is responsible for repairs. In case the supplier performs repairs, the requirements influence the cost and profitability of the repair operation.

12.8.1 Identify Support Needs

Three cases need to be considered: for products and systems, maintenance may be performed by the customer (owner/purchaser) of the equipment, by the supplier, or by a third party under contract. A service is normally maintained by the service provider.

12.8.1.1 *Maintenance by the customer/owner/purchaser*

Better supportability may be used to improve the value proposition for a prospective system purchaser when the purchaser will be responsible for support. This is common in the defense and similar industries in which standards prevail and requirements are negotiated between customers and suppliers. Support needs in this case include assistance with various support factors, such as

- sale of spare parts and subassemblies (field-replaceable units) needed according to the system maintenance concept,
- transfer or sale of documentation, tools, etc., needed for repairs,
- assistance with (sale of) intellectual property factors such as spares inventory sizing and management, repair facility design and optimization, etc., and
- assistance with (sale of) training materials and classes to familiarize customer personnel with repair procedures.

12.8.1.2 *Maintenance by the supplier*

When the supplier performs maintenance, there is incentive for cutting cost while remaining effective. Good supportability is needed here so that repair operations, if not actually turning a profit, remain low in cost so they do not degrade the profitability of the system. Because repair operations are under control of the supplier, they have the opportunity to optimize all facets of repair operations, including a robust FRACAS implementation that offers the opportunity to learn from failures with a short feedback loop time. A supplier in this case should carefully consider the factors covered in Chapter 13 and implement optimization models based on the introductory material in Chapter 13 to promote a low-cost and effective repair enterprise.

12.8.1.3 Maintenance by a third party

In case maintenance is performed by a third party unrelated to the supplier or to the customer, contractual provisions determine how effective and efficient the operation can be for the customer (system purchaser). The third party also becomes a potential customer of the supplier for materials, tools, spare parts, etc., needed to carry out repairs. Quality of repairs may be more difficult to control unless incentives are arranged so that the third party has a stake in not only speed and low cost but also quality (error-free repairs).

12.8.2 Balance Support with Economics

It may seem that only in the case of maintenance performed under contract by a supplier should cost and labor-hour factors be significant. However, even in the case where maintenance is to be performed by the customer, it is in the supplier's interest to understand and factor into support requirements the cost and labor-hour burdens created by maintenance procedures. Customers want to see that the supplier understands their needs and is acting to help them succeed. When possible, the supplier should work with customers to understand their processes and arrange the system to better align with those processes to improve supportability.

The business case for the system should be considered when developing the system or service support concept. Support for low-value products or products that become obsolete quickly may not need to be as extensive as support for systems with long useful lives or for high-consequence systems. It may be possible to contemplate building an optimization model to help decide the appropriate level of support. Such a model would require specific information on costs, profitability objectives, projected sales, etc., that might be known only probabilistically, so stochastic optimization might be required to carry out models like these. The mathematics and even the application of these models are beyond the scope of this book. Suffice it to say that even informal consideration of these factors will add value to system or service development.

12.8.3 Use Quantitative Modeling to Promote Rationally Based Support Requirements

Many of the factors affecting supportability may be fruitfully studied quantitatively with a variety of operations research models. Logistics, transportation, inventory sizing and management, staff sizing, and other factors bearing on the ability to repair a system or service effectively and efficiently are widely studied, and results for even quite complicated scenarios can be found in the literature. In this book, Chapter 13 discusses some of these in some detail. In keeping with the spirit of the book, not all topics are covered in detail, and references are provided for you to start your own exploration of these models where desired. Topics included here are discussed because systems engineers need to know the kinds of things that need to be done in the sustainability disciplines without necessarily needing to know the details needed to carry out the studies themselves. In practice, this means that support requirements need to be guided by quantitative models of relevant support operations, and the systems engineer will normally be a supplier and a customer for the specialists who develop and use the models. As a supplier, be prepared to communicate customer needs for support that were determined from a formal or informal balance between those needs and the business case for the system or service. As a customer, be prepared to use the results of quantitative modeling to develop requirements on a rational basis.

12.8.4 Manage Supportability by Fact

We have recommended writing requirements in quantitative form so that it is possible to collect data and verify whether they are being met. Routine verification using a systematic, repeatable process approach is recommended so that a realistic understanding of realized supportability may be acquired. As we discussed for maintainability, we do not want to take action every time we see a requirement not being met because measurements on any process subject to noise factors will exhibit some degree of fluctuation. It's important not to waste resources responding to every fluctuation in measurement—you want to reserve corrective action for those cases where a real change in the process is indicated. Maintenance requirements and support requirements are similar in this regard. They both concern time durations that are relatively short and count events that happen fairly frequently. So many of the same ideas concerning management by fact are similar for support requirements as they are for maintenance requirements. The ideas discussed in Section 10.8.4 may be applied here equally well.

12.9 CHAPTER SUMMARY

This chapter is concerned with the creation of effective support requirements. The property of the system that comprehends its readiness for rapid, low-cost, and error-free repair is called supportability. The chapter stresses beginning to create a support plan as soon as a design concept is developed and continually updating the support plan as the design concept matures. Several support effectiveness criteria and figures of merit are reviewed, including continuum variables and count variables. Interpretation and verification of support requirements are discussed through the use of statistical sampling and analysis techniques for support-related data. The chapter prepares for the discussion of design for supportability to be covered in Chapter 13.

12.10 EXERCISES

1. Why is "degree of supportability" redundant?
2. For each of the items listed in Section 12.3, find and critique examples of relevant requirements in systems or services from your own experience.
3. Devise suitable requirements for each of the items listed in Section 12.3. Review your results for clarity and completeness. What data should be collected to enable the determination of whether these requirements are satisfied?
4. Is the mean support time or the median support time more appropriate for characterizing the duration of maintenance preparation? (Hint: determine who or what organization is going to use the information, and for what purpose).
5. Consider the supportability requirements examples shown in Sections 12.5.1 and 12.5.2, or some supportability requirements from your own experience. Do these, as written, conform to the guidelines listed in Section 12.5? Rewrite your examples so that they conform to these guidelines.
6. When repairs are performed by a third party (other than the system supplier or system purchaser), what kinds of contract provisions should be implemented so that the third party adheres to standards of speed, cost, and quality needed by the customer? What stake does the system supplier have in this process? Give some examples of successful and fraught third-party repair scenarios.

REFERENCES

1. Moore DS, McCabe GP. *Introduction to the Practice of Statistics*. New York: Freeman and Co.; 1993.
2. Wadsworth HM, Stephens KS, Godfrey AB. *Modern Methods for Quality Control and Improvement*. New York: John Wiley & Sons, Inc; 2002.

13

Design for Supportability

13.1 WHAT TO EXPECT FROM THIS CHAPTER

The material we present in this chapter supports execution of the prescription we consistently advocate: pay attention to sustainability engineering at the early stages of system design so that better results may be achieved at lower cost. Good supportability promotes customer satisfaction and supplier profitability by decreasing the amount of time it takes to recover from failures, decreasing the burden on maintenance staff, and increasing system availability. Some important factors influencing supportability were reviewed in Section 12.2.2. This chapter discusses several useful practices that provide a quantitative foundation for enhancing supportability. These practical techniques form the core of design for supportability. Many of these quantitative techniques may also be extended to optimize their application. The decision whether to take the extra time and resources to carry out this optimization rests, as usual, on a balance of prevention and external failure costs.

Coverage of modeling and optimization for all relevant supportability techniques is beyond the scope of this book, but we use this chapter to show how some important supportability issues may be addressed with quantitative modeling. As always, the depth to which techniques like these are applied is dictated by the economics of the system and its total life cycle cost picture.

Reliability, Maintainability, and Supportability: Best Practices for Systems Engineers, First Edition. Michael Tortorella.

Products that become obsolete very quickly, are low value, or otherwise economically less consequential may receive less supportability attention, and some of the ideas in this chapter may be incorporated informally or possibly not at all. However, for expensive, complex, high-value, or high-consequence systems, the quantitative techniques introduced here are valuable in getting optimized supportability built into the system effectively. High-value and high-consequence systems justify more prevention cost and additional resources expended on design for supportability. Techniques presented here are not meant to be all-encompassing but rather are meant to provide an introduction to the thought process used in dealing with supportability needs quantitatively. It is unlikely that you will find a model here that exactly matches your situation. Rather, the models should serve as a source of ideas that may be adaptable to your needs. More complex models tailorable to more comprehensive needs, as well as models for other supportability practices not covered in this chapter, are available in the literature. The exercises also offer some practice in building quantitative models for supportability practices.

13.2 SUPPORTABILITY ASSESSMENT

13.2.1 Quantitative Supportability Assessment

As with reliability, it is useful for product and service designers to have a way to determine the supportability of the design as it progresses. Supportability is many-faceted, so different assessment models are needed for different facets. Aspects of supportability that lend themselves to quantitative modeling include

- inventory management for spare parts, repair parts, and consumables,
- logistics management for the transportation of required goods around the various locations specified in the level of repair analysis (LoRA) (Chapter 11),
- facility location,
- facility layout, and
- staffing levels.

While a major purpose of quantitative supportability modeling is to enable creation of suitable supportability requirements and their implementation, quantitative modeling may also be used to compare an existing supportability plan to existing requirements to determine whether the system design is capable of meeting those requirements. It is not the purpose of this book to show all the quantitative models possible for these factors, but we will discuss in this section a simple inventory model and a simple facility location model as illustrations of the relevant thought process.

13.2.1.1 A simple inventory management model

Two aspects of inventory management come into play in supportability: determination of a correct inventory size and continuing maintenance of appropriate inventory levels throughout operation. A model to determine correct initial inventory size is described in Section 13.3.7. In this section, we discuss a simple inventory management model useful for maintaining a suitable level of inventory to support the system's maintainability needs.

We consider ongoing operation of an inventory of spare field-replaceable units[1] (FRUs) that was designed using a procedure like that in Section 13.3.7. Denote by S the number of FRUs of a single specified type to be stocked in this inventory.[2] S is an output of the design procedure for the inventory (Section 13.3.7 and similar). There is also given a number $s<S$, which is a reorder threshold ($s>0$). From time to time, an FRU will be removed from the inventory to repair a failed system. The inventory manager checks the stock at the end of each month, and if the number of FRUs in the inventory at the time of checking is s or greater, the manager does not order any units. However, if the inventory at the time of checking falls to $a<s$, then enough FRUs (namely, $S-a$) are ordered from the supplier to return the inventory to size S. This is called an (s, S) inventory management policy. Some of the quantities that may be of interest to the maintenance manager include

- the probability that the inventory is depleted before the restocking parts arrive (this is called the "stockout probability") and
- the number of units that need to be ordered to bring the stock level back up to S.

Set the clock so that the inventory process starts at time 0. The demand in the nth month is $D_n \geq 0$, $n=1, 2,\ldots$. For the purposes of this model, we assume that $D_1, D_2,\ldots$ are identically distributed with $P\{D_n=k\}=\pi_k, k=1,2,\ldots,$ for all n. This is reasonable if the number of systems being serviced by the inventory does not change from month to month because the demand for spare FRUs is determined by the number of failures each month of the FRU over all the systems being serviced by the inventory. If we let Z_n denote the number of FRUs in the inventory just before the end of month n (which is to say, at any time after the last FRU is drawn from the inventory before the end of month n), then the stock size Z_n may take values $S, S-1, S-2,\ldots, 1, 0, -1, -2,\ldots,$ where negative numbers represent unfulfilled requests from the inventory (i.e., system failures due to failure of this FRU type) that could not be restored because there were not enough FRUs in the inventory. Then, the (s, S) rule shows that we may write a recursive expression for Z_n as follows:

$$Z_{n+1} = \begin{cases} Z_n - D_{n+1} & \text{if } s < Z_n \leq S \\ S - D_{n+1} & \text{if } Z_n \leq s \end{cases}.$$

[1] We introduce this term as synonymous with line-replaceable unit (LRU).

[2] In this formulation, a separate inventory management model is used for each FRU type.

If we also assume the demands are mutually independent, then $\{Z_1, Z_2, \ldots\}$ is a Markov chain with transition probabilities

$$p_{ij} = P\{Z_{n+1} = j \mid Z_n = i\} = \begin{cases} \pi_{i-j} & \text{if } s < i \leq S \\ \pi_{S-j} & \text{if } i \leq s \end{cases}.$$

The inventory manager's analyst uses knowledge about the distribution of demands and the Markov chain formulation to compute the stockout probability and the distribution of (or the expected value of) the number of FRUs that need to be ordered at the end of each month. An example of the kind of analysis needed to do this can be found in Chapter 3 of Ref. 22. See also Exercise 2.

This example is possibly the simplest inventory management scheme that has been studied quantitatively [17, 18, 22]. One clear disadvantage of this model is that it does not account for growth or shrinkage of the population of systems the inventory serves. Other variations may be needed to accommodate particular conditions in realistic inventory management applications. Fortunately, inventory management is one of the most widely studied operations research disciplines, and many models have been developed to enable quantitative inventory management under a dizzying variety of different operational conditions. Many of these models have been developed into open-source and commercial software. If you are considering using software for this task, the same review of its underlying model is needed as in any other sustainability engineering software to make sure that the assumptions the developers of the software made are close enough to your conditions that the results obtained from using the software will be relevant.

Again, systems engineers are not likely to be undertaking detailed work in support of a particular inventory management implementation, but they are more likely to be involved in determining requirements for the inventory management and in periodic reviews of data collection and analysis to verify continued satisfactory operation of the inventory. Seek help from operations research professionals, particularly if the operation you are studying has features that are not found in standard models.

13.2.1.2 A simple facility location model

Imagine that you are considering a repair scheme that incorporates an intermediate level of repair. The location of the intermediate repair facility or facilities has a great deal to do with the cost of the scheme. Ideally, one would like to locate the intermediate repair facility so that the sum of all relevant costs is as small as possible. Obviously, there is interplay between the facility location problem and the LoRA (Section 11.4.2) when the site of the intermediate repair facility is still to be determined. In this section, we will introduce a simple facility location model, again, not so much as a comprehensive model that you can use in many circumstances but more as an illustration of the relevant thought process.

Suppose that there are n systems in service, located at $(x_1, y_1),\ldots, (x_n, y_n)$ in the plane,[3] that are to subtend one intermediate maintenance facility whose location (x, y) is currently undetermined. Its location will be determined by minimizing the total transportation costs from (x, y) to the n individual system locations, weighted by the amount of traffic expected to flow to and from each location. The transportation cost from (x, y) to (z, w) is given by a nonnegative real function $C((x, y), (z, w))$. The proportion of demands on the intermediate facility from system i is $\alpha_i \geq 0$ with $\alpha_1 + \cdots + \alpha_n = 1$. These may be unequal because there may be different numbers and types of systems at each site, and the demand for replacement FRUs is determined by the frequency of failures of each FRU type in each month and the total number of FRUs at the site. Then, the location of the intermediate facility may be determined by finding the (x, y) that minimizes

$$\sum_{i=1}^{n} \alpha_i C((x,y),(x_i,y_i)).$$

See Exercise 3. The cost function may also include a factor relating to delay. One of the reasons for carefully considering the location of an intermediate repair facility is to minimize logistics delay time (Section 12.4).

Again, this is possibly the simplest facility location model capturing the essential ideas. It is unlikely that it contains enough detail to be useful for realistic facility location problems, but sometimes a simple model is all that can be justified by the quality of the input information. In any case, it can provide some guidance even when all factors known to be important cannot be explicitly incorporated. There is an extensive research and pedagogical literature on facility location problems. Ref. 12 provides a way in.

13.2.2 Qualitative Supportability Assessment

Supportability assessment includes qualitative as well as quantitative techniques. Qualitative supportability assessment, which may also be used for developments in which the additional expense and time required to apply quantitative assessments via modeling may not be justified, or for which quantitative models don't make sense, may be implemented in checklists for the adequate provision of

- comprehensive supportability requirements,
- diagnostic procedures,
- documentation,
- staff training, and
- test equipment,

and other related needs.

[3] Two-dimensional Euclidean space with (x, y)-coordinates.

13.3 IMPLEMENTATION OF FACTORS PROMOTING SUPPORTABILITY

Several factors promoting supportability were listed in Section 12.2.2. Here we consider each of these in more detail and discuss their implementation and how they may lead to improved supportability.

13.3.1 Diagnostics and Fault Location

One of the most important factors in preparing for repair is the ability to determine speedily just what needs to be repaired. Obviously, repair can't begin until the source of the failure is identified and located, and the longer it takes to do this, the greater the outage time and the worse it is for supportability and system availability. This section discusses some principles useful in designing systems so that faults can be speedily identified and located, paving the way for rapid initiation of repair.

When a failure occurs, time spent on diagnosing and locating the cause of the failure adds to outage duration. Supportability is improved by providing means for rapid diagnosis (what function failed and which part of the system is responsible) and fault location (identification of the specific subassembly or FRU that contains the fault). Online techniques for rapid diagnosis and fault location are procedures that run continually or periodically while the system is operating and include

- built-in test (BIT) facilities (also known as built-in self-test): BIT comprises means for ascertaining whether the system is producing proper outputs. Not only the final output but also intermediate outputs may be included. The diagnostic ability of BIT stems from an understanding of what a proper output should be for the stage under test, what the current output for that stage is, and how any differences indicate fault(s) that may have occurred in that stage. BIT may be implemented at the system level overall to help identify faulty FRUs and may also be implemented within an FRU to assist repair personnel if the FRU is repairable, either on-site or at a remote repair facility as dictated by the LoRA. See Ref. 30 for more information.
- parity checking and use of error-detecting (EDC) or error-correcting codes (ECC): these are a more rudimentary form of BIT because parity checking and EDC indicate only the presence of a fault but may not pinpoint the location of the fault. ECC provides additional robustness by correcting as well as detecting errors, but imposes a small penalty in throughput. The mathematical theory of error-correcting codes can be found in Ref. 25. Engineering applications are covered in Ref. 26.
- Diagnostic processes that run in background while the system is performing its functions.

Once the system has failed and is in an outage condition, off-line procedures may be applied. Off-line diagnostic and fault location procedures are tests and routines run using specialized test equipment and tools specifically designed to aid in determining the type and location of the fault(s) causing the failure. For example, a transmission test set for a specific digital communications mode (e.g., radioteletype) creates a signal of known integrity that is inserted into a transmitter input. The waveform is measured by the test set at predetermined test points in the transmitter and compared with the input. Deviations from known good signal quality at a certain test point indicate a difficulty with the transmitter stage(s) monitored by that test point. Physical design of the system can promote or impede supportability by making it easier or harder to access the required test points. Tests of this kind may also be automated if test points are brought out to a single external interface facilitating connection to a test set. These are not new ideas but are included here to provide examples of the design for supportability process.

13.3.2 Tools and Equipment

Some equipment is designed to be disassembled without tools. These systems, typically lower-value consumer products such as printers, implement snap fasteners that can be rapidly undone without any special tools (although specialized knowledge about the location and operation of the fasteners sometimes is required). But most more complex, higher-value systems intended to be used in more challenging environments usually require tools to undo fasteners, unseat circuit cards from connectors, etc. All such tools should be provided within easy reach so that time is not wasted searching for the proper tool.[4] It is possible to write a contract calling for any special tools to be provided as part of the system's physical design and to be shipped with the product. Some older examples include the R-390A/URR and Collins 51S-1 HF receivers, which were shipped with tools for disassembly and alignment incorporated into a compartment as an integral part of the receiver's physical design. Ideally, the use of special tools should be minimized because ready availability of a common tool saves time should a tool be misplaced.

13.3.3 Documentation and Workflow Management

Maintenance staff may work with several different parts of a system if not different systems entirely. While training is necessary, the press of time sometimes means it is not sufficient, and staff may need to consult documentation to refresh their ability to execute required procedures. Time spent searching for and through documentation adds to outage duration. Each repair should be studied for proper workflow, and documentation should be integrated into the workflow so that it is readily available when needed. A disciplined process management approach [13, 28] should be followed so that inefficiencies are rooted out while opportunities for error are minimized.

[4] Time so spent is tallied as part of inefficiency delay time (Section 12.4).

Workflow management software provides step-by-step directions for carrying out particular repair tasks. It may also be useful for in-process quality control. Particularly complex or tedious procedures may benefit from support with workflow management software, increasing convenience and minimizing the chance for error. Workflow management software may also be of benefit to inexperienced maintenance staff by providing both real-time instruction and a supplement to training.

13.3.4 Staff Training

When repair personnel have not been trained in the procedures required for repairs carried out in their facility, excess time is consumed by various inefficiencies: looking up the proper execution of a step in the repair process, asking a colleague for help, figuring out the correct sequence of operations, etc. These add time to repair execution and decrease maintainability. The possibility of errors increases. Many studies show the positive return on investment from training [2, 6, 29] in a variety of areas. The value of training in repair operations, where actions are repetitive and errors have serious consequences, should not be underestimated.

13.3.5 Layout of Repair Facility and Workstation Design

While on-site servicing and repair is often performed under *ad hoc* circumstances, intermediate and higher-level repair is performed in dedicated facilities whose design and layout may promote or impede efficient execution of repair tasks. A permanent installation offers an opportunity to maximize efficiency and throughput by good design of the facility overall and of the individual workstations in the facility. A facility in which more time is consumed than necessary in repair suffers from reduced throughput, unwarranted delays in returning repaired items to useful service, and additional costs due to overstaffing.

The stochastic network flow model provides a fruitful approach to modeling layout of a repair facility. It requires understanding the mix of repair jobs to be performed in the facility as well as the individual steps needed for each repair job type. From the required steps, one can

- plan the layout of the facility and
- gather information about how long a job may spend at each workstation in the facility.

The latter is facilitated by maintenance task analysis (Section 11.3.2.2). Once these are known, a network flow model may be implemented. In this context, where the network usually has only one input node (a location in the facility where repair jobs enter the facility) and one output node (a location in the facility where repair jobs leave the facility), the network flow model resembles the precedence diagram method or critical path method. We present an example of a stochastic network flow model for a repair facility in the context of a performance (throughput and delay) analysis for the facility in Section 13.4.1.

Design of a maintenance facility should also include provisions for data collection and archiving to support a FRACAS as described in Chapter 5.

13.3.6 Design of Maintenance Procedures

While the layout of the repair facility and the flow of material through it greatly influence the time needed to complete a repair job, the procedures used for each step of the job are no less important. Facility layout covers the sequencing and placement of workstations. Design of maintenance procedures considers specification of the sequence of operations at each individual workstation. Design of maintenance procedures comprehends specification of

- test equipment needed to carry out the functions of the workstation,
- tools needed for disassembly, repair, alignment, reassembly, etc., to be accomplished at the workstation,
- a sequence of operations that will accomplish the tasks required at the workstation quickly while minimizing the chance for error,
- staffing of the workstation, including how many technicians are required and the duties of each,
- documentation to support the workstation's operations, including overall process descriptions and detailed documentation for particularities of the system under repair,
- training for workstation operators, including periodic refresher training for experienced operators and introductory training for new operators,
- procedures for gathering management feedback from workstation operators to identify opportunities for greater efficiency, minimizing errors, and other factors that may become apparent only after experience is gained with existing operations, and
- continual improvement via periodic reassessment of workstation performance and redesign based on documented areas for improvement.

Workflow management software may also be useful in cases of a large number of inexperienced operators when there are few experienced operators they can learn from, and in cases where traceability of operations or material is required.

13.3.7 Spare Parts, Repair Parts, and Consumables Inventory

In any repair scheme, multilevel or not, in which repairs are performed by replacing modules or subassemblies (referred to as line-replaceable units or FRUs), a stock of known-good units is needed that is readily accessible to repair staff. Usually, this means some number of spare units will be kept at each site where repair using those units will be performed. How is that number chosen?

Support management tries to balance the costs of acquiring and carrying an inventory of spare units against the costs incurred when a repair requiring a certain type of unit cannot be completed in a timely fashion because the inventory of that type of unit on the site is depleted. Most early inventory optimization models involved finding an optimal value for the stockout probability (Section 12.4). When the stockout probability is low, a larger number of spare units are required, and acquisition and carrying costs are high. When the stockout probability is high, uncompleted repairs will occur more frequently, and the costs due to emergency acquisition of spare units and to longer outages increase. The sum of these costs should have a U-shaped graph, and from this formalism, an optimal value of the stockout probability, depending on the particular values of the costs involved, may be selected. Again, this is likely the simplest inventory optimization model that has been studied quantitatively, and it is not likely to apply directly in many particular cases. Many realistic inventory optimization models have been considered in an extensive operations research literature. See also Ref. 11 for an overview.

Failures and outages are not divorced from availability, however, and most systems have some availability objective. So, for instance, if an outage persists for a longer time because a spare unit is not present in the on-site inventory, then the system availability (any of inherent, operational, or achieved, see Section 10.6.4) will decrease. In a sense, stockout probability is a proxy for system availability: the reason one wants to keep the stockout probability low is so that the system availability will be high. The presence of an availability requirement means that this may be addressed directly by making the system availability requirement a constraint in formulating the inventory sizing optimization. In this way, system supportability is arranged so that the system availability requirement is directly considered—a more holistic approach integrating supportability and reliability. Some examples of this approach include Refs. 1, 7, 23.

A great variety of models have been developed to cover the many different operational possibilities in use: spares kept on-site only, spares kept on-site with backups kept at an intermediate serving location that provides backup spares for several sites, spares ordered directly from the manufacturer, etc. Many of these can be found in Ref. 27.

In practice, most enterprise management software contains inventory optimization and management features, and it would be rare to need to develop a new model from scratch. As with all contemplated software applications, it is worthwhile to check the assumptions used by the software's provider to make sure that they are compatible with the operation and results needed. Tuning after implementation may be desirable if experience indicates that the numbers suggested by the software are too large or too small.

13.3.8 Transportation and Logistics

A multilevel repair scheme requires extensive transportation of items to be repaired, repaired items, spare parts, etc. The more levels and locations in the scheme, the more transportation required. Transportation and logistics typically is the largest component of cost in these schemes, after acquisition costs and labor costs. Optimal facility location (Section 13.2.1.2) is one way to minimize transportation costs. However, it is often the case that facilities cannot be located at will, for example, in cases where one may wish to exploit legacy infrastructure already in place. In such cases, it is still important to try to minimize transportation costs and delays. When locations (system installations, intermediate and higher-level repair locations) are fixed, some of the factors that influence transportation and logistics costs include

- the mix of transportation modalities used for different routes in the repair scheme,
- batching and staging of jobs and material for transport, and
- internal versus subcontracted transportation.

Formal optimization models may be devised for cost minimization considering these factors. In many cases, simpler accounting models similar to the LoRA may be adequate to provide a good starting point for the operation, which may then be tuned after some experience is gained with its operational properties.

13.4 QUANTITATIVE DESIGN FOR SUPPORTABILITY TECHNIQUES

Most supportability engineering concerns actions taken by support staff working in a system whose goal is the promotion of speedier, less costly, and less error-prone repair. Section 13.3 discussed some important factors whose proper implementation helps promote better supportability. Some of these factors may be dealt with quantitatively, and so they may be settled by optimization. Earlier in this chapter, we considered some elementary quantitative developments for inventory management and facility location, showing the most basic models that apply and pointing to the literature for further developments. In this section, we will consider some more detailed models for the layout of a maintenance facility and staff sizing for individual workstations so that these may be designed to achieve stated supportability requirements at minimum cost through optimization of the control parameters available.

13.4.1 Performance Analysis of a Maintenance Facility

Some maintenance facilities are designed to support many system types, and many different types of repairs may be performed there. For instance, in the multilevel model studies in Section 11.4, more complicated repair demands are aggregated from field deployment sites and are serviced according to some planned scheme. Supportability concerns itself with enabling speedier repairs, so it is of interest to know how the design of the maintenance facility may promote or inhibit speedy repairs. This Section discusses a simple model that can be used to study the performance of a maintenance facility. Commonly used performance effectiveness criteria include throughput (the number of units flowing from the entrance of the facility to the exit of the facility per hour) and delay (the amount of time a unit spends in the facility waiting and undergoing service). The model may be used as a basis for optimization of the facility by adjusting the quantities and arrangement of workstations so that throughput and delay requirements are met at lowest cost.

Maintenance facilities commonly consist of some number of workstations that may be all the same (if there is only one type of repair being performed there) or different (if the facility handles more than one type of repair). Units undergoing maintenance enter the facility and routed to an incoming inspection and sorting activity, from where they are sent to the proper workstation. They then may be routed to other workstations if additional maintenance is needed and are routed to an exit when all maintenance is complete. It is common to refer to units requiring maintenance as "jobs." So one can picture a maintenance facility as a collection of workstations and streams of jobs flowing around the workstations. A convenient quantitative model for this activity is a *flow network* [14] in which an exogenous demand of jobs enters the network, spend some time at various workstations in the network, and leave the network when their required maintenance is complete.

In addition, in most cases,

- the number of persons staffing each workstation is limited, and
- the amount of time a job spends at a workstation is not predictable in advance.

So the network is actually a network of queues [3, 8, 19], where by a "queue," we mean a service system subject to random demands, and the time required to service each demand is random too. Each workstation is represented as a queue in which the number of servers is the number of operators attending that workstation, and the service times are the (random) times it takes to complete maintenance at that workstation (see maintenance task analysis, Section 11.3.2.2). The flow of jobs in the network is described by a Markov process having a transition matrix whose (i, j) entry is the conditional probability that a job will travel next to workstation j given that it is now at workstation i. This routing model also accommodates fixed, deterministic routing in which the path of a given job type through the facility is fixed and determined in advance. Routing is determined by the sequence of maintenance operations required by a unit, the capabilities of each workstation, and, in case there is more than one workstation having a given capability, the random completion times of jobs at workstations. Figure 13.1 shows a typical model of this sort of operation.

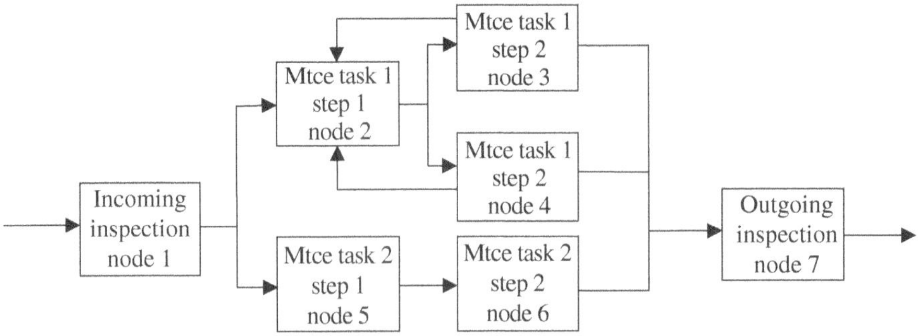

Figure 13.1 *A maintenance facility flow network.*

In this example, there are two different maintenance tasks because there are two types of units being serviced at this facility. Each task has two steps. The second step of Task 1 is more time consuming than the others, so two workstations are provided for this step.[5] The path from Task 1 step 2 back to Task 1 step 1 represents possible rework due to erroneous execution of Task 1. Jobs arrive from outside the network (call this node 0) and exit to outside the network when they are completed. Each workstation may be thought of as a queue with an arrival process composed of the jobs waiting to be worked on at the workstation, the number of servers equal to the number of people staffing the workstation, and the service times equal to the time required to complete a job at that workstation. A routing matrix for this example network is $(r_{ij} : i, j = 0, 1, \ldots, 7)$ (Figure 13.2).

This matrix represents that approximately 60% of the jobs entering the facility are of type 1 and 40% are of type 2. Half the completed jobs from step 1 of Task 1 (node 2) are sent to the workstation at node 3, and the other half are sent to node 4. The diagram and matrix represent that 2% of the jobs leaving nodes 3 and 4 need to return to node 2 because Task 1 was not completed correctly. All jobs receive an outgoing inspection before leaving the facility. This network is considered *open* because there is at least one node where jobs may enter or leave the network (in a *closed* network, a fixed number of jobs circulate around the network, no new jobs may enter, and no jobs leave the network; a closed network is usually not appropriate for modeling a repair facility because jobs are intended to be completed and leave the facility).

The simplest queuing network model is the *Jackson network* [19, 20] in which each individual workstation is an M/M/*c* FCFS[6] queue [15], and the routing is Markovian (where a job goes next depends only on where it is now, and not on any of its prior location(s)). The Jackson network allows for only one

[5] In the example, this is presented as an *a priori* judgment, but a major purpose of this kind of performance modeling is optimization of the facility, i.e., choosing the number of workstations for each task and/or the routing scheme to optimize some system effectiveness measure such as total operational cost of the facility or total time from entry to exit of the facility.

[6] First-come-first-served.

$$\begin{pmatrix} 0 & 1 & 0 & 0 & 0 & 0 & 0 & 0 \\ 0 & 0 & 0.6 & 0 & 0 & 0.4 & 0 & 0 \\ 0 & 0 & 0 & 0.5 & 0.5 & 0 & 0 & 0 \\ 0 & 0 & 0.02 & 0 & 0 & 0 & 0 & 0.98 \\ 0 & 0 & 0.02 & 0 & 0 & 0 & 0 & 0.98 \\ 0 & 0 & 0 & 0 & 0 & 0 & 1 & 0 \\ 0 & 0 & 0 & 0 & 0 & 0 & 0 & 1 \\ 1 & 0 & 0 & 0 & 0 & 0 & 0 & 0 \end{pmatrix}$$

Figure 13.2 *Example routing matrix.*

job class,[7] but Poisson traffic may arrive from outside the network at any node, jobs may leave the network at any node, and load-dependent arrival and service rates are supported. The key feature of the Jackson network is that it has what is called a *product-form solution*. That is, the joint distribution of the number of jobs at each workstation may be written as the product of the individual distributions of the number of jobs at each workstation. In other words, the network behaves as though the individual workstations were stochastically independent. This result depends crucially on Burke's theorem [5], which asserts that the departure process from an M/M/c queue is also a Poisson process. Formally, let J denote the number of workstations in the Jackson network, let N_i denote the number of jobs in service at workstation i in equilibrium,[8] and Q_i denote the (equilibrium) total number of jobs at workstation i ($i=1,\ldots,J$) (i.e., in service and waiting). Then the product-form solution asserts that

$$P\{N_1 = n_1,\ldots,N_J = n_J\} = \prod_{j=1}^{J} P\{N_j = n_j\}$$

and

$$P\{Q_1 = q_1,\ldots,Q_J = q_J\} = \prod_{j=1}^{J} P\{Q_j = q_j\}.$$

Denote the service rate for a single server at workstation i by μ_i and the number of servers at workstation i by c_i ($i=1,\ldots,J$). Then, the total service rate at workstation i is $c_i\mu_i$. The number of servers c_i at workstation i is a control

[7] This is not an insurmountable restriction. It comes into play only when more than one repair type (job class) is accommodated at one or more workstations. Extensions of the Jackson model to accommodate multiple job classes have been developed; perhaps the most well known is [3]. See also [4, 9] for approaches to computation in queueing network models.

[8] "Equilibrium" in this context means that enough time has passed so that any transient effects due to initial variations in the conditions of the queue have damped out, and a "steady-state" operation prevails. Formal definition of "equilibrium" requires consideration of ergodicity, which is beyond the scope of this book. It is usually enough to postulate that equilibrium is achieved after some long period of stable operation (stable meaning that the characteristics of the arrival process and service time distribution do not change).

variable in optimizing the facility.[9] The number of operators to be assigned to each workstation changes the service time for jobs at that workstation and also affects waiting times and buffer occupancies. Typically in maintenance operations, each workstation has only a finite buffer space, or "waiting room," and the buffer size may be a control variable in optimization, but the Jackson network model allows only infinite buffers. While this is clearly only an approximation to the real maintenance facility design, in applications, one would choose a buffer size large enough to accommodate most anticipated demand because one would not allow materials waiting for service to be ignored, discarded, or otherwise leave the system without receiving attention.

The first step in solving a Jackson network problem is to determine the composite arrival rates at each node when both the exogenous arrivals and the arrivals routed from other network nodes are included. This is done by solving the *traffic equation*

$$\lambda_i = \lambda_{0i} + \sum_{j=1}^{J} \lambda_j r_{ji}$$

where λ_i is the composite arrival rate at node i, λ_{0i} is the exogenous arrival rate at node i, and r_{ij} is the (i, j) entry in the routing matrix R ($r_{ij} = P\{$job next visits workstation j | job is currently at workstation $i\}$). Let λ and λ_0 be row vectors containing the individual λ_i and λ_{0i} values. Writing the traffic equation in matrix form as $\lambda(I - R^T) = \lambda_0$, we readily obtain the composite arrival rates $\lambda = \lambda_0(I - R^T)^{-1}$. $I - R^T$ is invertible because the network is open [8]. Knowing each composite arrival rate allows the individual workstations to be analyzed as M/M/c_i queues. We write $\rho_i = \lambda_i / c_i\mu_i$, $i = 1, \ldots, J$. A sufficient condition for the existence of an equilibrium solution for the M/M/c_i queue is $\rho_i < 1$. The probability that there are no jobs (either waiting or in service) at workstation i is

$$p_{0i} = \left[\sum_{n=0}^{c_i - 1} \frac{1}{n!} \left(\frac{\lambda_i}{\mu_i} \right)^n + \frac{1}{c_i!} \left(\frac{\lambda_i}{\mu_i} \right)^{c_i} \left(\frac{c_i\mu_i}{c_i\mu_i - \lambda_i} \right) \right]^{-1}, \quad \rho_i < 1.$$

Some performance variables of interest at workstation i are [15]

- The expected number of jobs in the buffer at the workstation is

$$\mathrm{E}Q_i = p_{0i} \left[\frac{(\lambda_i / \mu_i)^{c_i} \lambda_i \mu_i}{(c_i - 1)!(c_i\mu_i - \lambda_i)^2} \right].$$

  You will want to be sure there are enough buffer spaces to accommodate at least this many queued jobs plus maybe some margin for those times when the buffer occupancy exceeds its mean.

[9] For some ideas on choosing the number of operators for a workstation without analysis of the entire facility, see Section 13.4.2.

- The expected total number of jobs at workstation i, including both waiting and in service, is

$$\mathrm{E}N_i = \frac{\lambda_i}{\mu_i} + \left[\frac{(\lambda_i/\mu_i)^{c_i}\lambda_i\mu_i}{(c_i-1)!(c_i\mu_i-\lambda_i)^2}\right]p_{0i} = \frac{\lambda_i}{\mu_i} + \mathrm{E}Q_i.$$

- W_i is the time a job spends waiting for service (in the buffer) at workstation i. Its expected value is

$$\mathrm{E}W_i = \left[\frac{(\lambda_i/\mu_i)^{c_i}\mu_i}{(c_i-1)!(c_i\mu_i-\lambda_i)^2}\right]p_{0i}.$$

- Let T_i denote the total time a job spends at workstation i, including waiting time and service time. Then, with S_i denoting the service time for a job at workstation i,

$$T_i = S_i + W_i \text{ and } \mathrm{E}T_i = \frac{1}{\mu_i} + \mathrm{E}W_i.$$

- Assuming that the maintenance facility has one node at which jobs enter (call it node a) and one node at which jobs leave (call it node b) (these are nodes 1 and 7, respectively, in the earlier example), the expected time it takes for a job to complete a trip through the maintenance facility is the (a, b) entry in the matrix

$$(I-R)^{-1}(S\#R)(I-R)^{-1}$$

  where I is the identity matrix, R is the routing matrix for the facility, S is a matrix whose (i, j) entry is $T_i + \tau_{ij}$ (τ_{ij} is the expected transit time from node i to node j; in most cases, this will be taken to be zero unless the transit time is not negligible compared to the average service and wait times), and $(S\#R)_{ij} = S_{ij} \cdot R_{ij}$ (this is called the Hadamard or direct product of S and R) [31].
- The throughput is the expected number of jobs leaving the exit node b per unit time. Node b is an $M/M/c_b$ FIFO queue with composite arrival rate λ_b (from the traffic equation) and service rate μ_b. By Burke's theorem, the departure rate from node b is also λ_b, and this is the throughput expressed in the time unit of the composite arrival rate.

The performance parameters of the queueing network model used for the maintenance facility can be used as variables in a scheme to optimize the performance of the facility. The objective for this optimization could be to minimize the total time a job spends in the facility, maximize the throughput, minimize the total cost of the operation, etc., as required by the economics of the system. Full exploration of maintenance facility optimization is beyond the scope of this book, but you can use the simple Jackson network analysis summarized earlier to get some guidance on an initial design. More elaborate or more detailed queueing network models are available and would be appropriate to use if the quality of your information about the input parameters (arrival processes, service time distributions, etc.) warrants this sharper pencil. Given the uncertainty in many systems like these in which human performance is a major factor, the Jackson network model is often all that can be justified. Related models have been considered in the literature, including Refs. 10, 16.

The model discussed in this section makes strong assumptions about the operation (Poisson arrivals, exponential service, etc.), which may or may not be valid in particular cases. As always, the need to use a sharper model (one that better matches known characteristics of the true maintenance situation) is guided by the economics of the system development. In most cases, the simple model can provide useful guidance for facility design and optimization, and you are better off using it rather than nothing. Spend additional resources in fine-tuning a more complicated model only if the economics justifies the additional cost.

13.4.2 Staff Sizing: The Machine Servicing Model

A key to planning for efficient maintenance is the choice of an appropriate number of personnel to carry out each maintenance task. In this section, we introduce a quantitative approach to staff sizing by adapting a standard industrial engineering queueing model, the machine servicing model, to this need. The number of technicians at workstation i is c_i in the notation of Section 13.4.1, so there is an interplay between this workstation staffing model and the performance analysis considered earlier. When optimizing a facility using the earlier performance analysis model, the model in this section provides a good starting point for c_i. The cumulative effect of these choices becomes visible when the whole network model is executed.

The machine servicing model is a finite-source queueing model. That is, there are only a finite number of sources generating jobs for the queue. These sources are the machines requiring service from time to time (when they break down). The servers are the repair personnel. In this adaptation, we take the number of sources, S, to be the number of products, systems, or services generating maintenance tasks to be performed at a given workstation and c, the number of servers, to be the number of technicians who carry out the tasks at that workstation. In the planning exercise, c is unknown and is to be chosen to minimize some operational effectiveness measure such as the total time required to carry out all required maintenance operations or the total cost of the maintenance operation at the workstation. The rate at which the sources demand service from the maintenance staff is related to the frequency of failures of the products, systems, or services covered by the maintenance facility and the number of products, systems, or services served by the maintenance facility. The service times in the queueing model are the outage durations. Our initial model treats these as given (having a known distribution) and uses c as a control variable. A more-advanced model can be constructed using also the distribution of the service times (the outage times) as a control parameter. In that case, there is an interplay between the number of servers and the service times that makes the model analytically more complicated, but simulation can be used to obtain results from this more realistic model.

In this section, we will consider a simple machine-servicing model that makes several simplifying assumptions so that you can get an idea of what this technology can do without getting bogged down in details. More realistic models can be developed, often requiring simulation for solution, when there is a need for greater precision. Accordingly, we assume that there are S products, systems, or services of the same type assigned to the workstation in question, and each is a repairable system whose failures appear in time as homogeneous Poisson processes [22] with rate λ. The arrival rate of jobs to the c-server queue that is the maintenance staff is $\lambda_n=(S-n)\lambda$ when n, the number of systems currently in service is $<S$, and $\lambda_n=0$ when $n\geq S$. We postulate that the repair time for one system is exponentially distributed with mean $1/\mu$, so overall the repair times for all the systems in the facility are exponentially distributed with mean $1/\mu_n=1/n\mu$ for $0\leq n<c$ and $1/\mu_n=1/c\mu$ for $n\geq c$. Write $\rho=\lambda/\mu$ and assume that $\rho<1$. Solution of this model uses the theory of birth-and-death processes [22], from which we may obtain useful equilibrium operating characteristics of the model [15]:

- The probability that there are i systems being serviced is

$$p_i=\begin{cases}\binom{S}{i}\rho^i p_0 & \text{for } 1\leq i<c\\ \binom{S}{i}\dfrac{i!}{c^{i-c}c!}\rho^i p_0 & \text{for } c\leq i\leq S\end{cases}$$

  where p_0 is the probability that the facility is idle, given by

$$p_0(c)=\left[\sum_{j=1}^{c-1}\binom{S}{j}\rho^j+\sum_{j=c}^{S}\binom{S}{j}\frac{j!}{c^{j-c}c!}\rho^j\right]^{-1}.$$

- The expected number of systems at the workstation, including those being serviced and those waiting, is given by

$$\text{E}N=\sum_{j=0}^{S}jp_j=p_0(c)\sum_{j=0}^{c-1}j\binom{S}{j}\rho^j+\sum_{j=c}^{S}j\binom{S}{j}\frac{j!}{c^{j-c}c!}\rho^j.$$

- The expected number of systems waiting for service is given by

$$EQ = EN - c + p_0(c)\sum_{j=0}^{c-1}(c-j)\binom{S}{j}\rho^j.$$

- The expected time that a system spends at the workstation is given by

$$ET = \frac{EN}{\lambda(S - EN)}$$

  and the expected time that a system spends in the buffer waiting for service is given by

$$EW = \frac{EQ}{\lambda(S - EN)}.$$

When using this model to plan the number of operators to staff a workstation, choose a supportability figure of merit (such as the expected number of systems waiting for service) and use c as a control variable in the appropriate expression (we have written the idle probability as $p_0(c)$ to emphasize this), minimizing it with respect to c within a given cost constraint.

Again, even though this may look complicated, it is possibly the simplest machine-servicing model that has been treated quantitatively, and it may not contain all the details of a particular operation. In practical cases, though, it provides better guidance than guesswork for a starting point, and it doesn't take long before any inadequacies in the operation become apparent so that adjustments may be made from a sound baseline.

13.5 CURRENT BEST PRACTICES IN DESIGN FOR SUPPORTABILITY

13.5.1 Customer Needs and Supportability Requirements

Supportability has a direct impact on outage durations, so it is important to understand the customer's needs for the speed of service restoration so that they can be folded into supportability requirements. For example, the customer may have a need for a failover time (Section 11.3.1) not exceeding 50 milliseconds. This can only be achieved with automated processes, so all support aspects of bringing a redundant unit online need to be worked out in advance. These aspects include diagnosis and fault location, test of the redundant unit (if necessary), and predetermination of what is to happen if the switchover to the redundant unit does not complete properly. Supportability requirements should address at least the aspects that bear directly on decreasing the length of outages. These include

- diagnosis and fault location time,
- on-site spares inventory management,
- dispatch time for technicians to reach the faulty replaceable unit, and
- procedures for error minimization during replacement.

13.5.2 Team Integration

As with maintainability engineering, a significant danger to be avoided with supportability engineering is beginning to consider supportability too late in the development process. For example, we have seen how an important supportability consideration, namely, the proper sizing of spares inventories, has an effect on system availability. Given an availability requirement, and many systems have these, it pays to begin an inventory optimization early in development so that information is available for downstream use, for example, in the LoRA.

It is not reasonable to expect every team member to be an expert in supportability, so "omnibus" development management meetings, in which representatives from all sustainability engineering areas are involved, help ensure that important supportability issues do not escape attention. When a team member hears a plan from another part of the team that has an impact on supportability, immediate coordination can take place, and better results obtained.

13.5.3 Modeling and Optimization

When there is an opportunity to use new facilities or processes in maintenance, these should be planned on the basis of some quantitative modeling, even if only simple modeling. This chapter has shown examples of modeling applied to inventory management, facility location, maintenance facility design, and maintenance workstation staffing. The models shown here cannot support every plan in these four areas, but they are intended to show the reasoning process used when using quantitative methods for these plans. Most enterprise management software contains integrated models for these kinds of plans, but not every organization uses enterprise management software, and even for those that do it pays to verify that the assumptions on which the software is predicated reasonably match the conditions of your applications. Models don't need to be (indeed, can never be) perfect, but they do provide guidance that is better than guesswork for initial design of supportability facilities and processes.

13.5.4 Continual Improvement

It is rare that the initial design for supportability produces stellar results. Tuning of the control parameters in any facility or process is almost always needed. But beyond that, continual improvement is always of value even when support processes are functioning well. Conditions change: failures occur more or less frequently, suppliers come and go, staff turnover increases or decreases, etc., so what worked well a few months ago may no longer be optimal. A healthy program of continual improvement helps keep support costs lower while keeping results where they need to be. Continual improvement is supported by adapting quality control and management tools to support process needs. For instance, facility throughput may be monitored with a control chart so that when throughput changes, it is possible to determine whether the change is due to some important factor in the facility's operation (a "special cause") or whether it is a reflection of the normal statistical fluctuation present in any process that is subject to random influences or "noise variables" (a "common cause").

13.6 CHAPTER SUMMARY

This chapter has been devoted to actions you can take to improve the supportability of a design. Design for supportability is similar to design for reliability and design for maintainability in that it attempts to anticipate the support environment a system or service will live in and optimize operational conditions to deliver a level of support consistent with customer needs and the system's or service's business case. Instead of trying to prevent failures or optimize maintenance, design for supportability develops actions that can be taken to improve the support environment so that the overall goal of speedy, low-cost, and error-free repair can be achieved. Some of the specific aspects of a system's support environment that are covered here include inventory management for spare parts and consumables, location of intermediate repair facilities, arrangement of a repair facility to optimize throughput and cost, and choosing a good staff size (number of operators) for a repair workstation. Other characteristics of the support environment that design for supportability covers include procedures and tools for failure diagnostics, fault location, logistics management, documentation and training, and design of maintenance procedures. The chapter is designed to help systems engineers become familiar with basic supportability principles and only in a few cases pursues development of models in detail. Textbooks for the details of supportability engineering include Refs. 21, 24.

13.7 EXERCISES

1. Suppose $X_1, X_2, \ldots$ are random variables having an exponential distribution with mean μ_1 and that $Y_1, Y_2, \ldots$ are random variables having an exponential distribution with mean μ_2. Show that the combined population $X_1, Y_1, X_2, Y_2, \ldots$ has an exponential distribution with mean $(\mu_1^{-1}+\mu_2^{-1})^{-1}$. Does this work for more than two populations?

2. Determine the stockout probability for the (6, 9) inventory management system when the demand distribution is Poisson with rate 4.
3. Consider the facility location model described in Section 13.2.1.2. Find the location of the intermediate facility when there are four systems located at (0, 0), (10, 0), (1, 11), and (17, 24); each system sends the same demand on the intermediate facility; and the cost function is $C((x,y),(z,w))=5+6\sqrt{(x-z)^2+(y-w)^2}$.

REFERENCES

1. Adams CM. Inventory optimization techniques, system vs. item level inventory analysis. *2004 Annual Reliability and Maintainability Symposium.* January 26–29; Piscataway, NJ: IEEE; 2004. p 55-60.
2. Bartel AP. Measuring the employer's return on investments in training: evidence from the literature. Ind Relat J Econ Soc 2000;39 (3):502–524.
3. Baskett F, Chandy KM, Muntz RR, Palacios F. Open, closed, and mixed networks of queues with different classes of customers. J ACM 1975;22:248–260.
4. Bolch G, Grenier S, de Meer H, Trivedi K. *Queueing Networks and Markov Chains: Modeling and Performance Evaluation with Computer Science Applications.* New York: John Wiley & Sons, Inc; 1998.
5. Burke PJ. The output of a queueing system. Oper Res 1956;4 (6):699–704.
6. Carnevale AP, Schulz ER. Return on investment: accounting for training. Train Dev J 1990;44 (7):S1–S32.
7. Chan CK, Tortorella M. Spares inventory sizing for end-to-end service availability. Proceedings of the Annual Reliability and Maintainability Symposium; January 22–25; Philadelphia, PA; 2001. p 98–102.
8. Chen H, Yao DD. *Fundamentals of Queueing Networks.* New York: Springer; 2001.
9. Conway AE, Georganas ND. *Queueing Networks—Exact Computational Algorithms.* Cambridge: MIT Press; 1989.
10. Crespo Marquez A, Sánchez Heguedas A. Models for maintenance optimization: a study for repairable systems and finite time periods. Reliab Eng Syst Saf 2002;75 (3):367–377.
11. Davis RA. *Demand-Driven Inventory Optimization and Replenishment: Creating a More Efficient Supply Chain.* Hoboken: John Wiley & Sons, Inc.; 2013.
12. Drezner Z, editor. *Facility Location: A Survey of Applications and Methods.* New York: Springer-Verlag; 1995.
13. Dumas M, LaRosa M, Mending J. *Fundamentals of Business Process Management.* New York: Springer-Verlag; 2013.
14. Ford LR, Fulkerson DR. *Flows in Networks.* Princeton: Princeton University Press; 1962.
15. Gross D, Harris CM. *Fundamentals of Queueing Theory.* New York: John Wiley & Sons; 1974.
16. Hani Y, Amodeo L, Yalaoui F, Chen H. Simulation based optimization of a train maintenance facility. J Intell Manuf 2008;19 (3):293–300.
17. Heyman DP, Sobel M. *Stochastic Models in Operations Research.* Mineola: Dover Publications; 2003.

18. Hillier FS, Lieberman GJ. *Introduction to Operations Research*. 8th ed. New York: McGraw-Hill; 2005.
19. Jackson JR. Networks of waiting lines. Oper Res 1957;5:518–521.
20. Jackson JR. Jobshop-like queueing systems. Manag Sci 1963;10:131–142.
21. Jones JV. *Supportability Engineering Handbook: Implementation, Measurement and Management*. New York: McGraw-Hill; 2006.
22. Karlin S, Taylor HM. *A First Course in Stochastic Processes*. 2nd ed. New York: Academic Press; 1975.
23. Kumar UD, Knezevic J. Availability based spare optimization using renewal process. Reliab Eng Syst Saf 1998;59 (2):217–223.
24. Kumar UD, Crocker J, Knezevich J. *Reliability, Maintenance and Logistic Support: A Life Cycle Approach*. Dordrecht: Kluwer Academic Publishers; 2000.
25. MacWilliams FJ, Sloane NJA. *The Theory of Error-Correcting Codes*. Amsterdam: North-Holland; 1977.
26. Michelson AM, Levesque AH. *Error Control Techniques for Digital Communications*. New York: John Wiley & Sons, Inc; 1985.
27. Muckstadt JA. *Analysis and Algorithms for Service Parts Supply Chains*. New York: Springer; 2005.
28. Sharp A, McDermott P. *Workflow Modeling: Tools for Process Improvement and Application Development*. 2nd ed. Norwood: Artech House; 2008.
29. Stolovitch HD, Maurice JG. Calculating the return on investment in training: a critical analysis and a case study. Perform Improv 1998;37 (8):9–20.
30. Stroud CE. *A Designer's Guide to Built-In Self-Test*. New York: Springer-Verlag; 2002.
31. Tortorella M. Path-additive functionals in stochastic flow networks with Markovian routing. Rutgers University Department of Industrial and Systems Engineering Working Paper #06-004; 2006.

Index

Reliability, Maintainability, and Supportability: Best Practices for Systems Engineers, First Edition. Michael Tortorella.

WILEY SERIES IN SYSTEMS ENGINEERING AND MANAGEMENT

Andrew P. Sage, Editor

ANDREW P. SAGE and JAMES D. PALMER
Software Systems Engineering

WILLIAM B. ROUSE
Design for Success: A Human-Centered Approach to Designing Successful Products and Systems

LEONARD ADELMAN
Evaluating Decision Support and Expert System Technology

ANDREW P. SAGE
Decision Support Systems Engineering

YEFIM FASSER and DONALD BRETTNER
Process Improvement in the Electronics Industry, Second Edition

WILLIAM B. ROUSE
Strategies for Innovation

ANDREW P. SAGE
Systems Engineering

HORST TEMPELMEIER and HEINRICH KUHN
Flexible Manufacturing Systems: Decision Support for Design and Operation

WILLIAM B. ROUSE
Catalysts for Change: Concepts and Principles for Enabling Innovation

LIPING FANG, KEITH W. HIPEL, and D. MARC KILGOUR
Interactive Decision Making: The Graph Model for Conflict Resolution

DAVID A. SCHUM
Evidential Foundations of Probabilistic Reasoning

JENS RASMUSSEN, ANNELISE MARK PEJTERSEN, and LEONARD P. GOODSTEIN
Cognitive Systems Engineering

ANDREW P. SAGE
Systems Management for Information Technology and Software Engineering

ALPHONSE CHAPANIS
Human Factors in Systems Engineering

YACOV Y. HAIMES
Risk Modeling, Assessment, and Management, Third Edition

DENNIS M. BUEDE
The Engineering Design of Systems: Models and Methods, Second Edition

ANDREW P. SAGE and JAMES E. ARMSTRONG, Jr.
Introduction to Systems Engineering

WILLIAM B. ROUSE
Essential Challenges of Strategic Management

YEFIM FASSER and DONALD BRETTNER
Management for Quality in High-Technology Enterprises

THOMAS B. SHERIDAN
Humans and Automation: System Design and Research Issues

ALEXANDER KOSSIAKOFF and WILLIAM N. SWEET
Systems Engineering Principles and Practice

HAROLD R. BOOHER
Handbook of Human Systems Integration

JEFFREY T. POLLOCK and RALPH HODGSON
Adaptive Information: Improving Business Through Semantic Interoperability, Grid Computing, and Enterprise Integration

ALAN L. PORTER and SCOTT W. CUNNINGHAM
Tech Mining: Exploiting New Technologies for Competitive Advantage

REX BROWN
Rational Choice and Judgment: Decision Analysis for the Decider

WILLIAM B. ROUSE and KENNETH R. BOFF (editors)
Organizational Simulation

HOWARD EISNER
Managing Complex Systems: Thinking Outside the Box

STEVE BELL
Lean Enterprise Systems: Using IT for Continuous Improvement

J. JERRY KAUFMAN and ROY WOODHEAD
Stimulating Innovation in Products and Services: With Function Analysis and Mapping

WILLIAM B. ROUSE
Enterprise Tranformation: Understanding and Enabling Fundamental Change

JOHN E. GIBSON, WILLIAM T. SCHERER, and WILLAM F. GIBSON
How to Do Systems Analysis

WILLIAM F. CHRISTOPHER

Holistic Management: Managing What Matters for Company Success

WILLIAM B. ROUSE

People and Organizations: Explorations of Human-Centered Design

MO JAMSHIDI

System of Systems Engineering: Innovations for the Twenty-First Century

ANDREW P. SAGE and WILLIAM B. ROUSE

Handbook of Systems Engineering and Management, Second Edition

JOHN R. CLYMER

Simulation-Based Engineering of Complex Systems, Second Edition

KRAG BROTBY

Information Security Governance: A Practical Development and Implementation Approach

JULIAN TALBOT and MILES JAKEMAN

Security Risk Management Body of Knowledge

SCOTT JACKSON

Architecting Resilient Systems: Accident Avoidance and Survival and Recovery from Disruptions

JAMES A. GEORGE and JAMES A. RODGER

Smart Data: Enterprise Performance Optimization Strategy

YORAM KOREN

The Global Manufacturing Revolution: Product-Process-Business Integration and Reconfigurable Systems

AVNER ENGEL

Verification, Validation, and Testing of Engineered Systems

WILLIAM B. ROUSE (editor)

The Economics of Human Systems Integration: Valuation of Investments in People's Training and Education, Safety and Health, and Work Productivity

ALEXANDER KOSSIAKOFF, WILLIAM N. SWEET, SAM SEYMOUR, and STEVEN M. BIEMER

Systems Engineering Principles and Practice, Second Edition

GREGORY S. PARNELL, PATRICK J. DRISCOLL, and DALE L. HENDERSON (editors)

Decision Making in Systems Engineering and Management, Second Edition

ANDREW P. SAGE and WILLIAM B. ROUSE

Economic Systems Analysis and Assessment: Intensive Systems, Organizations, and Enterprises

BOHDAN W. OPPENHEIM
Lean for Systems Engineering with Lean Enablers for Systems Engineering

LEV M. KLYATIS
Accelerated Reliability and Durability Testing Technology

BJOERN BARTELS , ULRICH ERMEL, MICHAEL PECHT, and PETER SANDBORN
Strategies to the Prediction, Mitigation, and Management of Product Obsolescence

LEVANT YILMAS and TUNCER ÖREN
Agent-Directed Simulation and Systems Engineering

ELSAYED A. ELSAYED
Reliability Engineering, Second Edition

BEHNAM MALAKOOTI
Operations and Production Systems with Multiple Objectives

MENG-LI SHIU, JUI-CHIN JIANG, and MAO-HSIUNG TU
Quality Strategy for Systems Engineering and Management

ANDREAS OPELT, BORIS GLOGER, WOLFGANG PFARL, and RALF MITTERMAYR
Agile Contracts: Creating and Managing Successful Projects with Scrum

KINJI MORI
Concept-Oriented Research and Development in Information Technology

KAILASH C. KAPUR and MICHAEL PECHT
Reliability Engineering

MICHAEL TORTORELLA
Reliability, Maintainability, and Supportability: Best Practices for Systems Engineers